## PART I

## STUDENT'S SOLUTIONS MANUAL

# CALCULUS

## FINNEY/THOMAS

PART I

STUDENT'S SOLUTIONS MANUAL

# CALCULUS

FINNEY/THOMAS

MICHAEL B. SCHNEIDER
THOMAS L. COCHRAN

BELLEVILLE AREA COLLEGE

ADDISON-WESLEY PUBLISHING COMPANY
Reading, Massachusetts • Menlo Park, California • New York
Don Mills, Ontario • Wokingham, England • Amsterdam • Bonn
Sydney • Singapore • Tokyo • Madrid • San Juan

Reproduced by Addison-Wesley from camera-ready copy supplied by the author.

ISBN 0-201-19345-0



BCDEFGHIJ-AL-943210

The authors would like to dedicate this book to their wives, Jane and Barbara, the two who have put up with so much for so long.

Michael Schneider
Tom Cochran

The authors have attempted to make this manual as error free as possible but nobody's "perfeck". If you find errors, we would appreciate knowing them. You are more than welcome to write us at:

Belleville Area College
2500 Carlyle Road
Belleville, Illinois 62221

# TABLE OF CONTENTS

PART I

STUDENT'S SOLUTIONS MANUAL

# CALCULUS

FINNEY/THOMAS

# CHAPTER 1

# PREREQUISITES FOR CALCULUS

## 1.1 COORDINATES AND GRAPHS IN THE PLANE

| | P(x,y) | Q(x,−y) | R(−x,y) | S(−x,−y) |
|---|---|---|---|---|
| 1. | (3,1) | (3,−1) | (−3,1) | (−3,−1) |
| 3. | (−2,2) | (−2,−2) | (2,2) | (2,−2) |
| 5. | (−1,−1) | (−1,1) | (1,−1) | (1,1) |
| 7. | (1,0) | (1,0) | (−1,0) | (−1,0) |
| 9. | $(\sqrt{3},\sqrt{3})$ | $(\sqrt{3},-\sqrt{3})$ | $(-\sqrt{3},\sqrt{3})$ | $(-\sqrt{3},-\sqrt{3})$ |
| 11. | $(0,\pi)$ | $(0,-\pi)$ | $(0,\pi)$ | $(0,-\pi)$ |

13. $d = \sqrt{(0-1)^2 + (1-0)^2} = \sqrt{2}$

15. $d = \sqrt{(2\sqrt{3} - (-\sqrt{3}))^2 + (4-1)^2}$

$= \sqrt{27+9} = 6$

17. $d = \sqrt{(0-a)^2 + (0-b)^2} = \sqrt{a^2+b^2}$

19. a) (x,y) on graph $\Rightarrow y = -x^2 \not\Rightarrow -y = -x^2$ $\therefore$ Not symmetric to x–axis

b) (x,y) on graph $\Rightarrow y = -x^2 \Rightarrow y = -(-x)^2 \Rightarrow$ (−x,y) on the graph $\therefore$ Symmetric to y–axis

c) (x,y) on graph $\Rightarrow y = -x^2 \not\Rightarrow -y = -(-x^2)$ $\therefore$ Not symmetric to origin

21. a) (x,y) on graph $\Rightarrow y = \frac{1}{x^2} \not\Rightarrow -y = \frac{1}{x^2}$ $\therefore$ Not symmetric to x–axis

b) (x,y) on graph $\Rightarrow y = \frac{1}{x^2} \Rightarrow y = \frac{1}{(-x)^2} \Rightarrow$ (−x,y) on graph $\Rightarrow$ Symmetric to y–axis

c) (x,y) on graph $\Rightarrow y = \frac{1}{x^2} \not\Rightarrow -y = \frac{1}{(-x)^2}$ $\therefore$ Not symmetric to origin

23. a) (x,y) on graph $\Rightarrow xy = 1 \not\Rightarrow x(-y) = 1$ $\therefore$ Not symmetric to x–axis

b) (x,y) on graph $\Rightarrow xy = 1 \not\Rightarrow (-x)y = 1$ $\therefore$ Not symmetric to y–axis

c) (x,y) on graph $\Rightarrow xy = 1 \Rightarrow (-x)(-y) = 1 \Rightarrow$ (−x,−y) on graph $\Rightarrow$ Symmetric to origin

25. a) (x,y) on graph $\Rightarrow x^2y^2 = 1 \Rightarrow x^2(-y)^2 = 1 \Rightarrow$ (x,−y) on graph $\Rightarrow$ Symmetric to x–axis

b) (x,y) on graph $\Rightarrow x^2y^2 = 1 \Rightarrow (-x)^2y^2 = 1 \Rightarrow$ (−x,y) on graph $\Rightarrow$ Symmetric to y–axis

c) Symmetry to x and y axes $\Rightarrow$ Symmetry to origin

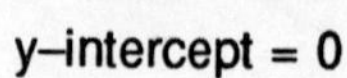

27. x–intercept = –1

y–intercept = 1

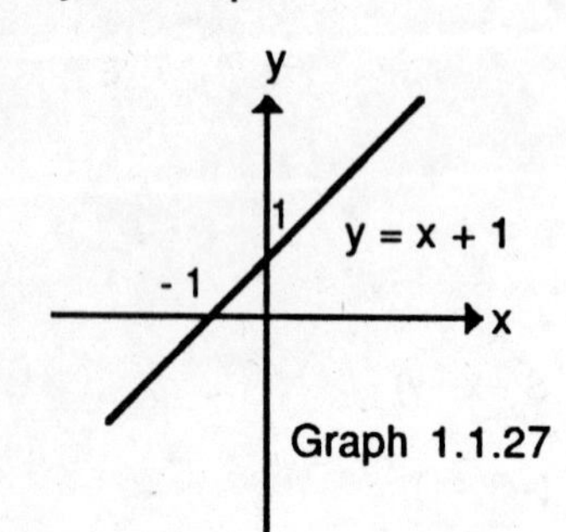

Graph 1.1.27

29. x–intercept = 0

y–intercept = 0

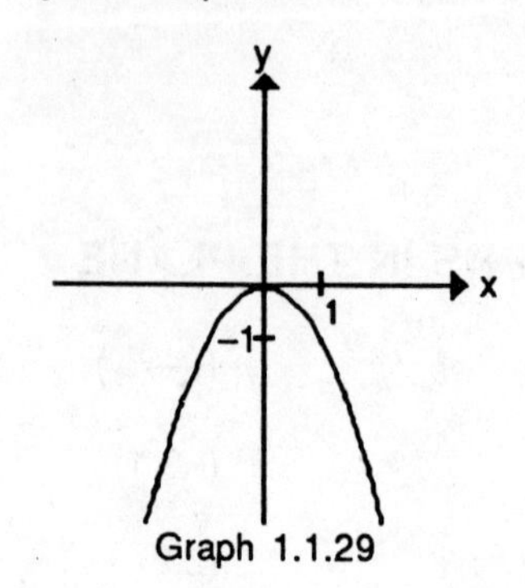

Graph 1.1.29

31. x–intercept = 0

y–intercept = 0

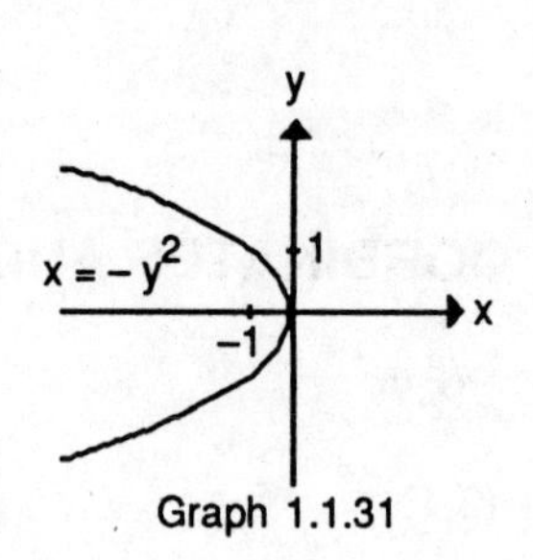

Graph 1.1.31

33. x–intercepts = ±2

y–intercepts = ±2

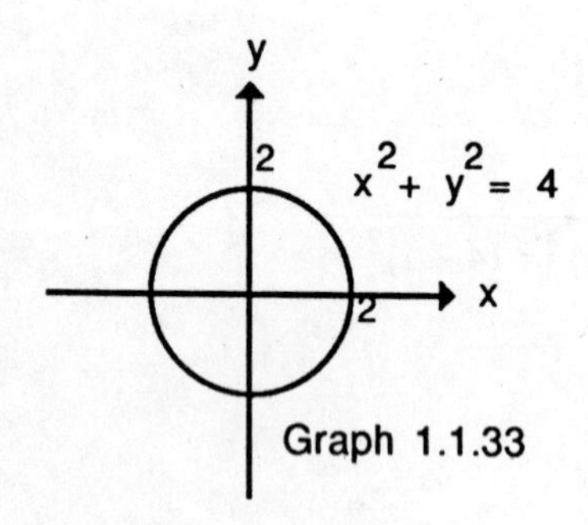

Graph 1.1.33

35. x–intercept = 0

y–intercept = 0

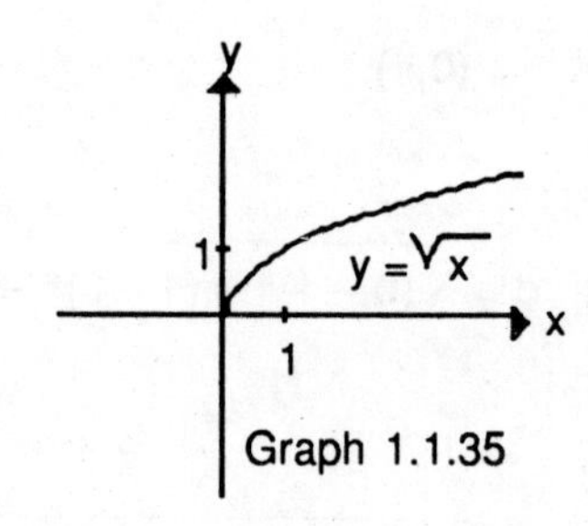

Graph 1.1.35

37. A 45° angle

39. AD = 3DC, P = 56 $\Rightarrow$ AD + DC + BC + AB = 56 $\Rightarrow$ 3DC + DC + 3DC + DC = 56 $\Rightarrow$ 8DC = 56. Let C = (9,y) $\Rightarrow$ m(DC) = 2 – y $\Rightarrow$ 8(2 – y) = 56 $\Rightarrow$ y = –5 $\Rightarrow$ C = (9,–5). m(DC) = 7 $\Rightarrow$ m(AD) = 21. Let A = (x,2) $\Rightarrow$ m(AD) = 9 – x = 21 $\Rightarrow$ x = –12 $\Rightarrow$ A = (–12,2) $\Rightarrow$ B = (–12,–5).

41. Let A = (2,0), B = (0,b), C = (1,1), D = (1,0), O = (0,0). Then $\Delta ADC \approx \Delta AOB \Rightarrow \frac{m(AD)}{m(CD)} = \frac{m(AO)}{m(BO)}$. m(AD) = 1; m(CD) = 1, m(AO) = 2; m(BO) = b $\Rightarrow$ $\frac{1}{1} = \frac{2}{b} \Rightarrow b = 2$.

43. Yes

## 1.2 SLOPE, AND EQUATIONS FOR LINES

1. $\Delta x = 1 - (-1) = 2$

   $\Delta y = 2 - 1 = 1$

3. $\Delta x = -1 - (-3) = 2$

   $\Delta y = -2 - 2 = -4$

5. $\Delta x = -8 - (-3) = -5$

   $\Delta y = 1 - 1 = 0$

7. $m = \frac{\Delta y}{\Delta x} = \frac{1-(-2)}{2-1} = 3$

   $m_{\perp} = -\frac{1}{3}$

9. $m = \frac{\Delta y}{\Delta x} = \frac{-2-(-1)}{1-(-2)} = -\frac{1}{3}$

   $m_{\perp} = 3$

11. $m = \frac{\Delta y}{\Delta x} = \frac{3-3}{-1-2} = 0$

    $m_{\perp}$ does not exist

13. a) $x = 2$

    b) $y = 3$

15. a) $x = -7$

    b) $y = -2$

17. a) $x = 0$

    b) $y = -\sqrt{2}$

19. $P(1,1),\ m = 1 \Rightarrow y - 1 = 1(x-1) \Rightarrow y = x$

21. $P(-1,1),\ m = 1 \Rightarrow y - 1 = 1(x-(-1)) \Rightarrow y = x + 2$

23. $P(0,b),\ m = 2 \Rightarrow y - b = 2(x-0) \Rightarrow y = 2x + b$

25. $(0,0),\ (2,3) \Rightarrow m = \frac{3}{2} \Rightarrow y - 0 = \frac{3}{2}(x-0) \Rightarrow y = \frac{3}{2}x$

27. $(1,1),\ (1,2) \Rightarrow$ m does not exist $\Rightarrow$ Vertical line $\Rightarrow x = 1$

29. $(-2,0),\ (-2,-2) \Rightarrow$ m does not exist $\Rightarrow$ Vertical line $\Rightarrow x = -2$

31. $y = 3x - 2$

33. $y = x + \sqrt{2}$

35. $y = -5x + 2.5$

37. x–intercept = 4

    y–intercept = 3

39. x–intercept = 3

    y–intercept = –4

41. x–intercept = –2

    y–intercept = 4

43. $P(0,0)$, $L: y = -x + 2 \Rightarrow m_L = -1 \Rightarrow m_{\perp} = 1 \Rightarrow y - 0 = 1(x-0) \Rightarrow y = x$ is $\perp$ line. $y = x$ and $y = -x + 2 \Rightarrow$

    $-x + 2 = x \Rightarrow x = 1 \Rightarrow y = 1 \Rightarrow$ Point of intersection, Q, is $(1,1) \Rightarrow d(P,Q) = \sqrt{(1-0)^2 + (1-0)^2} = \sqrt{2}$

45. $P(1,2)$, $L: x + 2y = 3 \Rightarrow L: y = -\frac{1}{2}x + \frac{3}{2} \Rightarrow m_L = -\frac{1}{2} \Rightarrow m_{\perp} = 2 \Rightarrow y - 2 = 2(x-1) \Rightarrow y = 2x$ is $\perp$ line.

    $x + 2y = 3$ and $y = 2x \Rightarrow x + 2(2x) = 3 \Rightarrow x = \frac{3}{5} \Rightarrow y = \frac{6}{5} \Rightarrow$ Point of intersection, Q, is $\left(\frac{3}{5}, \frac{6}{5}\right) \Rightarrow d(P,Q) =$

    $\sqrt{\left(\frac{3}{5} - 1\right)^2 + \left(\frac{6}{5} - 2\right)^2} = \frac{2\sqrt{5}}{5}$

47. $P(3,6)$, $L: x + y = 3 \Rightarrow y = -x + 3 \Rightarrow m_L = -1 \Rightarrow m_\perp = 1 \Rightarrow y - 6 = 1(x - 3) \Rightarrow y = x + 3$.

$y = x + 3$ and $x + y = 3 \Rightarrow x + x + 3 = 3 \Rightarrow x = 0 \Rightarrow y = 3 \Rightarrow$ Point of intersection, Q, is $(0,3) \Rightarrow d(P,Q)$ $= \sqrt{(0-3)^2 + (3-6)^2} = 3\sqrt{2}$

49. $P(2,1)$, $L: y = x + 2 \Rightarrow m_L = 1 \Rightarrow //$ line is $y - 1 = 1(x - 2) \Rightarrow y = x - 1$

51. $P(1,0)$, $L: 2x + y = -2 \Rightarrow m_L = -2 \Rightarrow //$ line is $y - 0 = -2(x - 1) \Rightarrow y = -2x + 2$

53. $A(-2,3)$, $\Delta x = 5$, $\Delta y = -6 \Rightarrow x_2 = 5 + (-2) = 3$ and $y_2 = -6 + 3 = -3 \Rightarrow (x_2, y_2) = (3,-3)$

55. $\Delta x = 5$, $\Delta y = 6$, $B(3,-3)$ Let $A = (x,y)$ Then $\Delta x = x_2 - x_1 \Rightarrow 5 = 3 - x \Rightarrow x = -2$. $\Delta y = y_2 - y_1 \Rightarrow$

$6 = -3 - y \Rightarrow y = -9 \therefore A = (-2,-9)$

57. a) $A \approx (69°, 0 \text{ in})$, $B \approx (68°, .4 \text{ in}) \Rightarrow m = \dfrac{68° - 69°}{.4 - 0} = -2.5°/\text{in}$

b) $A \approx (68°, .4 \text{ in})$, $B \approx (10°, 4 \text{ in}) \Rightarrow m = \dfrac{10° - 68°}{4 - .4} = -16.1°/\text{in}$

c) $A \approx (10°, 4 \text{ in})$, $B \approx (5°, 4.6 \text{ in}) \Rightarrow m = \dfrac{5° - 10°}{4.6 - 4} = -8.3°/\text{in}$

59. $p = kd + 1$ $(d_1, p_1) = (0,1)$, $(d_2, p_2) = (100, 10.94) \Rightarrow \dfrac{\Delta p}{\Delta d} = \dfrac{10.94 - 1}{100 - 0} = .0944$ atm/m. If $d = 50$, then $\dfrac{\Delta p}{\Delta d}$ $= \dfrac{p - 1}{50 - 0} = .0944 \Rightarrow p - 1 = 50(.0944) \Rightarrow p = 5.97$ atm

61. $F = \frac{9}{5}C + 32$. Let $F = C$. Then $C = \frac{9}{5}C + 32 \Rightarrow C = -40° = F$

63. Let $A = (-1,1)$, $B = (2,3)$, $C = (2,0)$. Since BC is vertical and $m(BC) = 3$, let $AD_1$ be vertical (upward) $\Rightarrow D_1 = (-1,4)$. Let $AD_2$ be vertical (downward) $\Rightarrow D_2 = (-1,-2)$. Let $D_3 = (x,y)$. m of AB = m of $CD_3 \Rightarrow$ $\dfrac{y-3}{x-2} = -\dfrac{1}{3}$ and m of AC = m of $BD_3 \Rightarrow \dfrac{y-0}{x-2} = \dfrac{2}{3}$ .Solve the system of equations to get $x = 5$, $y = 2 \Rightarrow$ $D_3 = (5,2)$

65. $2x + ky = 3 \Rightarrow ky = -2x + 3 \Rightarrow y = -\frac{2}{k}x + \frac{3}{k} \Rightarrow m = -\frac{2}{k}$. $x + y = 1 \Rightarrow y = -x + 1 \Rightarrow m = -1$.

$\therefore m_{//} = -1 \Rightarrow -\frac{2}{k} = -1 \Rightarrow k = 2$. $m_\perp = 1 \Rightarrow -\frac{2}{k} = 1 \Rightarrow k = -2$.

## SECTION 1.3 FUNCTIONS AND THEIR GRAPHS

1. D: $x \geq 0$

   R: $y \geq 0$

3. D: $x \geq 0$

   R: $y \leq 0$

5. D: $x + 4 \geq 0 \Rightarrow x \geq -4$

   R: $y \geq 0$

7. D: $x - 2 \neq 0 \Rightarrow x \neq 2$

   R: $y \neq 0$

9. Odd

11. Odd

13. Neither

15. Even

17. Even

19. Odd

21. D: $-\infty < x < \infty$

    R: $y \geq 0$

    Symmetric to y axis

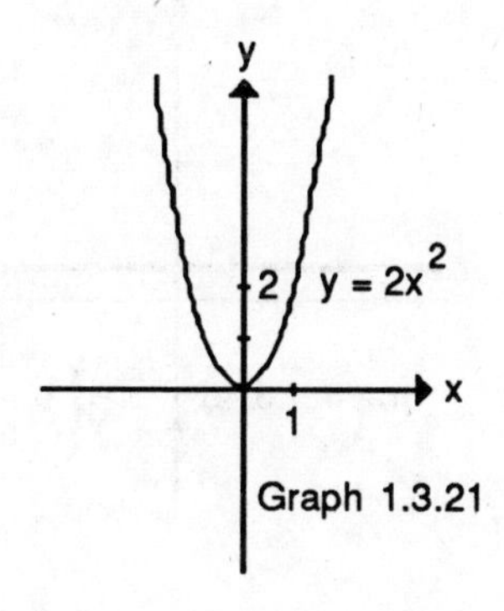


Graph 1.3.21

23. D: $-\infty < x < \infty$

    R: $y \geq -9$

    Symmetric to y axis

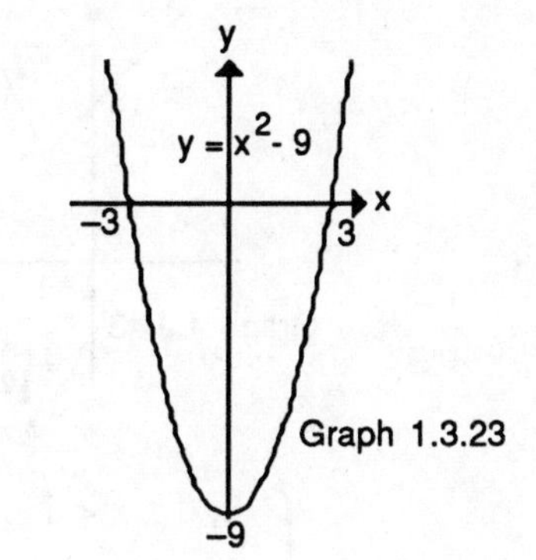


Graph 1.3.23

25. D: $-\infty < x < \infty$

    R: $-\infty < y < \infty$

    Symmetric to origin

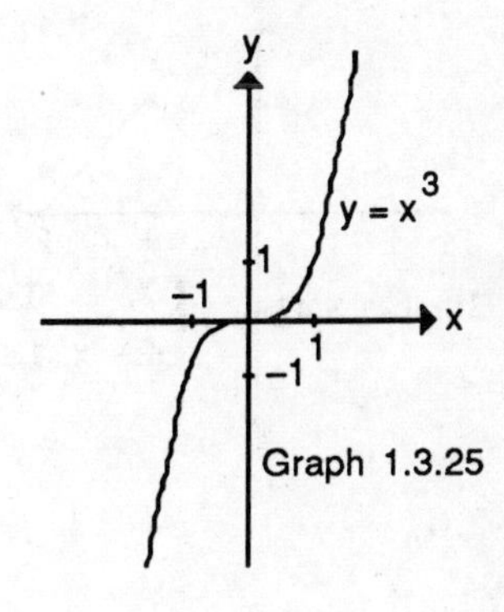


Graph 1.3.25

27. D: $x \geq -1$

    R: $y \geq 0$

    No symmetry

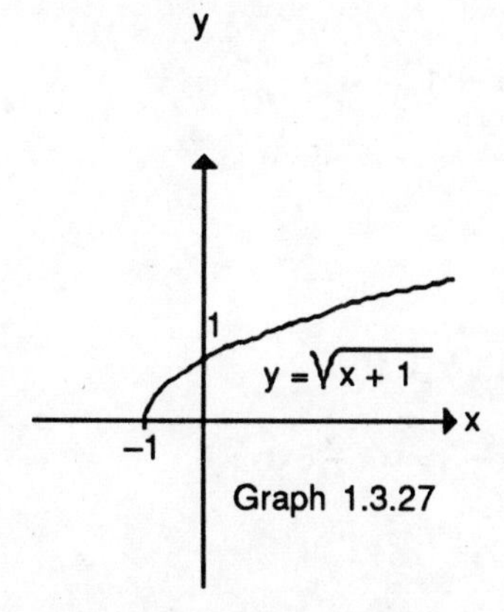


Graph 1.3.27

29. D: $x \neq 0$

    R: $y \neq 0$

    Symmetric to origin

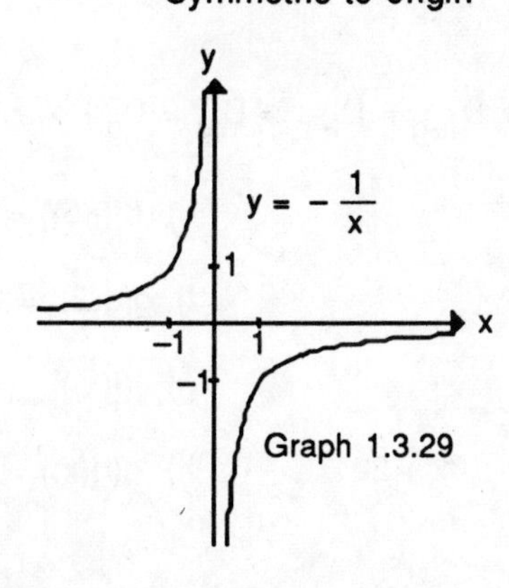


Graph 1.3.29

31. D: $x \neq 0$

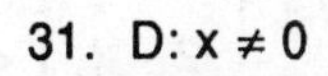

    R: $y \neq 1$

    No symmetry

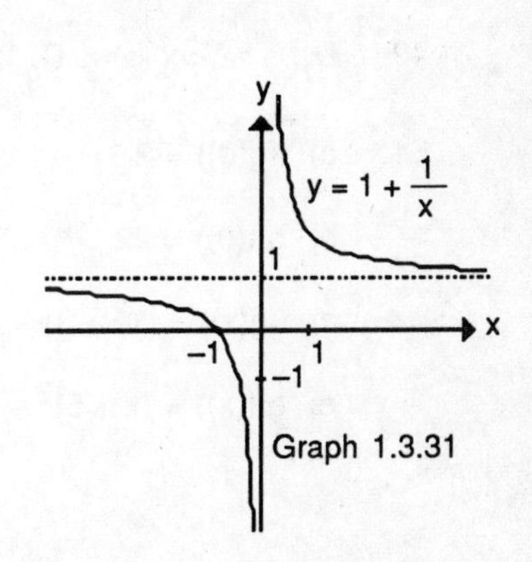


Graph 1.3.31

33. a) No. $\sqrt{x}$ not a Real Number if $x < 0$ b) No. Division by 0 is undefined c) D: $x > 0$

35. a) i, $y = x^2 - 1$ symmetric to y axis and R: $y \geq -1$ b) iv, R: $y \geq 0$ and $x = 1 \Rightarrow y = 0$

37. a) $0 \leq x < 1$ b) $-1 < x \leq 0$

39. a)

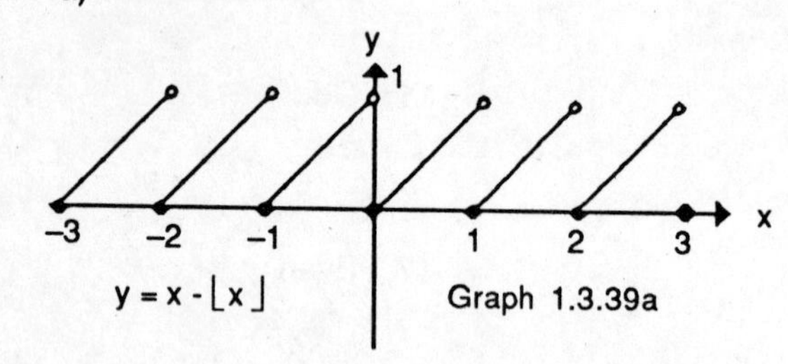


39. b)

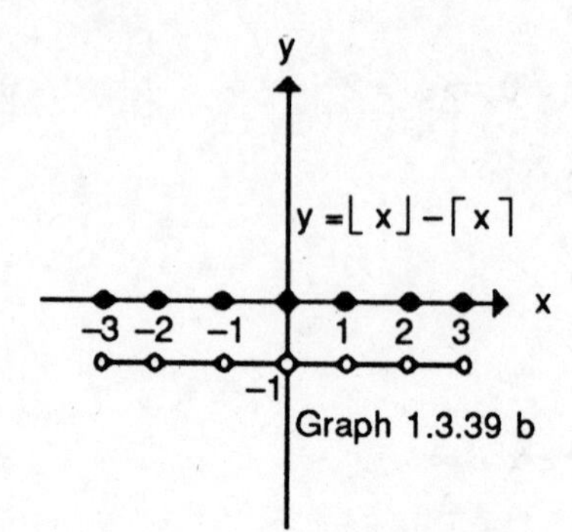


41. $y = \begin{cases} x, & 0 \leq x \leq 1 \\ 2 - x, & 1 \leq x \leq 2 \end{cases}$

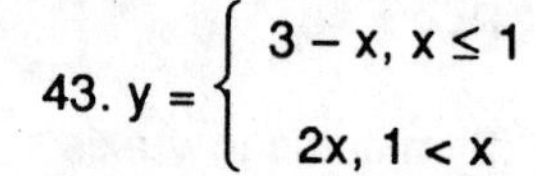

43. $y = \begin{cases} 3 - x, & x \leq 1 \\ 2x, & 1 < x \end{cases}$

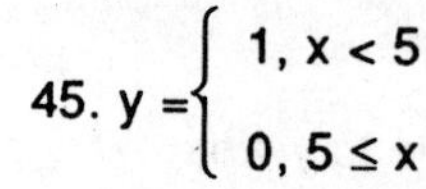

45. $y = \begin{cases} 1, & x < 5 \\ 0, & 5 \leq x \end{cases}$

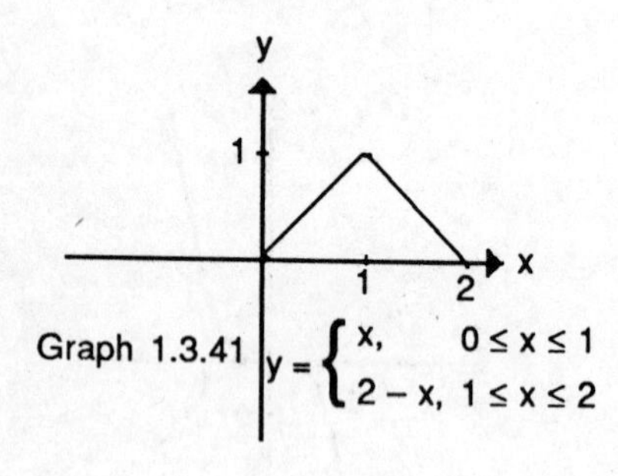


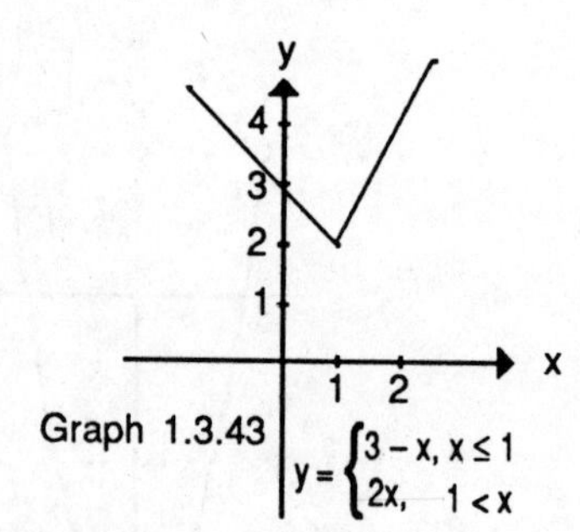


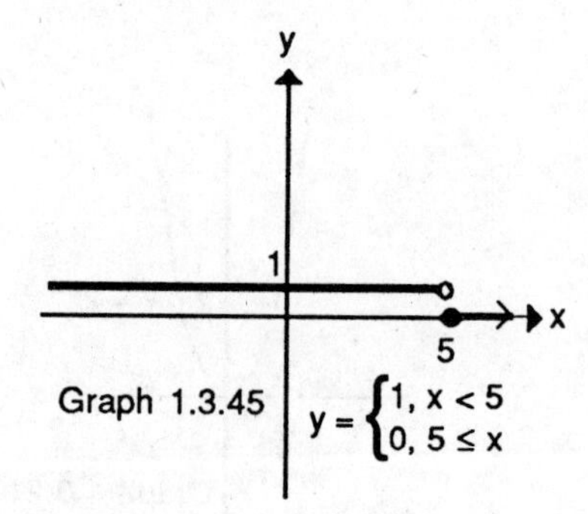


47. a) $y = \begin{cases} x, & 0 \leq x \leq 1 \\ 2 - x, & 1 < x \leq 2 \end{cases}$ b) $y = \begin{cases} 2, & 0 \leq x < 1 \\ 0, & 1 \leq x < 2 \\ 2, & 2 \leq x < 3 \\ 0, & 3 \leq x \leq 4 \end{cases}$

49. $D_f: -\infty < x < \infty$, $D_g: x \geq 1 \Rightarrow D_{f+g} = D_{f-g} = D_{fg} = D_{g/f}: x \geq 1$; $D_{f/g}: x > 1$

51. a) $f(g(0)) = 2$

b) $g(f(0)) = 22$

c) $f(g(x)) = (x^2 - 3) = x^2 + 2$

d) $g(f(x)) = (x + 5)^2 - 3 = x^2 + 10x + 22$

e) $f(f(-5)) = 5$

f) $g(g(2)) = -2$

g) $f(f(x)) = (x + 5) + 5 = x + 10$

h) $g(g(x)) = (x^2 - 3)^2 - 3 = x^4 - 6x^2 + 6$

53.

| | $g(x)$ | $f(x)$ | $(f \circ g)(x)$ |
|---|---|---|---|
| a) | $x - 7$ | $\sqrt{x}$ | $\sqrt{x - 7}$ |
| b) | $x + 2$ | $3x$ | $3(x + 2) = 3x + 6$ |
| c) | $x^2$ | $\sqrt{x - 5}$ | $\sqrt{x^2 - 5}$ |
| d) | $\frac{x}{x - 1}$ | $\frac{x}{x - 1}$ | $\frac{\frac{x}{x-1}}{\frac{x}{x-1} - 1} = \frac{x}{x - (x - 1)} = x$ |
| e) | $\frac{1}{x - 1}$ | $1 + \frac{1}{x}$ | $1 + \frac{1}{\frac{1}{x-1}} = 1 + (x - 1) = x$ |
| f) | $\frac{1}{x}$ | $\frac{1}{x}$ | $\frac{1}{\frac{1}{x}} = x$ |

55. Don't look it up in here! Try it for yourself.

## SECTION 1.4 SHIFTS, CIRCLES, AND PARABOLAS

1. a) $y = (x + 4)^2$

   b) $y = (x - 7)^2$

3. a) Position 4

   b) Position 1

   c) Position 2

   d) Position 3

5. $C(0,2)$, $a = 2 \Rightarrow x^2 + (y - 2)^2 = 4$

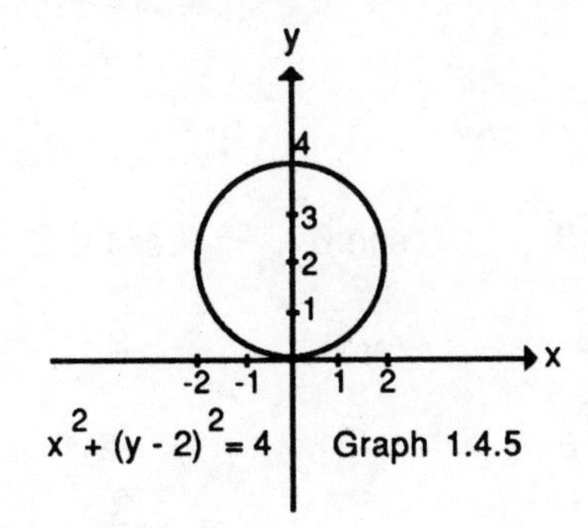

Graph 1.4.5

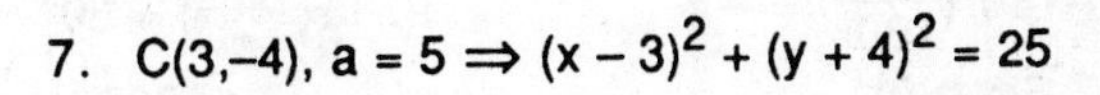

7. $C(3,-4)$, $a = 5 \Rightarrow (x - 3)^2 + (y + 4)^2 = 25$

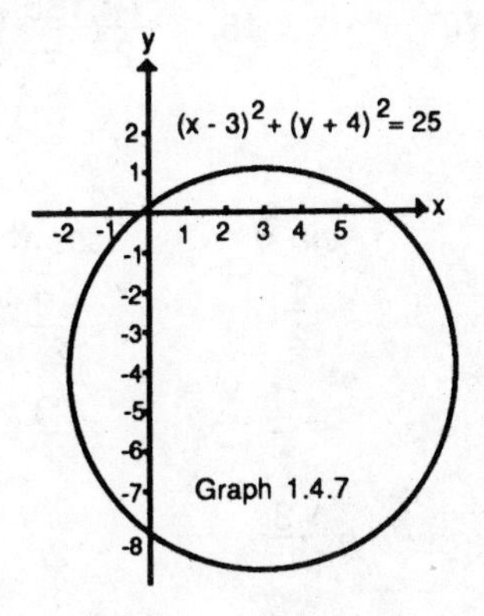

Graph 1.4.7

9. $x^2 + y^2 = 4$

11. $(x - 3)^2 + (y - 3)^2 = 9$

13. a) Exterior of circle with center (0,0) and radius = 1

    b) Interior of circle with center (0,0) and radius = 2

    c) Interior of concentric ring centered at (0,0) with interior radius = 1 and exterior radius = 2

15. $(x+2)^2 + (y+1)^2 < 6$

17. F(0,4), Directrix: $y = -4 \Rightarrow$
$y = \frac{x^2}{4(4)} \Rightarrow y = \frac{1}{16}x^2$

19. F(0,–3), Directrix: $y = 3 \Rightarrow$
$y = -\frac{x^2}{4(3)} \Rightarrow y = -\frac{1}{12}x^2$

21. $\frac{1}{4p} = 4 \Rightarrow p = \frac{1}{16} \Rightarrow$
$F(0,\frac{1}{16})$, Directrix: $y = -\frac{1}{16}$

23. $\frac{1}{4p} = -3 \Rightarrow p = -\frac{1}{12} \Rightarrow$
$F(0,-\frac{1}{12})$, Directrix: $y = \frac{1}{12}$

25. $y + 3 = (x+2)^2$

27. $y = x^2$

29. $y = \sqrt{x+4}$

31. $y = \frac{1}{2}x$

33. $y = \sqrt{-(x-9)}$

35. $y - y_0 = m(x - x_0)$

37. $y = x^2$, $y = 2 - x^2 \Rightarrow x^2 = 2 - x^2 \Rightarrow x^2 = 1 \Rightarrow x = \pm 1$. $x = 1 \Rightarrow y = 1$, $x = -1 \Rightarrow y = 1 \therefore$ (1,1) and (–1,1)

39. $x = y^2$, $x - y = 2 \Rightarrow y^2 - y = 2 \Rightarrow y^2 - y - 2 = 0 \Rightarrow y = 2$ or $y = -1$. $y = 2 \Rightarrow x = 4$, $y = -1 \Rightarrow x = 1 \Rightarrow$
(4,2) and (1,–1)

## SECTION 1.5 A REVIEW OF TRIGONOMETRIC FUNCTIONS

1. $120°(\frac{\pi}{180°}) = \frac{2\pi}{3}$

3. $270°(\frac{\pi}{180°}) = \frac{3\pi}{2}$

5. $-45°(\frac{\pi}{180°}) = -\frac{\pi}{4}$

7. $-180°(\frac{\pi}{180°}) = -\pi$

9. $\frac{\pi}{6}(\frac{180°}{\pi}) = 30°$

11. $\frac{5\pi}{4}(\frac{180°}{\pi}) = 225°$

13. $-\frac{3\pi}{2}(\frac{180°}{\pi}) = -270°$

15. $-\frac{7\pi}{2}(\frac{180°}{\pi}) = -630°$

| | $\theta$ | $\sin\theta$ | $\cos\theta$ | $\tan\theta$ | $\cot\theta$ | $\sec\theta$ | $\csc\theta$ |
|---|---|---|---|---|---|---|---|
| 17. a) | $\frac{\pi}{6}$ | $\frac{1}{2}$ | $\frac{\sqrt{3}}{2}$ | $\frac{1}{\sqrt{3}}$ | $\sqrt{3}$ | $\frac{2}{\sqrt{3}}$ | 2 |
| b) | $-\frac{\pi}{6}$ | $-\frac{1}{2}$ | $\frac{\sqrt{3}}{2}$ | $-\frac{1}{\sqrt{3}}$ | $-\sqrt{3}$ | $\frac{2}{\sqrt{3}}$ | –2 |
| 19. a) | $\frac{\pi}{3}$ | $\frac{\sqrt{3}}{2}$ | $\frac{1}{2}$ | $\sqrt{3}$ | $\frac{1}{\sqrt{3}}$ | 2 | $\frac{2}{\sqrt{3}}$ |
| b) | $-\frac{\pi}{3}$ | $-\frac{\sqrt{3}}{2}$ | $\frac{1}{2}$ | $-\sqrt{3}$ | $-\frac{1}{\sqrt{3}}$ | 2 | $-\frac{2}{\sqrt{3}}$ |
| 21. a) | $\frac{3\pi}{4}$ | $\frac{1}{\sqrt{2}}$ | $-\frac{1}{\sqrt{2}}$ | –1 | –1 | $-\sqrt{2}$ | $\sqrt{2}$ |
| b) | $-\frac{3\pi}{4}$ | $-\frac{1}{\sqrt{2}}$ | $-\frac{1}{\sqrt{2}}$ | 1 | 1 | $-\sqrt{2}$ | $-\sqrt{2}$ |
| 23. a) | 0 | 0 | 1 | 0 | undefined | 1 | undefined |
| b) | $\pi$ | 0 | –1 | 0 | undefined | –1 | undefined |

25.

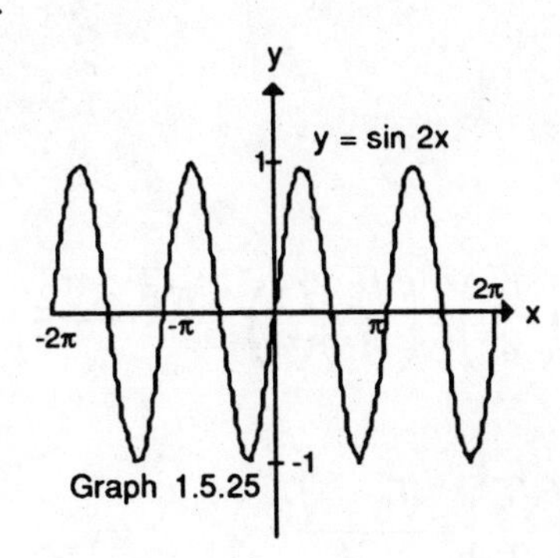

Graph 1.5.25

27.

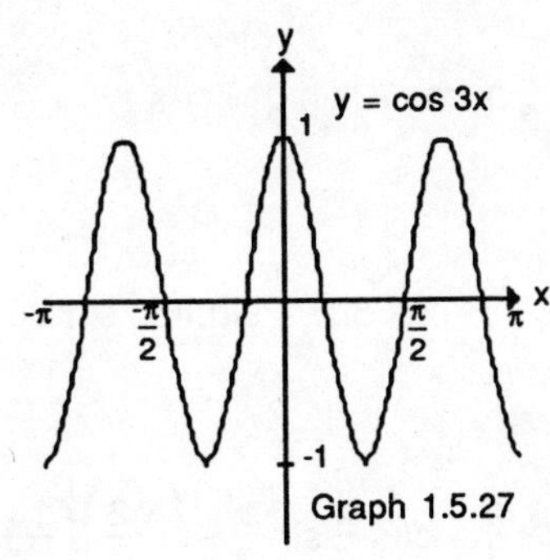

Graph 1.5.27

29.

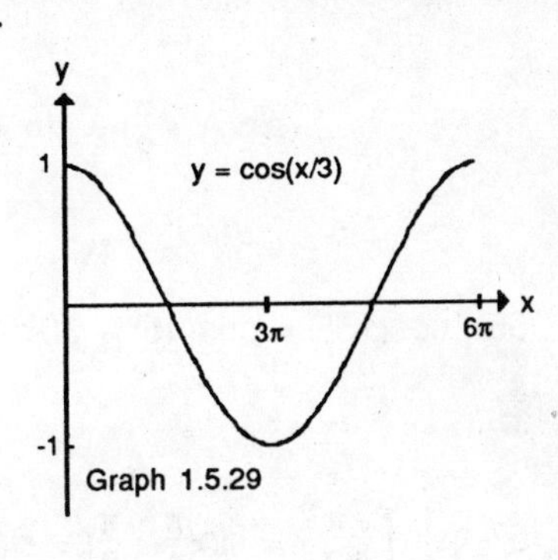

Graph 1.5.29

31.

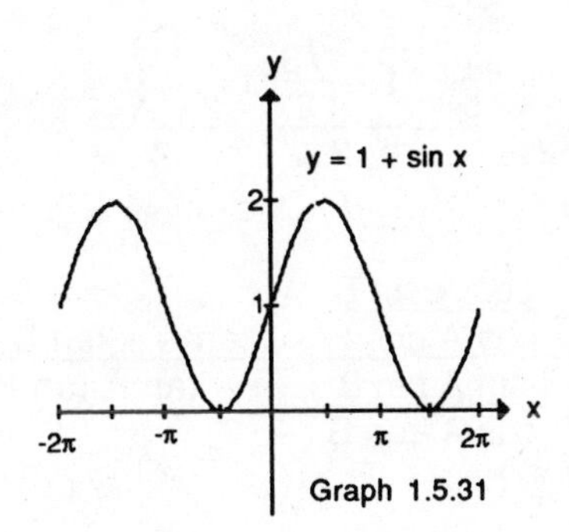

Graph 1.5.31

33.

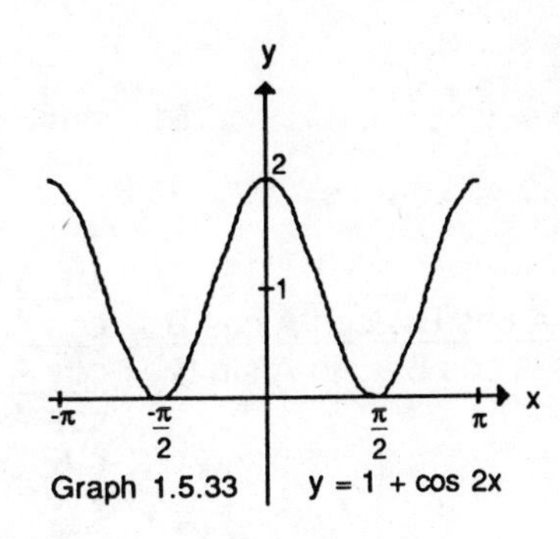

Graph 1.5.33

35.

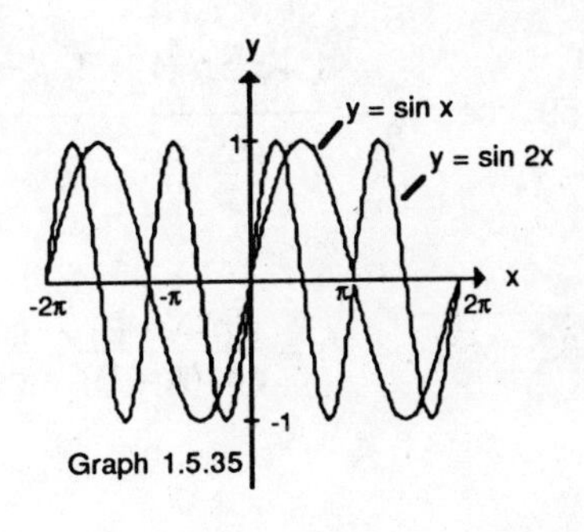

Graph 1.5.35

37.

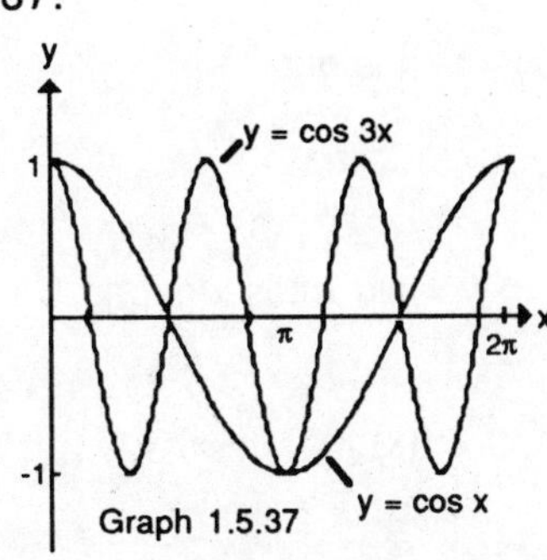

Graph 1.5.37

39.

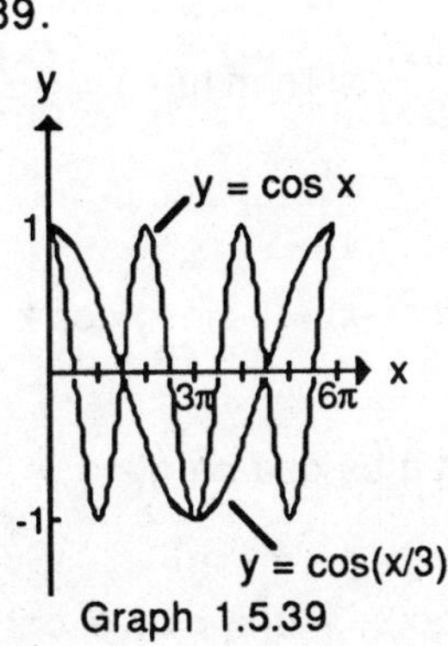

Graph 1.5.39

41.

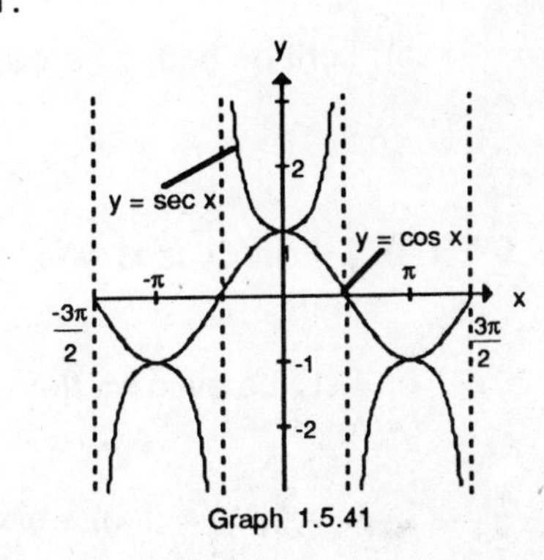

Graph 1.5.41

43. a,j ; b,d,g ; e,c,i,m,n ; f,k,l,h

45. Cosine: Symmetric to y axis only; Sine: Symmetric to origin only; Tangent: Symmetric to origin only

47. a) Yes, since $-1 \le \cos 2x \le 1 \Rightarrow 0 \le 1 + \cos 2x \le 2 \Rightarrow 0 \le \frac{1 + \cos 2x}{2} \le 1$ b) 1 and –1

c) 1 and 0 d) D: $-\infty < x < \infty$ R: $0 \le y \le 1$

49. a) 37 b) Period: $\frac{2\pi}{\frac{2\pi}{365}} = 365$ c) Horizontal Shift: Right 101 d) Vertical Shift: Up 25

51. If A = B, $\cos(0) = \cos A \cos A + \sin A \sin A = \cos^2 A + \sin^2 A = 1$

53. If $B = \frac{\pi}{2}$, $\cos(A + \frac{\pi}{2}) = \cos A \cos \frac{\pi}{2} - \sin A \sin \frac{\pi}{2} = -\sin A$

$$\sin(A + \tfrac{\pi}{2}) = \sin A \cos \tfrac{\pi}{2} + \cos A \sin \tfrac{\pi}{2} = \cos A$$

55. $\cos(15°) = \cos(45° - 30°) = \cos 45° \cos 30° + \sin 45° \sin 30° = \left(\frac{\sqrt{2}}{2}\right)\left(\frac{\sqrt{3}}{2}\right) + \left(\frac{\sqrt{2}}{2}\right)\left(\frac{1}{2}\right) = \frac{\sqrt{6} + \sqrt{2}}{4}$

57. $\sin\left(\frac{7\pi}{12}\right) = \sin(\frac{\pi}{4} + \frac{\pi}{3}) = \sin \frac{\pi}{4} \cos \frac{\pi}{3} + \cos \frac{\pi}{4} \sin \frac{\pi}{3} = \left(\frac{\sqrt{2}}{2}\right)\left(\frac{1}{2}\right) + \left(\frac{\sqrt{2}}{2}\right)\left(\frac{\sqrt{3}}{2}\right) = \frac{\sqrt{6} + \sqrt{2}}{4}$

59. $\cos^2 \frac{\pi}{8} = \frac{1 + \cos(\frac{\pi}{4})}{2} = \frac{1 + \frac{\sqrt{2}}{2}}{2} = \frac{2 + \sqrt{2}}{4}$

61. $\sin^2 \frac{\pi}{12} = \frac{1 - \cos \frac{\pi}{6}}{2} = \frac{1 - \frac{\sqrt{3}}{2}}{2} = \frac{2 - \sqrt{3}}{4}$

63. $\tan(A + B) = \frac{\sin(A + B)}{\cos(A + B)} = \frac{\sin A \cos B + \cos A \cos B}{\cos A \cos B - \sin A \sin B} = \frac{\frac{\sin A \cos B}{\cos A \cos B} + \frac{\cos A \sin B}{\cos A \cos B}}{\frac{\cos A \cos B}{\cos A \cos B} - \frac{\sin A \sin B}{\cos A \cos B}} = \frac{\tan A + \tan B}{1 - \tan A \tan B}$

65. a) $\cot x = \frac{\cos x}{\sin x}$ $\therefore \cot(-x) = \frac{\cos(-x)}{\sin(-x)} = \frac{\cos x}{-\sin x} = -\cot x \Rightarrow f(x) = \cot x$ is odd

b) Let f be odd, g be even, $h(x) = \frac{f(x)}{g(x)}$. Then $h(-x) = \frac{f(-x)}{g(-x)} = \frac{-f(x)}{g(x)} = -h(x) \Rightarrow$ h is odd.

67. a) $y = \sin x \cos x$: $y(-x) = \sin(-x) \cos(-x) = (-\sin x)(\cos x) = -\sin x \cos x \Rightarrow y = \sin x \cos x$ is odd

b) Let f be even $\Rightarrow f(-x) = f(x)$. Let g be odd $\Rightarrow g(-x) = -g(x)$. Then, if $h(x) = f(x)g(x)$, $h(-x) =$

$f(-x)g(-x) = (f(x))(-g(x)) = -f(x)g(x) = -h(x) \Rightarrow$ h is odd.

## SECTION 1.6 ABSOLUTE VALUE OR MAGNITUDE, AND TARGET VALUES

1. a) False

   b) True

   c) True

   d) True

   e) True $2 < x < 6 \Rightarrow \frac{1}{3} < \frac{x}{6} < 1 \Rightarrow 3 > \frac{6}{x} > 1$

   f) True $|x - 4| < 2 \Rightarrow -2 < x - 4 < 2 \Rightarrow 2 < x < 6$

   g) True $2 < x < 6 \Rightarrow -2 > -x > -6 \Rightarrow -6 < -x < -2 \Rightarrow -6 < -x < 2$

   h) True

3. 3

5. 5

7. $x = \pm 2$

9. $|2x + 5| = 4 \Rightarrow$

   $2x + 5 = 4$ or $2x + 5 = -4$

   $\Rightarrow x = -\frac{1}{2}$ or $x = -\frac{9}{2}$

11. $|8 - 3x| = 9 \Rightarrow$

    $8 - 3x = 9$ or $8 - 3x = -9 \Rightarrow$

    $x = -\frac{1}{3}$ or $x = \frac{17}{3}$

13. $|x + 3| < 1 \Rightarrow$

    $-1 < x + 3 < 1$

    $\Rightarrow -4 < x < -2 \Rightarrow$ g

15. $\left|\frac{x}{2}\right| < 1 \Rightarrow$

    $-1 < \frac{x}{2} < 1 \Rightarrow$

    $-2 < x < 2 \Rightarrow$ e

17. $|2x - 5| \le 1 \Rightarrow$

    $-1 \le 2x - 5 \le 1 \Rightarrow$

    $2 \le x \le 3 \Rightarrow$ h

19. $\left|\frac{x-1}{2}\right| < 1 \Rightarrow$

    $-1 < \frac{x-1}{2} < 1 \Rightarrow$

    $-1 < x < 3 \Rightarrow$ b

21. $-2 < y < 2$

23. $|y - 1| \le 2 \Rightarrow$

    $-2 \le y - 1 \le 2 \Rightarrow$

    $-1 \le y \le 3$

25. $|3y - 7| < 2 \Rightarrow$

    $-2 < 3y - 7 < 2 \Rightarrow$

    $\frac{5}{3} < y < 3$

27. $\left|\frac{y}{3}\right| \le 10 \Rightarrow$

    $-10 \le \frac{y}{3} \le 10 \Rightarrow$

    $-30 \le y \le 30$

29. $\left|\frac{y}{2} - 1\right| \le 1 \Rightarrow$

    $-1 \le \frac{y}{2} - 1 \le 1 \Rightarrow$

    $0 \le y \le 4$

31. $|1 - y| < \frac{1}{10} \Rightarrow$

    $-\frac{1}{10} < 1 - y < \frac{1}{10} \Rightarrow$

    $\frac{11}{10} > y > \frac{9}{10}$

33. $3 < x < 9 \Rightarrow$

    $-3 < x - 6 < 3 \Rightarrow$

    $|x - 6| < 3$

35. $-5 < x < 3 \Rightarrow$

    $-4 < x + 1 < 4 \Rightarrow$

    $|x + 1| < 4$

37. $|x^2 - 100| < 0.1 \Rightarrow$

    $-0.1 < x^2 - 100 < 0.1 \Rightarrow$

    $99.9 < x^2 < 100.1 \Rightarrow$

    $\sqrt{99.9} < x < \sqrt{100.1}$

39. $\left|\sqrt{x-7} - 4\right| < 0.1 \Rightarrow$

    $-0.1 < \sqrt{x-7} - 4 < 0.1 \Rightarrow$

    $3.9 < \sqrt{x-7} < 4.1 \Rightarrow$

    $15.21 < x - 7 < 16.81 \Rightarrow$

    $15.21 < x - 7 < 16.81 \Rightarrow$

    $22.21 < x < 23.81$

41. $\left|\frac{120}{x} - 5\right| < 1 \Rightarrow$

$-1 < \frac{120}{x} - 5 < 1 \Rightarrow$

$4 < \frac{120}{x} < 6 \Rightarrow$

$\frac{1}{4} > \frac{x}{120} > \frac{1}{6} \Rightarrow$

$30 > x > 20$

43. $|(x + 1) - 4| < 0.5 \Rightarrow$

$|x - 3| < 0.5$

45. $\left|\left(-\frac{x}{2} + 1\right) - (-2)\right| < \frac{1}{2} \Rightarrow$

$\left|-\frac{x}{2} + 3\right| < \frac{1}{2} \Rightarrow -\frac{1}{2} < -\frac{x}{2} + 3 < \frac{1}{2} \Rightarrow$

$-\frac{7}{2} < -\frac{x}{2} < -\frac{5}{2} \Rightarrow -7 < -x < -5 \Rightarrow$

$7 > x > 5 \Rightarrow 1 > x - 6 > -1 \Rightarrow |x - 6| < 1$

47. $|A - 9| \le 0.01 \Rightarrow \left|\pi\left(\frac{x}{2}\right)^2 - 9\right| \le 0.01 \Rightarrow$

$-0.01 \le \pi\left(\frac{x}{2}\right)^2 - 9 \le 0.01 \Rightarrow 8.99 \le \pi\left(\frac{x}{2}\right)^2 \le 9.01 \Rightarrow$

$\frac{8.99}{\pi} \le \left(\frac{x}{2}\right)^2 \le \frac{9.01}{\pi} \Rightarrow \sqrt{\frac{8.99}{\pi}} \le \frac{x}{2} \le \sqrt{\frac{9.01}{\pi}} \Rightarrow$

$2\sqrt{\frac{8.99}{\pi}} \le x \le 2\sqrt{\frac{9.01}{\pi}} \Rightarrow 3.383 \le x \le 3.387 \Rightarrow$

$-0.002 \le x - 3.385 \le 0.002 \Rightarrow |x - 3.385| \le 0.002$

49. a) iii b) iv c) i d) ii

51.

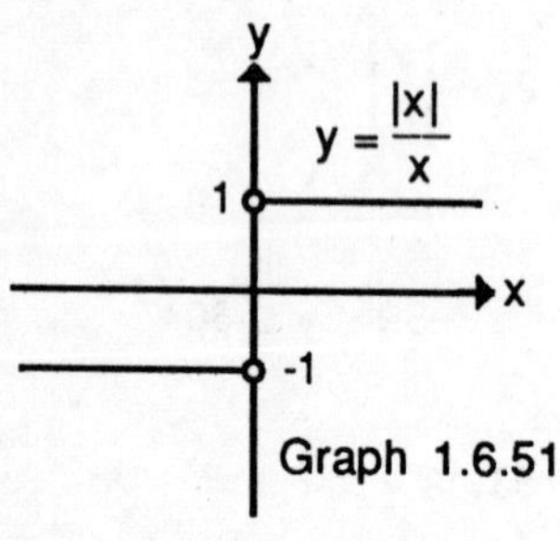


Graph 1.6.51

53.

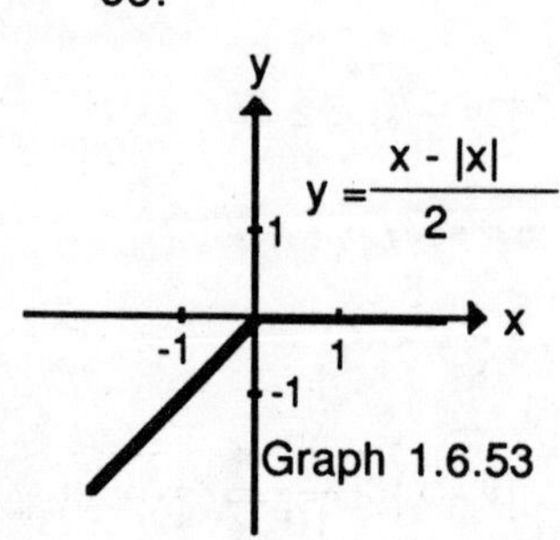


Graph 1.6.53

55. a) a any negative Real

b) $a \ge 0$

57. $y = \sqrt{x^2} \Rightarrow$ D: $-\infty < x < \infty$ and

R: $y \ge 0$

$y = \left(\sqrt{x}\right)^2 \Rightarrow$ D: $x \ge 0$ and R: $y \ge 0$

59. $f(x) = x^2 + 2x = 1$ and $(g \circ f)(x) = |x + 1| \Rightarrow (g \circ f)(x) = \sqrt{(x + 1)^2} = \sqrt{x^2 + 2x + 1} = \sqrt{f(x)} \Rightarrow g(x) = \sqrt{x}$

61. Use $x < -1$, $-1 \le x < 3$, and $x \ge 3$

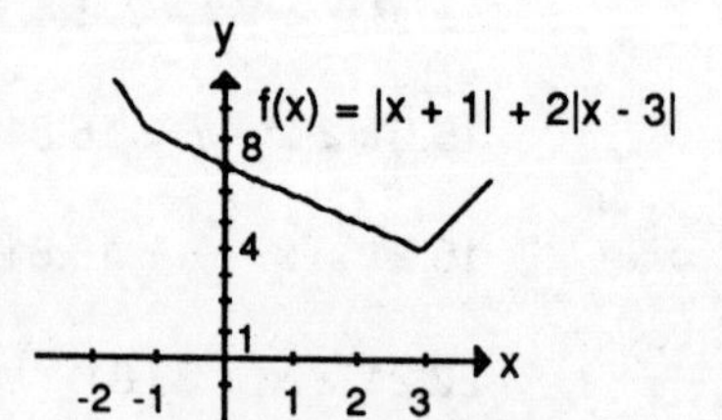


Graph 1.6.61

63. Use $x < 0$, $0 \le x < 1$, $1 \le x < 3$, and $x \ge 3$

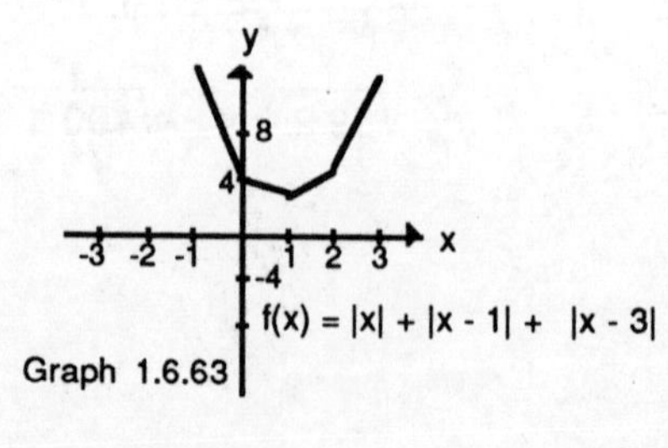


Graph 1.6.63

65. Use $x < -3$, $-3 \le x < 2$, $2 \le x < 4$, $2 \le x < 4$, and $x \ge 4$.
d(x) at minimum when $x = 2$

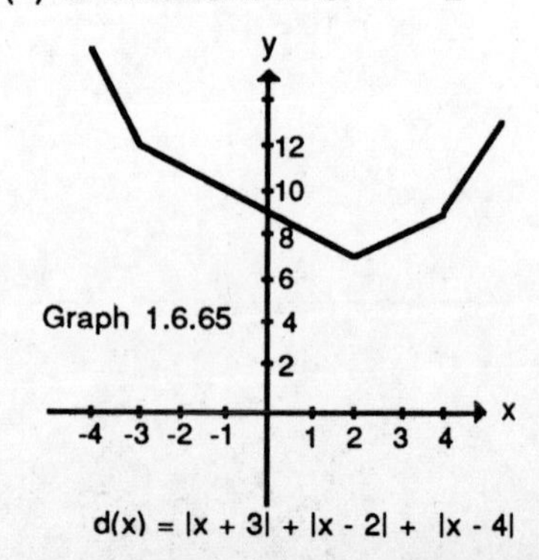


Graph 1.6.65

67. Use $x < -3$, $-3 \le x < -1$, $-1 \le x < 2$, $2 \le x < 6$, $6 \le x$. When $x < -3$, $d(x) = -7x + 5 \Rightarrow$ minimum value is close to 26 when $x \approx -3$; when $-3 \le x < -1$, $d(x) = -3x + 17 \Rightarrow$ minimum value is close to 20 when $x \approx -1$; when $-1 \le x < 2$, $d(x) = -x + 19 \Rightarrow$ minimum value is close to 17 when $x \approx 2$; when $2 \le x < 6$, $d(x) = 5x + 7 \Rightarrow$ minimum value is 17 when $x = 2$; when $x \ge 6$, $d(x) = 7x - 5 \Rightarrow$ minimum value is 37 when $x = 6$. $\therefore$ minimum value is 17 when $x = 2$, place the table at 2.

## SECTION 1.7 CALCULATORS AND CALCULUS

1. a) Answers vary

   b) $e^{-1} = 0.367879441$

   $e^{-10} = 0.000045399$

   $e^{-100} = 3.7200759 \times 10^{-44}$

   $e^{-1000} = 0$ (Answers vary)

3. Answers vary

   $x^* = \pi$

5. $y = x$

7. $\frac{x}{x+1} - \left(\frac{-1}{x+1}\right) = c \Rightarrow \frac{x+1}{x+1} = c \Rightarrow$

   $c = 1$ if $x \ne -1$

9. $\tan x \sin 2x - (-2\cos^2 x) = \tan x \sin 2x + 2\cos^2 x = (\tan x)(2 \sin x \cos x) + 2\cos^2 x = 2\sin^2 x + 2\cos^2 x = 2$ for every $x \ne \frac{n\pi}{2}$, n an odd integer

11. $\ln(2x) - \ln x = \ln\left(\frac{2x}{x}\right) = \ln 2$ $\therefore$ Constant for $x > 0$

13. a) $\sqrt{3} = 1.732050808$

    $\sqrt{1.732050808} = 1.316074013$

    $\sqrt{1.316074013} = 1.14720269$, etc

    b) $\sqrt{5} = 2.236067978$

    $\sqrt{2.236067978} = 1.495348781$

    $\sqrt{1.495348781} = 1.222844545$, etc

15. $\sqrt[10]{2} = 1.071773463$; $\sqrt[10]{1.071773463} = 1.00695555$

    $\sqrt[10]{1.00695555} = 1.000693387$; $\sqrt[10]{1.000693387} = 1.000069317$, etc

17. $x_0 = 1$

    $x_1 = 1.540302306$

    $x_2 = 1.570791601$

    $x_3 = 1.570796327$

    $x_4 = 1.570796327$, etc

19.

| $\Delta x$ | .1 | .01 | .001 | .00001 |
|---|---|---|---|---|
| $\frac{\sin \Delta x}{\Delta x}$ | .017453283 | .017453292 | .017453292 | .017453292 |

etc.

21. $m = \tan 40° = .8391$

23. min $\angle = \arctan \frac{5}{16} = 17.35°$ and max $\angle = \arctan \frac{9}{8} = 48.37°$

## CHAPTER 1 PRACTICE EXERCISES

1. a) (1,–4)
   b) (–1,4)
   c) (–1,–4)

3. a) (–4,–2)
   b) (4,2)
   c) (4,–2)

5. a) Origin
   b) y–axis

7. a) $(x,y)$ on graph $\Rightarrow x^2 - y^2 = 4 \Rightarrow x^2 - (-y)^2 = 4 \Rightarrow (x,-y)$ on graph $\Rightarrow$ Symmetric to x–axis
   $(x,y)$ on graph $\Rightarrow x^2 - y^2 = 4 \Rightarrow (-x)^2 - y^2 = 4 \Rightarrow (-x,y)$ on graph $\Rightarrow$ Symmetric to y–axis
   Symmetry to x and y axes $\Rightarrow$ Symmetry to origin

   b) $(x,y)$ on graph $\Rightarrow x - y = 4 \not\Rightarrow x - (-y) = 4 \Rightarrow (x,-y)$ not on graph $\Rightarrow$ Not symmetric to x–axis
   $(x,y)$ on graph $\Rightarrow x - y = 4 \not\Rightarrow -x - y = 4 \Rightarrow (-x,y)$ not on graph $\Rightarrow$ Not symmetric to y–axis
   $(x,y)$ on graph $\Rightarrow x - y = 4 \not\Rightarrow -x - (-y) = 4 \Rightarrow (-x,-y)$ not on graph $\Rightarrow$ Not symmetric to origin

9. $x = 1$ Vertical
   $y = 3$ Horizontal

11. $x = 0$ Vertical
    $y = -3$ Horizontal

13. $P(2,3)$, $m = 2 \Rightarrow y - 3 = 2(x - 2) \Rightarrow y = 2x - 1$
    x–intercept $= \frac{1}{2}$
    y – intercept $= -1$

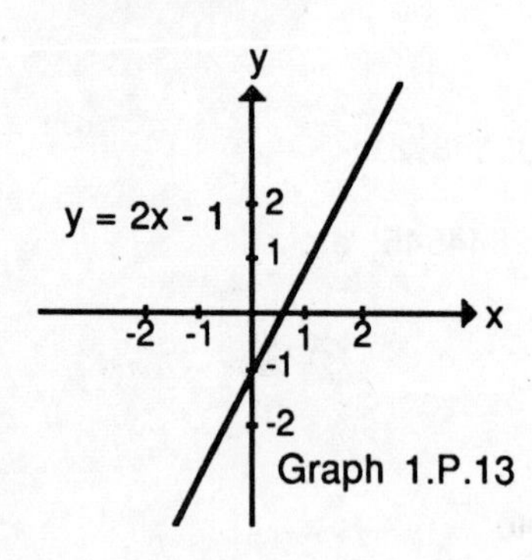


15. $P(1,0)$, $m = -1 \Rightarrow y - 0 = -1(x - 1) \Rightarrow y = -x + 11$
    x–intercept = 1
    y–intercept = 1

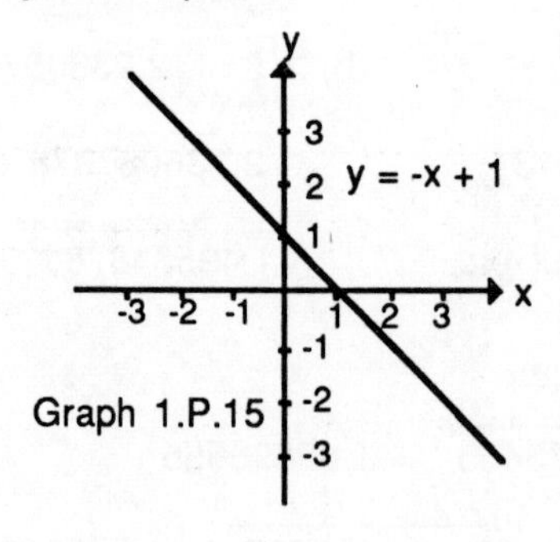


17. $P(1,-6)$, $m = 3 \Rightarrow y - (-6) = 3(x - 1) \Rightarrow y = 3x - 9$
    x–intercept = 3
    y–intercept = –9

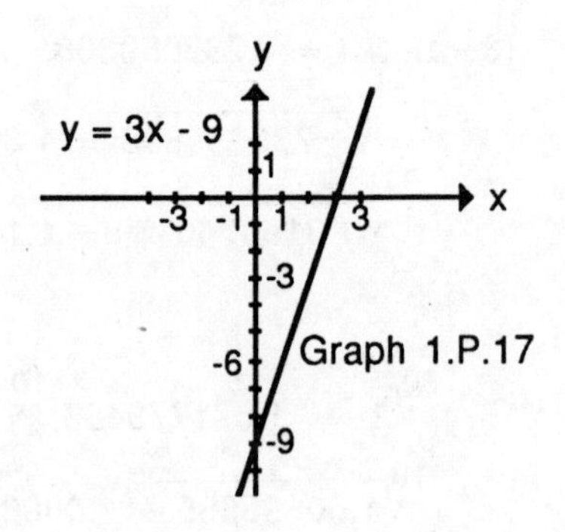


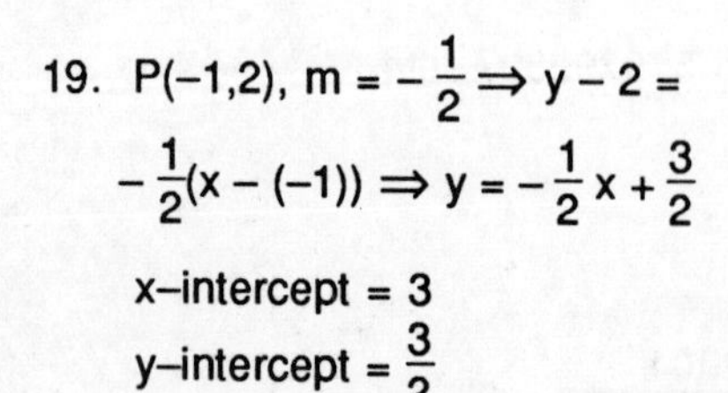

19. $P(-1,2)$, $m = -\frac{1}{2} \Rightarrow y - 2 = -\frac{1}{2}(x - (-1)) \Rightarrow y = -\frac{1}{2}x + \frac{3}{2}$
    x–intercept = 3
    y–intercept $= \frac{3}{2}$

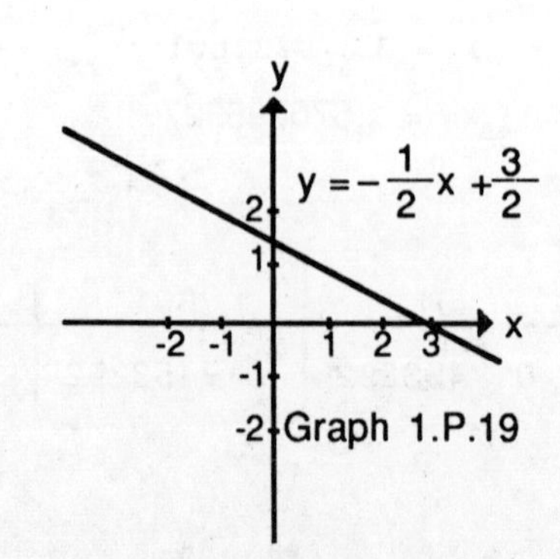

21. $(-2,-2)$ and $(1,3) \Rightarrow m = \frac{3-(-2)}{1-(-2)} = \frac{5}{3}$

$y - 3 = \frac{5}{3}(x-1) \Rightarrow y = \frac{5}{3}x + \frac{4}{3}$

23. $(2,-1)$ and $(4,4) \Rightarrow m = \frac{4-(-1)}{4-2} = \frac{5}{2}$

$y - 4 = \frac{5}{2}(x-4) \Rightarrow y = \frac{5}{2}x - 6$

25. $y = \frac{1}{2}x + 2$

27. $y = -2x - 1$

29. a) $P(6,0)$, $L: 2x - y = -2 \Rightarrow L: y = 2x + 2 \Rightarrow$ $m = 2 \Rightarrow y - 0 = 2(x-6) \Rightarrow y = 2x - 12$ is parallel line

b) $m_L = 2 \Rightarrow m_\perp = -\frac{1}{2} \Rightarrow y - 0 = -\frac{1}{2}(x-6) \Rightarrow y = -\frac{1}{2}x + 3$ is $\perp$ line. $y = 2x + 2$ and $y = -\frac{1}{2}x + 3 \Rightarrow$ $2x + 2 = -\frac{1}{2}x + 3 \Rightarrow x = \frac{2}{5} \Rightarrow y = \frac{14}{5}$.

$\therefore d = \sqrt{(\frac{2}{5}-6)^2 + (\frac{14}{5}-0)^2} = \sqrt{\frac{980}{25}} = \frac{14\sqrt{5}}{5}$

31. a) $P(4,-12)$, $L: 4x + 3y = 12 \Rightarrow L: y = -\frac{4}{3}x + 4 \Rightarrow m = -\frac{4}{3}$. $y - (-12) = -\frac{4}{3}(x-4)$ $\Rightarrow y = -\frac{4}{3}x - \frac{20}{3}$ is parallel line

b) $m_L = -\frac{4}{3} \Rightarrow m_\perp = \frac{3}{4} \Rightarrow y - (-12) = \frac{3}{4}(x-4) \Rightarrow y = \frac{3}{4}x - 15$ is $\perp$ line. $y = -\frac{4}{3}x + 4$ and $y = \frac{3}{4}x - 15 \Rightarrow$ $-\frac{4}{3}x + 4 = \frac{3}{4}x - 15 \Rightarrow x = \frac{228}{25} \Rightarrow y = \frac{-204}{25}$.

$\therefore d = \sqrt{\left(\frac{228}{25}-4\right)^2 + \left(\frac{-204}{25}-(-12)\right)^2} = \sqrt{\frac{1024}{25}} = \frac{32}{5}$

33. D: $-\infty < x < \infty$
R: $y \ge 0$


Graph 1.P.33

35. D: $-\infty < x < \infty$
R: $y \ge 0$

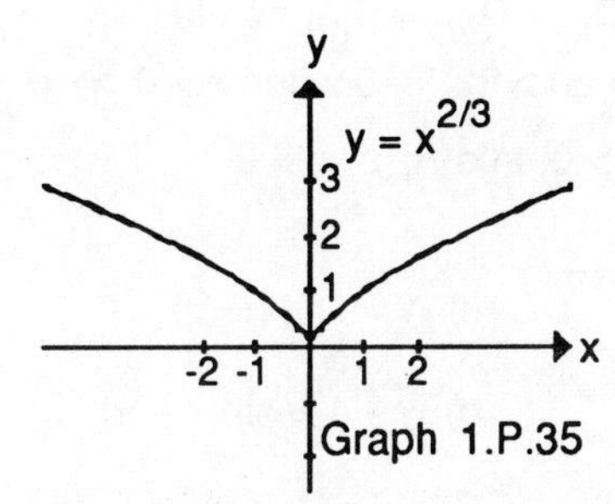

Graph 1.P.35

37. D: $-\infty < x < \infty$
R: $-\infty < y < \infty$


Graph 1.P.37

39. D: $x \ne 0$
R: $y > 0$

41. D: $-\infty < x < \infty$

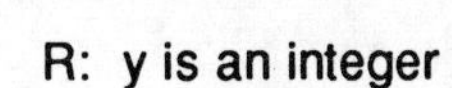

R: y is an integer

43. D: $-\infty < x < \infty$

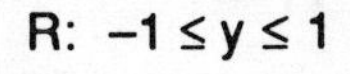

R: $-1 \le y \le 1$

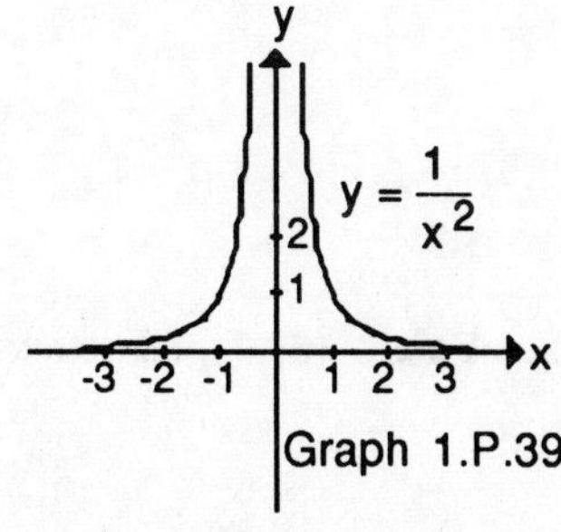

Graph 1.P.39

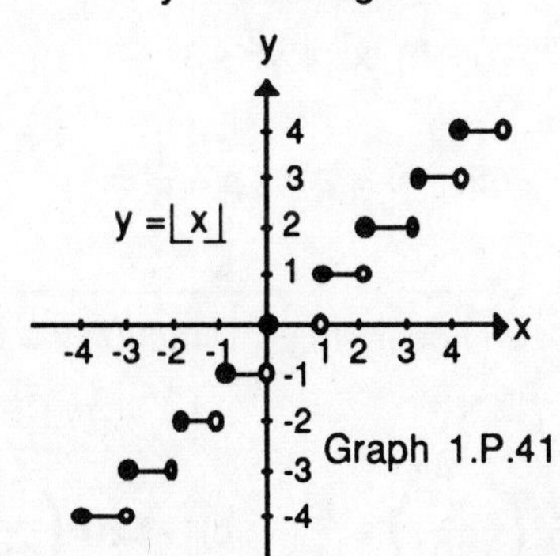

Graph 1.P.41

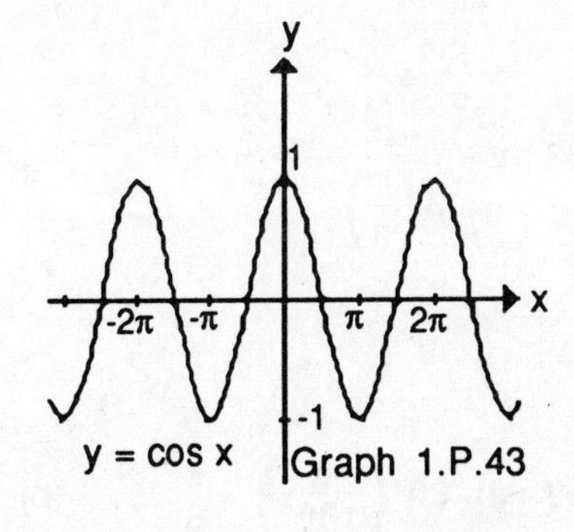

Graph 1.P.43

45. D: $x \neq \frac{n\pi}{2}$, n an odd interger

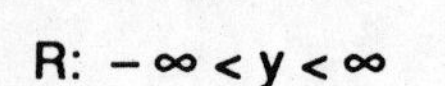

R: $-\infty < y < \infty$

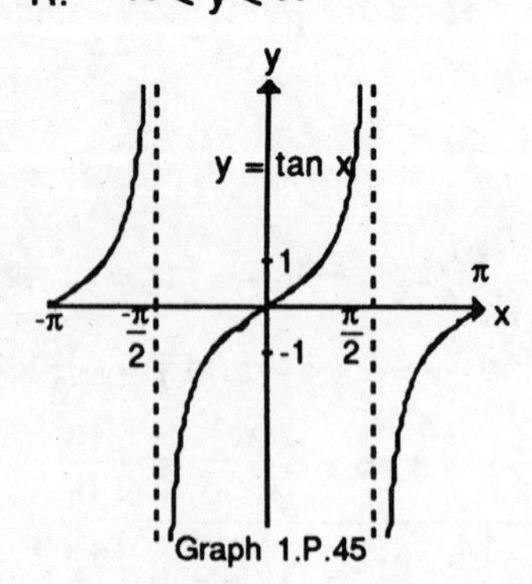


Graph 1.P.45

47. D: $x \neq \frac{n\pi}{2}$, n an odd integer

R: $y \leq -1$ or $y \geq 1$

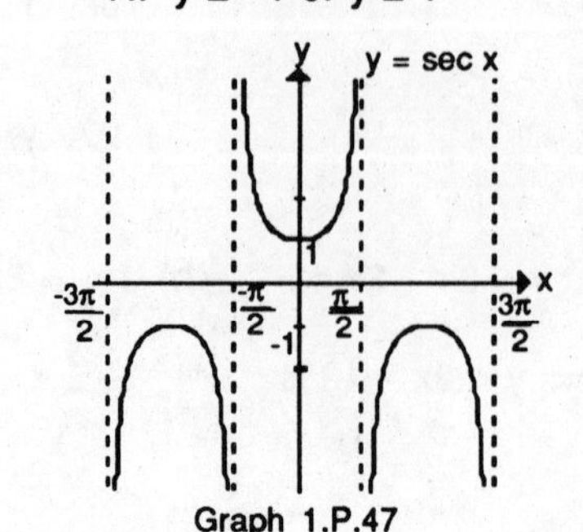


Graph 1.P.47

49. a) Even
b) Even
c) Even

51. a) Even
b) Odd
c) Odd

53. a) Even
b) Odd
c) Odd

55. Yes, period = 1 (See Problem 39, Section 1.3, Graph 1.3.39)

57.

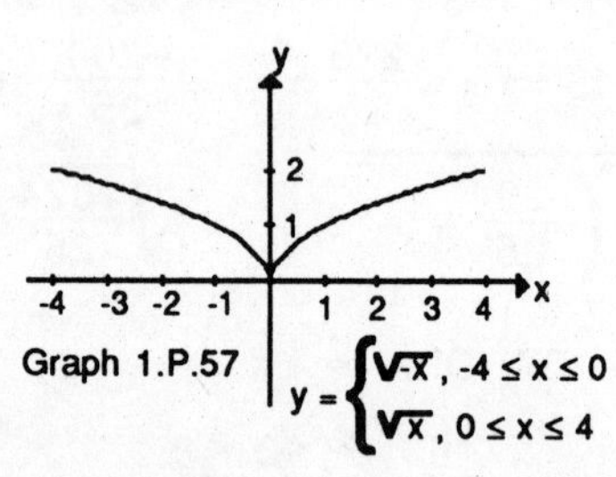


Graph 1.P.57

59.

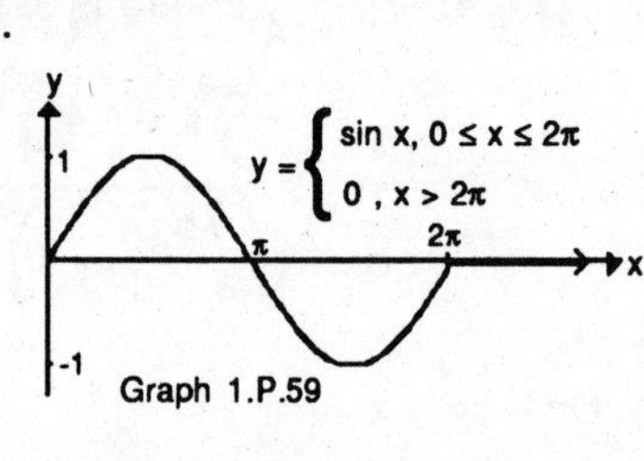


Graph 1.P.59

61. $y = \begin{cases} -x + 1, & 0 \leq x < 1 \\ -x + 2, & 1 \leq x \leq 2 \end{cases}$

63. $D_f: x \neq 0$, $D_g: x > 0 \Rightarrow D_{f+g} = D_{fg} = D_{f/g} = D_{g/f}: x > 0$ $R_f: y \neq 0$, $R_g: y > 0 \Rightarrow R_{f+g} = R_{fg} = R_{f/g} = R_{g/f}; y > 0$. $D_f: x \neq 0$ and $f(x) > 0$ when $x > 0 \Rightarrow D_{g \circ f}: x > 0$ and $R_{g \circ f}: y > 0$. $D_g: x > 0$ and $g(x) > 0$ for $x > 0 \Rightarrow D_{f \circ g}: x > 0$ and $R_{f \circ g}: y > 0$

65. a) $y + 1 = (x - 2)^2$

b) $y = (x + 2)^2$

67. a) $y = \sin(x - \frac{\pi}{2})$

b) $y - 1 = \sin(x - \pi)$

69. a) $y = \sqrt{x - 2}$

b) $y = \sqrt{x + 2}$

71. $(x - 1)^2 + (y - 1)^2 = 1$

73. $(x - 2)^2 + (y + 3)^2 = \frac{1}{4}$

75. C(3,–5), a = 4

77. C(–1,7), a = 11

79. a) $x^2 + y^2 < 1$

b) $x^2 + y^2 \leq 1$

81. $y = \frac{x^2}{4(2)} = \frac{x^2}{8}$

83. $y = \frac{-x^2}{4\left(\frac{1}{4}\right)} = -x^2$

85. $4p = 2 \Rightarrow p = \frac{1}{2} \Rightarrow$

Focus: $(0,\frac{1}{2})$, Directrix: $y = -\frac{1}{2}$

87. $4p = 4 \Rightarrow p = 1 \Rightarrow$

Focus: (0,–1), Directrix: y = 1

89. a) $30°\left(\frac{\pi}{180°}\right) = \frac{\pi}{6}$ b) $45°\left(\frac{\pi}{180°}\right) = \frac{\pi}{4}$ c) $-120°\left(\frac{\pi}{180°}\right) = -\frac{2\pi}{3}$ d) $-150°\left(\frac{\pi}{180°}\right) = -\frac{5\pi}{6}$

| | $\theta$ | $\sin\theta$ | $\cos\theta$ | $\tan\theta$ | $\cot\theta$ | $\sec\theta$ | $\csc\theta$ |
|---|---|---|---|---|---|---|---|
| 91. a) | $\frac{\pi}{3}$ | $\frac{\sqrt{3}}{2}$ | $\frac{1}{2}$ | $\sqrt{3}$ | $\frac{1}{\sqrt{3}}$ | 2 | $\frac{2}{\sqrt{3}}$ |
| b) | $-\frac{\pi}{3}$ | $-\frac{\sqrt{3}}{2}$ | $\frac{1}{2}$ | $-\sqrt{3}$ | $-\frac{1}{\sqrt{3}}$ | 2 | $-\frac{2}{\sqrt{3}}$ |
| c) | $\frac{2\pi}{3}$ | $\frac{\sqrt{3}}{2}$ | $-\frac{1}{2}$ | $-\sqrt{3}$ | $-\frac{1}{\sqrt{3}}$ | −2 | $\frac{2}{\sqrt{3}}$ |
| d) | $-\frac{2\pi}{3}$ | $-\frac{\sqrt{3}}{2}$ | $-\frac{1}{2}$ | $\sqrt{3}$ | $\frac{1}{\sqrt{3}}$ | −2 | $-\frac{2}{\sqrt{3}}$ |

93. a)

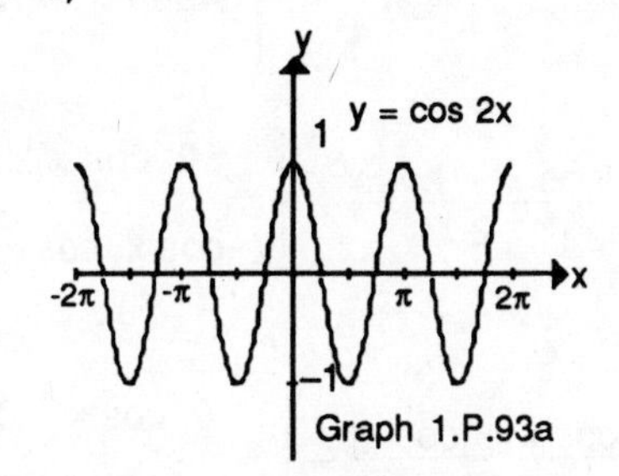


Graph 1.P.93a

b)

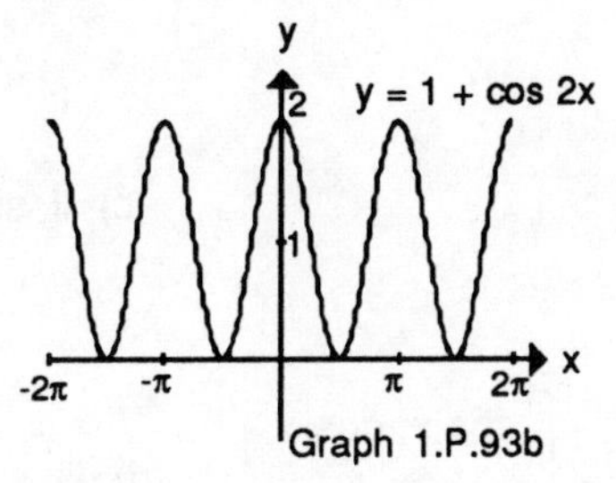


Graph 1.P.93b

c)

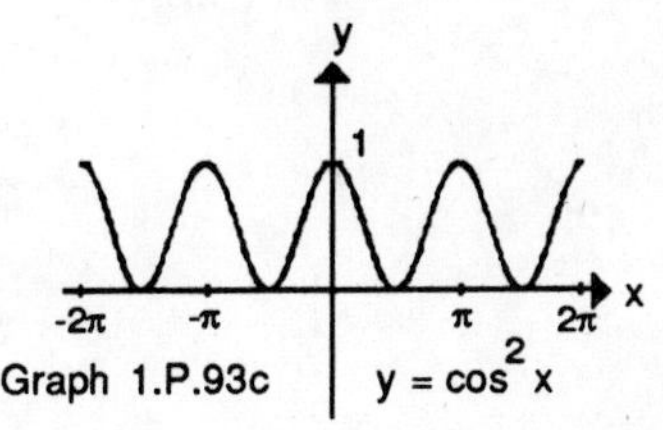


Graph 1.P.93c $y = \cos^2 x$

95. $\cos 15° = \cos(60° - 45°) = \cos 60° \cos 45° + \sin 60° \sin 45° =$
$\left(\frac{1}{2}\right)\left(\frac{\sqrt{2}}{2}\right) + \left(\frac{\sqrt{3}}{2}\right)\left(\frac{\sqrt{2}}{2}\right) = \frac{\sqrt{2} + \sqrt{6}}{4}$

97. $\sin\left(\frac{\pi}{12}\right) = \sin\left(\frac{\pi}{3} - \frac{\pi}{4}\right) = \sin\left(\frac{\pi}{3}\right)\cos\left(\frac{\pi}{4}\right) - \cos\left(\frac{\pi}{3}\right)\sin\left(\frac{\pi}{4}\right) = \left(\frac{\sqrt{3}}{2}\right)\left(\frac{\sqrt{2}}{2}\right) - \left(\frac{1}{2}\right)\left(\frac{\sqrt{2}}{2}\right) = \frac{\sqrt{6} - \sqrt{2}}{4}$

99. a) $\cos^2 \frac{\pi}{6} = \left(\frac{\sqrt{3}}{2}\right)^2 = \frac{3}{4}$ b) $\cos^2 \frac{\pi}{6} = \frac{\cos\frac{\pi}{3} + 1}{2} = \frac{\frac{1}{2} + 1}{2} = \frac{3}{4}$

101. $|x - 1| = \frac{1}{2} \Rightarrow x - 1 = \frac{1}{2}$ or $x - 1 = -\frac{1}{2} \Rightarrow x = \frac{3}{2}$ or $x = \frac{1}{2}$

103. $\left|\frac{2x}{5} + 1\right| = 7 \Rightarrow \frac{2x}{5} + 1 = 7$ or $\frac{2x}{5} + 1 = -7 \Rightarrow 2x + 5 = 35$ or $2x + 5 = -35 \Rightarrow x = 15$ or $x = -20$

105. $|x + 2| \le \frac{1}{2} \Rightarrow -\frac{1}{2} \le x + 2 \le \frac{1}{2} \Rightarrow -1 \le 2x + 4 \le 1 \Rightarrow -\frac{5}{2} \le x \le -\frac{3}{2}$

107. $\left|y - \frac{2}{5}\right| < \frac{3}{5} \Rightarrow -\frac{3}{5} < y - \frac{2}{5} < \frac{3}{5} \Rightarrow -\frac{1}{5} < y < 1$

109. $3 < x < 11 \Rightarrow 3 - 7 < x - 7 < 11 - 7 \Rightarrow -4 < x - 7 < 4 \Rightarrow |x - 7| \le 4$

111. $-1 < y < 7 \Rightarrow -1 - 3 < y - 3 < 7 - 3 \Rightarrow -4 < y - 3 < 4 \Rightarrow |y - 3| < 4$

113. $|(2x - 3) - 1| < 2 \Rightarrow |2x - 4| < 2 \Rightarrow 2|x - 2| < 2 \Rightarrow |x - 2| < 1$

115. $|\sqrt{x + 2} - 4| < 1 \Rightarrow -1 < \sqrt{x + 2} - 4 < 1 \Rightarrow 3 < \sqrt{x + 2} < 5 \Rightarrow 9 < x + 2 < 25 \Rightarrow 7 < x < 23 \Rightarrow$
$7 - 15 < x - 15 < 23 - 15 \Rightarrow -8 < x - 15 < 8 \Rightarrow |x - 15| < 8$

117.

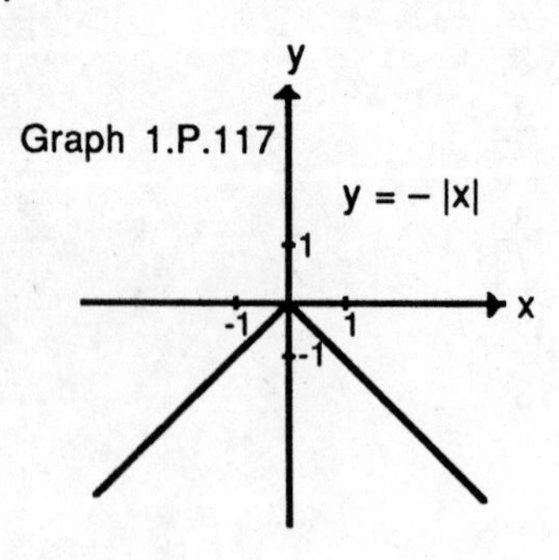

Graph 1.P.117

119.

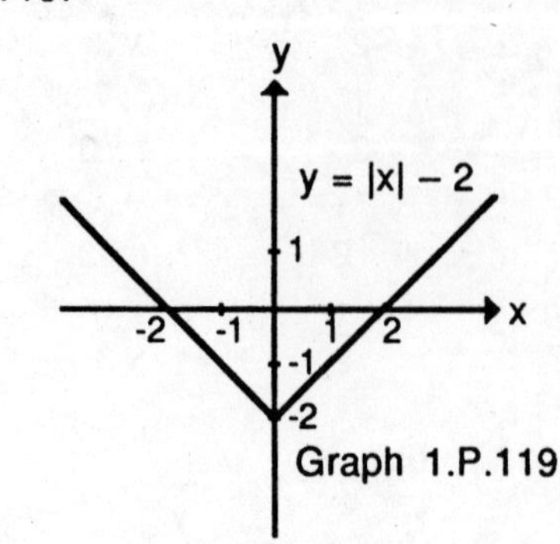

Graph 1.P.119

121.

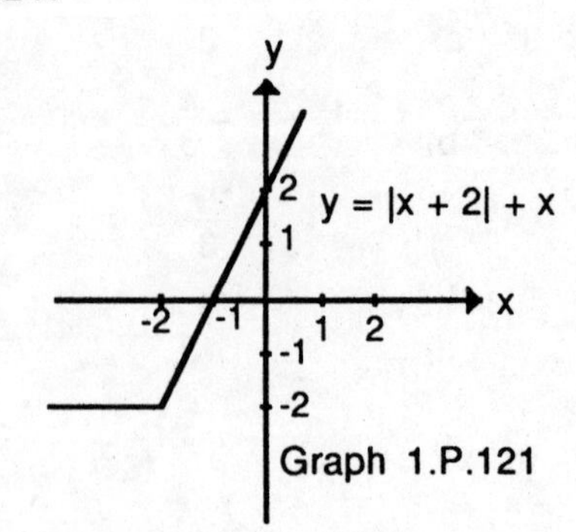

Graph 1.P.121

123. a) iv

b) i, since $y = \dfrac{\cos x + |\cos x|}{2} = \begin{cases} \cos x, & \cos x \ge 0 \\ 0, & \cos x < 0 \end{cases}$

c) ii, since $y = \dfrac{|\cos x| - \cos x}{2} = \begin{cases} 0, & \cos x \ge 0 \\ -\cos x, & \cos x < 0 \end{cases}$

d) iii, since $y = \dfrac{\cos x - |\cos x|}{2} = \begin{cases} 0, & \cos x \ge 0 \\ \cos x, & \cos x < 0 \end{cases}$

125. $\left|\dfrac{\sqrt{x}}{2} - 1\right| < 0.2 \Rightarrow -0.2 < \dfrac{\sqrt{x}}{2} - 1 < 0.2 \Rightarrow 0.8 < \dfrac{\sqrt{x}}{2} < 1.2 \Rightarrow$

$1.6 < \sqrt{x} < 2.4 \Rightarrow 2.56 < x < 5.76$

$\left|\dfrac{\sqrt{x}}{2} - 1\right| < 0.1 \Rightarrow -0.1 < \dfrac{\sqrt{x}}{2} - 1 < 0.1 \Rightarrow 0.9 < \dfrac{\sqrt{x}}{2} < 1.1 \Rightarrow$

$1.8 < \sqrt{x} < 2.2 \Rightarrow 3.24 < x < 4.84$

# CHAPTER 2

# LIMITS AND CONTINUITY

## 2.1 LIMITS

1. 4
3. 4
5. 2
7. 25
9. −5
11. 9
13. 1
15. 45
17. 4
19. −15
21. $\frac{5}{8}$
23. −2
25. 5
27. 0
29. $\frac{11}{4}$

31. $\lim_{x\to 1} \frac{x-1}{x^2-1} = \lim_{x\to 1} \frac{x-1}{(x+1)(x-1)} =$
$\lim_{x\to 1} \frac{1}{x+1} = \frac{1}{2}$

33. $\lim_{x\to -5} \frac{x^2+3x-10}{x+5} = \lim_{x\to -5} \frac{(x+5)(x-2)}{x+5}$
$= \lim_{x\to -5} (x-2) = -7$

35. $\lim_{x\to -2} \frac{x^2+x-2}{x^2-4} = \lim_{x\to -2} \frac{(x+2)(x-1)}{(x+2)(x-2)}$
$= \lim_{x\to -2} \frac{x-1}{x-2} = \frac{3}{4}$

37. $\lim_{t\to 1} \frac{t^2-3t+2}{t^2-1} = \lim_{t\to 1} \frac{(t-1)(t-2)}{(t-1)(t+1)} =$
$\lim_{t\to 1} \frac{t-2}{t+1} = -\frac{1}{2}$

39. $\lim_{x\to 5} \frac{x-5}{x^2-25} = \lim_{x\to 5} \frac{x-5}{(x-5)(x+5)} =$
$\lim_{x\to 5} \frac{1}{x+5} = \frac{1}{10}$

41. $\lim_{x\to 2} \frac{2x-4}{x^3-2x^2} = \lim_{x\to 2} \frac{2(x-2)}{x^2(x-2)} =$
$\lim_{x\to 2} \frac{2}{x^2} = \frac{1}{2}$

43. $\lim_{t\to 1} \frac{t^3-1}{t-1} = \lim_{t\to 1} \frac{(t-1)(t^2+t+1)}{t-1} =$
$\lim_{t\to 1} (t^2+t+1) = 3$

45. a) $\lim_{x\to 2^+} f(x) = 2$
$\lim_{x\to 2^-} f(x) = 1$
b) No, $\lim_{x\to 2^+} f(x) \neq \lim_{x\to 2^-} f(x)$

47. a) False
b) True
c) False
d) True
e) True
f) True
g) False
h) False
i) False
j) False

49. a)

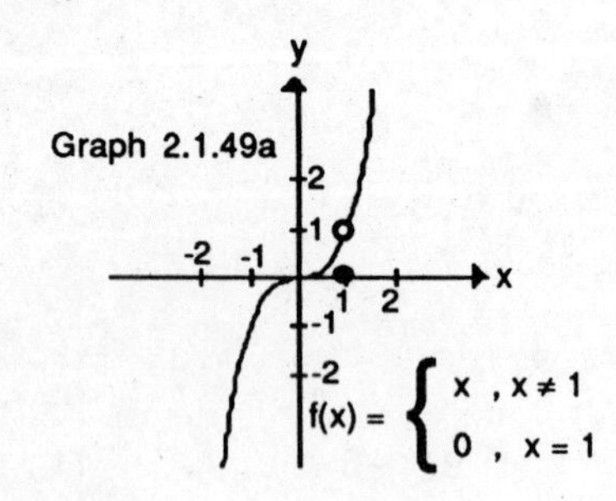


b) $\lim_{x\to1^+} f(x) = 1$

$\lim_{x\to1^-} f(x) = 1$

c) Yes, $\lim_{x\to1^-} f(x) = \lim_{x\to1^+} f(x)$

51.

Graph 2.1.51

$f(x) = \begin{cases} \sqrt{1-x^2}, & 0 \le x < 1 \\ 1, & 1 \le x < 2 \\ 2, & x = 2 \end{cases}$

a) $(0,1) \cup (1,2)$

b) $x = 2$

c) $x = 0$

53. a) $\lim_{x\to0^+} f(x)$ does not exist

b) $\lim_{x\to0^-} f(x) = 0$

c) $\lim_{x\to0} f(x)$ does not exist because $\lim_{x\to0^+} f(x)$ does not exist.

55. $\lim_{x\to0^+} \lfloor x \rfloor = 0$

57. $\lim_{x\to.5} \lfloor x \rfloor = 0$

59. $\lim_{x\to0^+} \frac{x}{|x|} = \lim_{x\to0^+} \frac{x}{x} = 1$

61. a) $-10$

b) $-20$

63. a) 4

b) $-21$

c) $-12$

d) $-\frac{7}{3}$

65. a)

| x | 1.1 | 1.01 | 1.001 |
|---|---|---|---|
| $\frac{x^2-1}{x-1}$ | 2.1 | 2.01 | 2.001 |

etc.

b) $\lim_{x\to1} \frac{x^2-1}{x-1} = \lim_{x\to1} \frac{(x-1)(x+1)}{x-1} = \lim_{x\to1} (x+1) = 2$

67.

| x | 1.1 | 1.01 | 1.001 |
|---|---|---|---|
| $\frac{\ln(x^2)}{\ln x}$ | 2 | 2 | 2 |

## SECTION 2.2 THE SANDWICH THEOREM AND (SIN θ)/θ

1. 1

3. $\frac{1}{2}$

5. $\lim_{x\to 0^+} \frac{x}{\sin x} = \lim_{x\to 0^+} \frac{1}{\frac{\sin x}{x}} = 1$

7. $\lim_{x\to 0} \frac{\sin 2x}{x} = \lim_{x\to 0} 2\left(\frac{\sin 2x}{2x}\right) = 2$

9. $\lim_{x\to 0} \frac{\tan 2x}{2x} = \lim_{x\to 0} \left(\frac{\sin 2x}{2x}\right)\left(\frac{1}{\cos 2x}\right)$
$= (1)(1) = 1$

11. $\lim_{x\to 0} \frac{\sin x}{2x^2 - x} = \lim_{x\to 0} \left(\frac{\sin x}{x}\right)\left(\frac{1}{2x-1}\right)$
$= (1)(-1) = -1$

13. If $1 - \frac{x^2}{6} < \frac{\sin x}{x} < 1$ when $-1 < x < 1$,
then since $\lim_{x\to 0} \left(1 - \frac{x^2}{6}\right) = 1$ and $\lim_{x\to 0} 1$
$= 1$, $\lim_{x\to 0} \frac{\sin x}{x} = 1$ by the sandwich theorem.

15. a)

| $x$ | .1 | .01 | .001 |
|---|---|---|---|
| $\frac{1-\cos x}{x^2}$ | .4995583474 | .4999958 | .5 |

b) If $\frac{1}{2} - \frac{x^2}{24} < \frac{1-\cos x}{x^2} < \frac{1}{2}$, then since
$\lim_{x\to 0} \left(\frac{1}{2} - \frac{x^2}{24}\right) = \frac{1}{2}$ and $\lim_{x\to 0} \frac{1}{2} = \frac{1}{2}$,
$\lim_{x\to 0} \frac{1-\cos x}{x^2} = \frac{1}{2}$ by the sandwich
theorem

## SECTION 2.3 LIMITS INVOLVING INFINITY

1. a) $\frac{2}{5}$
   b) $\frac{2}{5}$

3. a) 0
   b) 0

5. a) $-3$
   b) $-3$

7. a) $+\infty$
   b) $-\infty$

9. a) 0
   b) 0

11. a) 7
    b) 7

13. a) $-\frac{2}{3}$
    b) $-\frac{2}{3}$

15. a) $(-1)(1) = -1$
    b) $(-1)(1) = -1$

17. a) $\left(-\frac{1}{2}\right)(2) = -1$
    b) $\left(-\frac{1}{2}\right)(2) = -1$

19. $+\infty$

21. $+\infty$

23. $+\infty$

25. $+\infty$

27. $-\infty$

29. $\frac{1}{2}$

31. 1

33. $+\infty$

35. $\frac{1}{3} - \frac{5}{6} = -\frac{1}{2}$

37. $\lim_{x\to 2^+} f(x) = +\infty$ $\quad \lim_{x\to -2^+} f(x) = -\infty$
$\lim_{x\to 2^-} f(x) = -\infty$ $\quad \lim_{x\to -2^-} f(x) = +\infty$

39. $\lim_{x\to -2^+} f(x) = +\infty$

$\lim_{x\to -2^-} f(x) = -\infty$

41. $\lim_{x\to 0^+} \frac{\lfloor x \rfloor}{x} = \lim_{x\to 0^+} \frac{0}{x}$

$= 0$

43. $\lim_{x\to \infty} \frac{|x|}{|x| + 1} = \lim_{x\to \infty} \frac{x}{x+1} = 1$

45. $\lim_{x\to 0^+} \frac{1}{\sin x} = +\infty$

47. $\lim_{x\to(\pi/2)^+} \frac{1}{\cos x} - \infty$

49. $\lim_{x\to -\infty} f(x) = \lim_{x\to -\infty} \frac{1}{x} = 0$

$\lim_{x\to 0^-} f(x) = \lim_{x\to 0^-} \frac{1}{x} = -\infty$

$\lim_{x\to 0^+} f(x) = \lim_{x\to 0^+} -1 = -1$

$\lim_{x\to \infty} f(x) = \lim_{x\to \infty} -1 = -1$

51. $\lim_{x\to\infty} \left(2 + \frac{\sin x}{x}\right) = 2 + 0 = 2$

53. $\lim_{x\to\infty} \left(1 + \cos \frac{1}{x}\right) = 1 + 1 = 2$

55. 0

57. If $\frac{2x-3}{x} < f(x) < \frac{2x^2 + 5x}{x^2}$, then since $\lim_{x\to\pm\infty} \frac{2x-3}{x} = 2$ and $\lim_{x\to\pm\infty} \frac{2x^2+5x}{x^2} = 2$, $\lim_{x\to\pm\infty} f(x) = 2$ by the sandwich theorem.

59. a)

| $x$ | 10 | 100 | 1000 | etc. |
|---|---|---|---|---|
| $\frac{\ln(x+1)}{\ln x}$ | 1.041392685 | 1.002160687 | 1.000144693 | |

b)

| $x$ | 10 | 100 | 1000 | 10000 | 100000 |
|---|---|---|---|---|---|
| $\frac{\ln(x+999)}{\ln x}$ | 3.003891166 | 1.520498846 | 1.100270931 | 1.010338301 | 1.000863415 |

etc.

c) $\lim_{x\to\infty} \frac{\ln x^2}{\ln x} \approx 2$ (In fact the limit is exactly 2, as you will see in Chapter 7)

d)

| $x$ | 10 | 100 | 1000 | etc. |
|---|---|---|---|---|
| $\frac{\ln x}{\log x}$ | 2.302585093 | 2.302585093 | 2.302585093 | |

## SECTION 2.4 CONTINUOUS FUNCTIONS

1. a) Yes
   b) $\lim_{x \to -1^+} f(x) = 1$, yes
   c) Yes
   d) Yes

3. a) No
   b) No

5. a) $\lim_{x \to 2} f(x) = 0$
   b) $f(2) = 0$

7. All except $x = 2$

9. $[-1,0) \cup (0,1) \cup (1,2]$

11. All except $x = 1$

13. At all x except 0 and 1

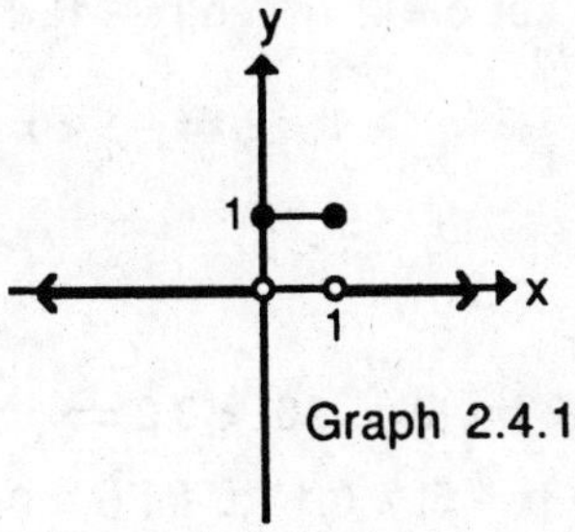

Graph 2.4.13

15. $x - 2 = 0 \Rightarrow x = 2$

17. $x + 1 = 0 \Rightarrow x = -1$

19. Continuous everywhere

21. $x^2 - 1 = 0 \Rightarrow x = \pm 1$

23. $x = 0$

25. Yes; $\lim_{x \to 1} \dfrac{x^2 - 1}{x - 1} = \lim_{x \to 1} \dfrac{(x + 1)(x - 1)}{x - 1} = \lim_{x \to 1} (x + 1) = 2$
    $f(1) = 2 \;\therefore\; \lim_{x \to 1} f(x) = f(1)$

27. $h(x) = \dfrac{x^2 + 3x - 10}{x - 2} = \dfrac{(x + 5)(x - 2)}{x - 2} = x + 5$ if $x \neq 2$. $\therefore$ let $h(2) = 7$

29. $g(x) = \dfrac{x^2 - 16}{x^2 - 3x - 4} = \dfrac{(x - 4)(x + 4)}{(x - 4)(x + 1)} = \dfrac{x + 4}{x + 1}$ if $x \neq 4$. $\therefore$ let $g(4) = \dfrac{8}{5}$

31. $f(x) \to 8$ when $x \to 3^-$
    $\Rightarrow 2ax = 8$ when $x = 3 \Rightarrow$
    $2a(3) = 8 \Rightarrow a = \dfrac{4}{3}$

33. $\lim_{x \to 0} \dfrac{1 + \cos x}{2} = \dfrac{1 + 1}{2} = 1$

35. $\lim_{x \to 0} \tan x = 0$

37. $x = 2$ and $3$. No minimum because $y \to 0$ as $x \to 1^-$ but $y > 0$ for any x.

39. No maximum because $y \to 1$ as $x \to 1^-$ or $x \to -1^+$ but $y < 1$ for any $x \in (-1,1)$.

41. Intermediate Value Theorem

## SECTION 2.5 DEFINING LIMITS FORMALLY WITH EPSILONS AND DELTAS

Note: On exercises 1 through 8, let $\delta = \min(|x_0 - a|, |x_0 - b|)$ (the smaller of the distances from $x_0$ to a and b)

1.

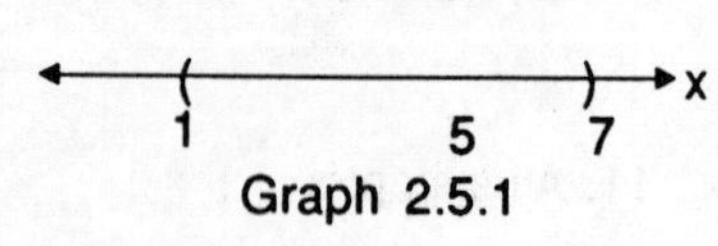


Graph 2.5.1

Let $\delta = 2$. Then $|x - 5| < 2 \Rightarrow$

$-2 < x - 5 < 2 \Rightarrow 3 < x < 7 \Rightarrow$

$1 < x < 7$

3.

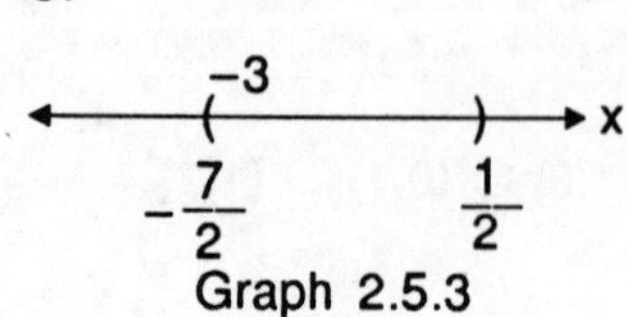


Graph 2.5.3

Let $\delta = \frac{1}{2}$. Then $|x - (-3)|$

$< \frac{1}{2} \Rightarrow -\frac{1}{2} < x + 3 < \frac{1}{2} \Rightarrow$

$1 < x < 7$

5.

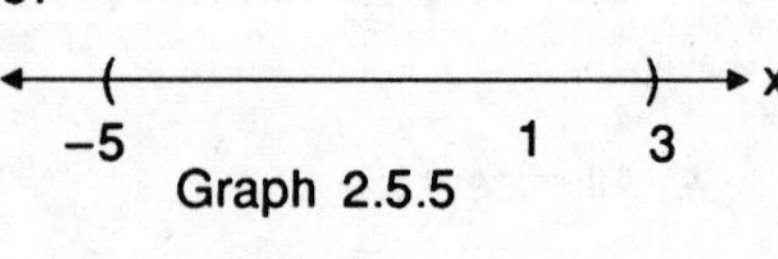


Graph 2.5.5

Let $\delta = 2$. Then $|x - 1| < 2 \Rightarrow$

$-2 < x - 1 < 2 \Rightarrow -1 < x < 3 \Rightarrow$

$-5 < x < 3$

7.

Graph 2.5.7

x

$\frac{4}{9}$ $\frac{1}{2}$ $\frac{4}{7}$

Let $\delta = \frac{1}{18}$. Then $|x - \frac{1}{2}| < \frac{1}{18} \Rightarrow$

$-\frac{1}{18} < x - \frac{1}{2} < \frac{1}{18} \Rightarrow \frac{4}{9} < x < \frac{5}{9} \Rightarrow$

$\frac{4}{9} < x < \frac{4}{7}$

9. $|(2x - 4) - 6| < 0.2 \Rightarrow |2x - 10| < 0.2 \Rightarrow$

$2|x - 5| < 0.2 \Rightarrow |x - 5| < 0.1 \quad \therefore$ let $\delta = 0.1$

11. $|x^2 - 4| < 1 \Rightarrow -1 < x^2 - 4 < 1 \Rightarrow$

$3 < x^2 < 5 \Rightarrow \sqrt{3} < x < \sqrt{5}$ Now

$\sqrt{5} - 2 \approx 0.236$ and $2 - \sqrt{3} \approx 0.268$.

13. $|\sqrt{x} - 1| < \frac{1}{4} \Rightarrow -\frac{1}{4} < \sqrt{x} - 1 < \frac{1}{4} \Rightarrow$

$\frac{3}{4} < \sqrt{x} < \frac{5}{4} \Rightarrow \frac{9}{16} < x < \frac{25}{16}$ Now $\frac{25}{16} - 1 = \frac{9}{16}$

and $1 - \frac{9}{16} = \frac{7}{16}$ $\therefore$ let $\delta = \frac{7}{16}$

15. $|(x + 1) - 4| < 0.01 \Rightarrow |x - 3| < 0.01 \Rightarrow$

$-0.01 < x - 3 < 0.01 \Rightarrow 2.99 < x < 3.01$

17. $|(x^2 - 5) - 4| < 0.05 \Rightarrow |x^2 - 9| < 0.05 \Rightarrow$

$-0.05 < x^2 - 9 < 0.05 \Rightarrow 8.95 < x^2 < 9.05$

$\Rightarrow \sqrt{8.95} < |x| < \sqrt{9.05}$

19. $|\sqrt{19 - x} - 4| < 0.03 \Rightarrow$

$-0.03 < \sqrt{19 - x} - 4 < 0.03 \Rightarrow$

$3.97 < \sqrt{19 - x} < 4.03 \Rightarrow$

$15.7609 < 19 - x < 16.2409 \Rightarrow$

$-3.2391 < -x < -2.7591 \Rightarrow 3.2391 > x > 2.7591$

21. $\left|\frac{1}{x} - 4\right| < 0.1 \Rightarrow -0.1 < \frac{1}{x} - 4 < 0.1 \Rightarrow$

$3.9 < \frac{1}{x} < 4.1 \Rightarrow \frac{1}{3.9} > x > \frac{1}{4.1} \Rightarrow$

$\frac{10}{39} > x > \frac{10}{41}$

23. $\lim_{x \to 1} (2x + 3) = 5$

$|(2x + 3) - 5| < 0.01 \Rightarrow |2x - 2| < 0.01 \Rightarrow$

$2|x - 1| < 0.01 \Rightarrow |x - 1| < 0.005$

$\therefore$ let $\delta \le 0.005$

25. $\lim_{x\to 1/2} (4x - 2) = 0$

$|(4x - 2) - 0| < 0.02 \Rightarrow |4x - 2| < 0.02 \Rightarrow$

$4\left|x - \frac{1}{2}\right| < 0.02 \Rightarrow \left|x - \frac{1}{2}\right| < 0.005$

$\therefore$ let $\delta \le 0.005$

27. $\lim_{x\to 2}\left(\frac{x^2 - 4}{x - 2}\right) = \lim_{x\to 2} (x + 2) = 4$

$\left|\frac{x^2 - 4}{x - 2} - 4\right| < 0.05 \Rightarrow |(x + 2) - 4| < 0.05$

$\Rightarrow |x - 2| < 0.05$

$\therefore$ let $\delta \le 0.05$

29. $\lim_{x\to 11} \sqrt{x - 7} = 2$

$\left|\sqrt{x - 7} - 2\right| < 0.01 \Rightarrow -0.01 < \sqrt{x - 7} - 2 < 0.01$

$\Rightarrow 1.99 < \sqrt{x - 7} < 2.01 \Rightarrow 3.9601 < x - 7 < 4.0401 \Rightarrow$

$10.9601 < x < 11.0401$ Now $11 - 10.9601 = 0.0399$ and

$11.0401 - 11 = 0.0401$ $\therefore$ let $\delta \le .0399$

31. $\lim_{x\to 2} \frac{4}{x} = 2$

$\left|\frac{4}{x} - 2\right| < 0.4 \Rightarrow -0.4 < \frac{4}{x} - 2 < 0.4 \Rightarrow$

$1.6 < \frac{4}{x} < 2.4 \Rightarrow \frac{1}{1.6} > \frac{x}{4} > \frac{1}{2.4} \Rightarrow$

$\frac{4}{1.6} > x > \frac{4}{2.4} \Rightarrow \frac{5}{2} > x > \frac{5}{3}$ Now $\frac{5}{2} - 2 = \frac{1}{2}$

and $2 - \frac{5}{3} = \frac{1}{3}$ $\therefore$ let $\delta \le \frac{1}{3}$

33. For $\varepsilon = 0.01$: $|(9 - x) - 5| < 0.01 \Rightarrow |4 - x| < 0.01$

$\Rightarrow |x - 4| < 0.01$

$\therefore$ let $\delta = 0.01$

For $\varepsilon = 0.0001$: $|(9 - x) - 5| < 0.0001 \Rightarrow$

$|4 - x| < 0.0001 \Rightarrow |x - 4| < 0.0001$

$\therefore$ let $\delta = 0.0001$

For $\varepsilon = 0.001$: $|(9 - x) - 5| < 0.001 \Rightarrow |4 - x| <$

$0.001 \Rightarrow |x - 4| < 0.001$

$\therefore$ let $\delta = 0.001$

For arbitrary $\varepsilon$: $|(9 - x) - 5| < \varepsilon \Rightarrow |4 - x| < \varepsilon$

$\Rightarrow |x - 4| < \varepsilon$

$\therefore$ let $\delta = \varepsilon$

35. Let $\sqrt{x - 5} < \varepsilon$ for $x > 5$. Then $x - 5 < \varepsilon^2$

$\Rightarrow x < 5 + \varepsilon^2 \Rightarrow 5 < x < 5 + \varepsilon^2$ $\therefore$ let $\delta = \varepsilon^2$

$\Rightarrow I = (5, 5 + \varepsilon^2)$

$\lim_{x\to 5^+} \sqrt{x - 5} = 0$

36. Let $\sqrt{4 - x} < \varepsilon$ for $x < 4$ Then $\sqrt{4 - x} < \varepsilon \Rightarrow$

$4 - x < \varepsilon^2 \Rightarrow -x < \varepsilon^2 - 4 \Rightarrow x > 4 - \varepsilon^2 \Rightarrow$

$4 > x > 4 - \varepsilon^2$ $\therefore$ let $\delta = \varepsilon^2 \Rightarrow I = (4 - \varepsilon^2, 4)$

$\lim_{x\to 4^-} \sqrt{4 - x} = 0$

37.

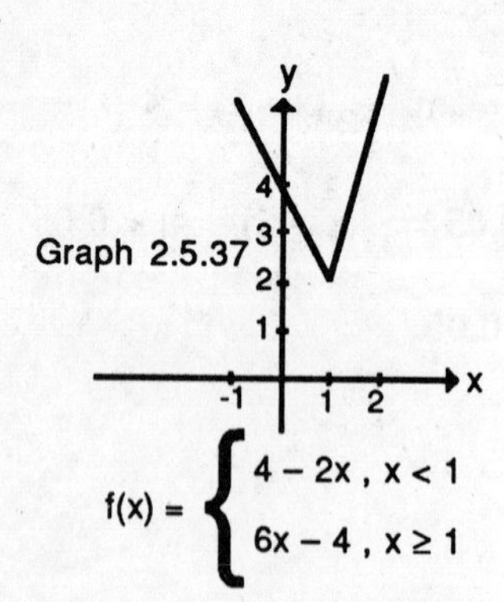


Graph 2.5.37

For $x \to 1^+$: $|f(x) - 2| = |(6x - 4) - 2| < \varepsilon \Rightarrow |6x - 6| < \varepsilon \Rightarrow$
$6|x - 1| < \varepsilon \Rightarrow |x - 1| < \frac{\varepsilon}{6}$

For $x \to 1^-$: $|f(x) - 2| = |(4 - 2x) - 2| < \varepsilon \Rightarrow |2 - 2x| < \varepsilon \Rightarrow$
$2|x - 1| < \varepsilon \Rightarrow |x - 1| < \frac{\varepsilon}{2}$ $\quad\therefore$ let $\delta = \min\left(\frac{\varepsilon}{6}, \frac{\varepsilon}{2}\right) = \frac{\varepsilon}{6}$

39. $\lim_{x \to 2} f(x) = 5$ if for every $\varepsilon > 0$, there exists a $\delta > 0$ so that $|f(x) - 5| < \varepsilon$ when $0 < |x - 2| < \delta$ if f is defined on an open interval about 2, except possibly at 2.

41. Given $0 < \varepsilon < 4$. $|x^2 - 4| < \varepsilon \Rightarrow -\varepsilon < x^2 - 4 < \varepsilon \Rightarrow 4 - \varepsilon < x^2 < 4 + \varepsilon$ Since $0 < \varepsilon < 4$, $4 - \varepsilon > 0 \Rightarrow$
$\sqrt{4 - \varepsilon} < x < \sqrt{4 + \varepsilon}$ $\;\therefore$ let $\delta = \min(\sqrt{4 + \varepsilon} - 2, 2 - \sqrt{4 - \varepsilon}) = \sqrt{4 + \varepsilon} - 2$ since $4 - \varepsilon < 4 + \varepsilon$

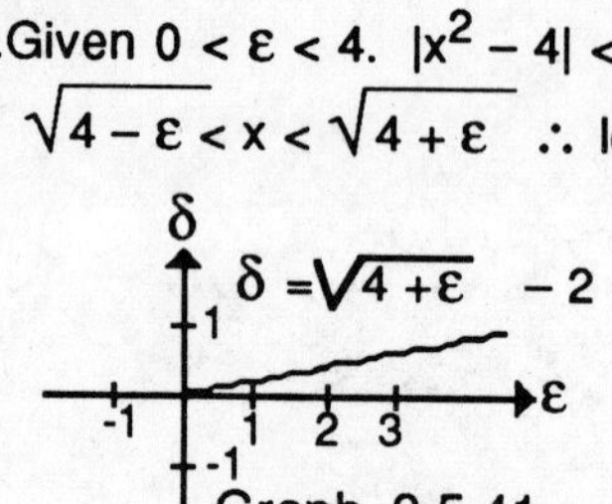

$\delta$
$\delta = \sqrt{4 + \varepsilon} - 2$
1, -1, 1 2 3, -1, $\varepsilon$

Graph 2.5.41

## PRACTICE EXERCISES, CHAPTER 2

1. 1 $\quad$ 3. $-4$ $\quad$ 5. $\frac{2}{3}$ $\quad$ 7. 0 $\quad$ 9. $-1$

11. $\lim_{x \to -1} \frac{x^2 - x - 2}{x + 1} = \lim_{x \to -1} \frac{(x - 2)(x + 1)}{x + 1} = \lim_{x \to -1} (x - 2) = -3$

13. $\lim_{x \to 1} \frac{x^2 - 1}{x - 1} = \lim_{x \to 1} \frac{(x - 1)(x + 1)}{x - 1} = \lim_{x \to 1} (x + 1) = 2$

15. $\lim_{x \to 2} \frac{x - 2}{x^2 + x - 6} = \lim_{x \to 2} \frac{x - 2}{(x - 2)(x + 3)} = \lim_{x \to 2} \frac{1}{x + 3} = \frac{1}{5}$

17. $\frac{2}{5}$ $\quad$ 19. 0 $\quad$ 21. $-\infty$ $\quad$ 23. $\infty$ $\quad$ 25. $\infty$

27. $\lim_{x \to 0} \frac{\sin 2x}{4x} = \lim_{x \to 0} \left(\frac{1}{2}\right)\left(\frac{\sin 2x}{2x}\right) = \frac{1}{2}$

29. $\lim_{x \to \pi} \sin\left(\frac{x + \pi}{4}\right) \cos\left(\frac{x - \pi}{x}\right) = 1 \cdot 1 = 1$

31. a) $\infty$
b) $-\infty$

33. a) $\lim_{x \to -1^+} f(x) = 1$ $\quad \lim_{x \to 1^+} f(x) = 1$

$\lim_{x \to -1^-} f(x) = 1$ $\quad \lim_{x \to 1^-} f(x) = -1$

$\lim_{x \to 0^+} f(x) = 0$

$\lim_{x \to 0^-} f(x) = 0$

b) $\lim_{x \to -1} f(x) = 1$

$\lim_{x \to 0} f(x) = 0$

$\lim_{x \to 1} f(x)$ does not exist since

$\lim_{x \to 1^+} f(x) \neq \lim_{x \to 1^-} f(x)$

c) f is continuous at x = −1 and x = 0.

35. a)

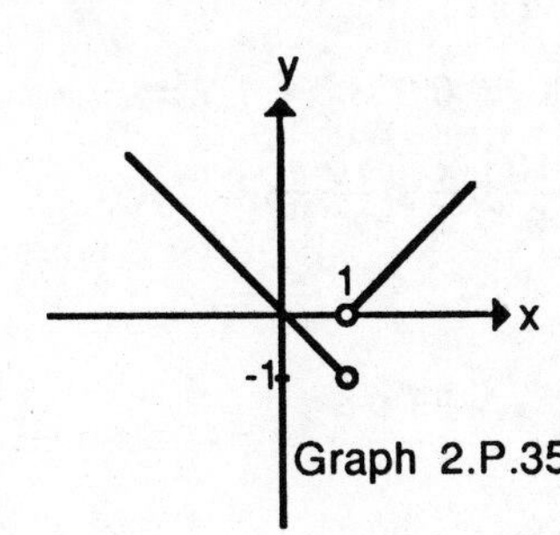

Graph 2.P.35a

b) $\lim_{x \to 1^+} f(x) = 0$

$\lim_{x \to 1^-} f(x) = -1$

c) None, not possible to make f continuous at x = 1.

37. a) $\lim_{x \to c} 3f(x) = (3)(-7) = -21$

b) $\lim_{x \to c} (f(x))^2 = (-7)^2 = 49$

c) $\lim_{x \to c} (f(x))(g(x)) = (-7)(0) = 0$

d) $\lim_{x \to c} \frac{f(x)}{g(x) - 7} = \frac{-7}{0 - 7} = 1$

e) $\lim_{x \to c} \cos(g(x)) = \cos 0 = 1$

f) $\lim_{x \to c} |f(x)| = |-7| = 7$

39. If $0 \le \left|\sqrt{x} \sin \frac{1}{x}\right| \le \sqrt{x}$ and since $\lim_{x \to 0^+} 0 = 0$ and $\lim_{x \to 0^+} \sqrt{x} = 0$, then $\lim_{x \to 0^+} \sqrt{x} \sin \frac{1}{x} = 0$ by the sandwich theorem.

41. Since $-1 \le \sin x \le 1$ for all x, $\frac{-1}{\sqrt{x}} \le \frac{\sin x}{\sqrt{x}} \le \frac{1}{\sqrt{x}}$ for all x > 0. Since $\lim_{x \to \infty} \frac{-1}{\sqrt{x}} = 0$ and $\lim_{x \to \infty} \frac{1}{\sqrt{x}} = 0$,

$\lim_{x \to \infty} \frac{\sin x}{\sqrt{x}} = 0$ by the sandwich theorem.

43. $\lim_{x \to \infty} \frac{x + \sin x}{x} = \lim_{x \to \infty} \left(1 + \frac{\sin x}{x}\right) = 1$

45. $\lim_{x \to 3} \frac{x^2 + 2x - 15}{x - 3} = \lim_{x \to 3} \frac{(x - 3)(x + 5)}{x - 3} = \lim_{x \to 3} (x + 5) = 8$. ∴ let k = 8

47. No, because the interval is not closed.

49. True. The Intermediate Value Theorem says since f is continuous on the interval [1,2] and since f(1) = 0, f(2) = 3, then f takes on all values between 0 and 3, including 2.5 for some $x \in (1,2)$.

51. $\lim_{x \to 1} f(x) = 3$ if for every $\varepsilon > 0$ there exists a $\delta > 0$ so that $|f(x) - 3| < \varepsilon$ whenever $0 < |x - 1| < \delta$ providing f is defined on an open interval about x = 1, except possibly at x = 1.

53. $\lim_{x \to 0} x^2 = 0$. But as $x \to 0$, $x^2$ gets closer to −1 (yes it does--think about it). But $L \neq -1$. (There are other answers to this exercise.)

55. $|f(x) - 1| < \varepsilon \Rightarrow |(2x - 3) - 1| < \varepsilon \Rightarrow$

$|2x - 4| < \varepsilon \Rightarrow 2|x - 2| < \varepsilon \Rightarrow$

$|x - 2| < \frac{\varepsilon}{2}$ ∴ let $\delta \le \frac{\varepsilon}{2}$

57. $\lim_{x\to 3}(5x - 10) = 5$

$|f(x) - 5| < 0.05 \Rightarrow |(5x - 10) - 5| < 0.05 \Rightarrow$

$|5x - 15| < 0.05 \Rightarrow 5|x - 3| < 0.05 \Rightarrow$

$|x - 3| < 0.01 \therefore$ let $\delta \le 0.01$

59. $\lim_{x\to 1}(5x - 10) = -5$

$|f(x) - (-5)| < 0.05 \Rightarrow |(5x - 10) - (-5)| < 0.05$

$|5x - 5| < 0.05 \Rightarrow 5|x - 1| < 0.05 \Rightarrow$

$|x - 1| < 0.01 \quad \therefore$ let $\delta \le 0.01$

61. $\lim_{x\to 9} \sqrt{x - 5} = 2$

$|f(x) - 2| < 1 \Rightarrow |\sqrt{x - 5} - 2| < 1 \Rightarrow$

$-1 < \sqrt{x - 5} - 2 < 1 \Rightarrow 1 < \sqrt{x - 5} < 3$

$\Rightarrow 1 < x - 5 < 9 \Rightarrow -3 < x - 9 < 5. \Rightarrow$

$-3 < x - 9 < 3 \therefore$ let $\delta = 3$

63. $\lim_{x\to 2} \frac{2}{x} = 1$

$|f(x) - 1| < 0.1 \Rightarrow \left|\frac{2}{x} - 1\right| < 0.1 \Rightarrow$

$-0.1 < \frac{2}{x} - 1 < 0.1 \Rightarrow 0.9 < \frac{2}{x} < 1.1 \Rightarrow$

$\frac{1}{0.9} > \frac{x}{2} > \frac{1}{1.1} \Rightarrow \frac{10}{9} > \frac{x}{2} > \frac{10}{11} \Rightarrow$

$\frac{20}{9} > x > \frac{20}{11} \Rightarrow \frac{2}{9} > x - 2 > -\frac{2}{11} \Rightarrow$

$\frac{2}{11} > x - 2 > -\frac{2}{11} \quad \therefore$ let $\delta \le \frac{2}{11}$

# CHAPTER 3

# DERIVATIVES

## 3.1 SLOPES, TANGENT LINES, AND DERIVATIVES

1. $\frac{f(x+h)-f(x)}{h} = \frac{3(x+h)^2 - x^2}{h} = \frac{3x^2 + 6xh + 3h^2 - 3x^2}{h} = 6x + 3h$. $f'(x) = \lim_{h \to 0} \frac{3(x+h)^2 - 3x^2}{h} = \lim_{h \to 0} 6x + h = 6x \Rightarrow m = 18$ when $x = 3$. $\therefore$ the tangent line $y - 27 = 18(x - 3) \Rightarrow y = 18x - 27$.

3. $f'(x) = \lim_{h \to 0} \frac{2(x+h)^2 - 5 - (2x^2 - 5)}{h} = 4x \Rightarrow m = 12$ when $x = 3$. $\therefore$ the tangent line $y - 13 = 12(x - 3) \Rightarrow y = 12x - 23$.

5. $\frac{f(x+h)-f(x)}{h} = \frac{(x+h)^2 - 6(x+h) - (x^2 - 6x)}{h} = \frac{x^2 + 2xh + h^2 - 6x - 6h - x^2 + 6x}{h} = 2x + h - 6$. $f'(x) = \lim_{h \to 0} \frac{f(x+h)-f(x)}{h} = 2x - 6 \Rightarrow m = 0$ when $x = 3$. $\therefore$ the tangent line $y + 9 = 0(x - 3) \Rightarrow y = -9$.

7. $\frac{f(x+h)-f(x)}{h} = \frac{2/(x+h) - 2/x}{h} = \frac{2x - 2x - 2h}{x(x+h)h} = \frac{-2}{x(x+h)}$. $f'(x) = \lim_{h \to 0} \frac{f(x+h)-f(x)}{h} = -2/x^2 \Rightarrow m = -2/9$ when $x = 3$. $\therefore$ the tangent line $y - 2/3 = (-2/9)(x - 3) \Rightarrow y = (-2/9)x + 4/3$.

9. $f'(x) = \lim_{h \to 0} \frac{(x+h)/(x+h+1) - x/(x+1)}{h} = 1/(x+1)^2 \Rightarrow m = 1/16$ when $x = 3$. $\therefore$ the tangent line $y - (3/4) = (1/16)(x - 3) \Rightarrow y = (1/16)x + 9/16$.

11. $\frac{f(x+h)-f(x)}{h} = \frac{x + h + \frac{9}{x+h} - x - \frac{9}{x}}{h} = \frac{hx^2 + h^2x + 9x - 9x - 9h}{h(x+h)x} = \frac{x^2 + hx - 9}{x^2}$. $f'(x) = \lim_{h \to 0} \frac{f(x+h)-f(x)}{h} = 1 - \frac{9}{x^2} \Rightarrow m = 0$ when $x = 3$. $\therefore$ the tangent line $y - 6 = 0(x - 3) \Rightarrow y = 6$.

13. $\frac{f(x+h)-f(x)}{h} = \frac{1 + \sqrt{x+h} - 1 - \sqrt{x}}{h} \cdot \frac{\sqrt{x+h} + \sqrt{x}}{\sqrt{x+h} + \sqrt{x}} = \frac{x + h - x}{h(\sqrt{x+h} + \sqrt{x})} = \frac{1}{\sqrt{x+h} + \sqrt{x}}$. $f'(x) = \lim_{h \to 0} \frac{f(x+h)-f(x)}{h} = \frac{1}{2\sqrt{x}} \Rightarrow m = \frac{1}{2\sqrt{3}}$ when $x = 3$. $\therefore$ the tangent line $y - (1 + \sqrt{3}) = \frac{1}{2\sqrt{3}}(x - 3) \Rightarrow$ $y = \frac{1}{2\sqrt{3}}x + 1 + \frac{\sqrt{3}}{2}$.

15. $f'(x) = \lim\limits_{h \to 0} \dfrac{\sqrt{2(x+h)} - \sqrt{2x}}{h} = \dfrac{1}{\sqrt{2x}} \Rightarrow m = \dfrac{1}{\sqrt{6}}$ when $x = 3$. $\therefore$ the tangent line $y - \sqrt{6} = \left(\dfrac{1}{\sqrt{6}}\right)(x-3) \Rightarrow y = \dfrac{1}{\sqrt{6}}x + \dfrac{3}{\sqrt{6}}$.

17. $\dfrac{f(x+h) - f(x)}{h} = \dfrac{1/\sqrt{x+h} - 1/\sqrt{x}}{h} = \dfrac{\sqrt{x} - \sqrt{x+h}}{h\sqrt{x}\sqrt{x+h}} \dfrac{\sqrt{x} + \sqrt{x+h}}{\sqrt{x} + \sqrt{x+h}} = \dfrac{x - x - h}{h\sqrt{x}\sqrt{x+h}(\sqrt{x} + \sqrt{x+h})} = \dfrac{-1}{\sqrt{x}\sqrt{x+h}(\sqrt{x} + \sqrt{x+h})}$. $f'(x) = \lim\limits_{h \to 0} \dfrac{f(x+h) - f(x)}{h} = \dfrac{-1}{2\sqrt{x^3}} \Rightarrow m = \dfrac{-1}{6\sqrt{3}}$ when $x = 3$. $\therefore$ the tangent line $y - \dfrac{1}{\sqrt{3}} = \left(\dfrac{-1}{6\sqrt{3}}\right)(x-3) \Rightarrow y = \dfrac{-1}{6\sqrt{3}}x + \dfrac{\sqrt{3}}{2}$.

19. $\dfrac{f(x+h) - f(x)}{h} = \dfrac{(x+h)^3 - x^3}{h} = \dfrac{x^3 + 3x^2h + 3xh^2 + h^3 - x^3}{h} = 3x^2 + 3xh + h^2$. $f'(x) = \lim\limits_{h \to 0} \dfrac{f(x+h) - f(x)}{h} = 3x^2 \Rightarrow m = 27$ when $x = 3$. $\therefore$ the tangent line $y - 27 = 27(x-3) \Rightarrow y = 27x - 54$.

21. 23. 25.

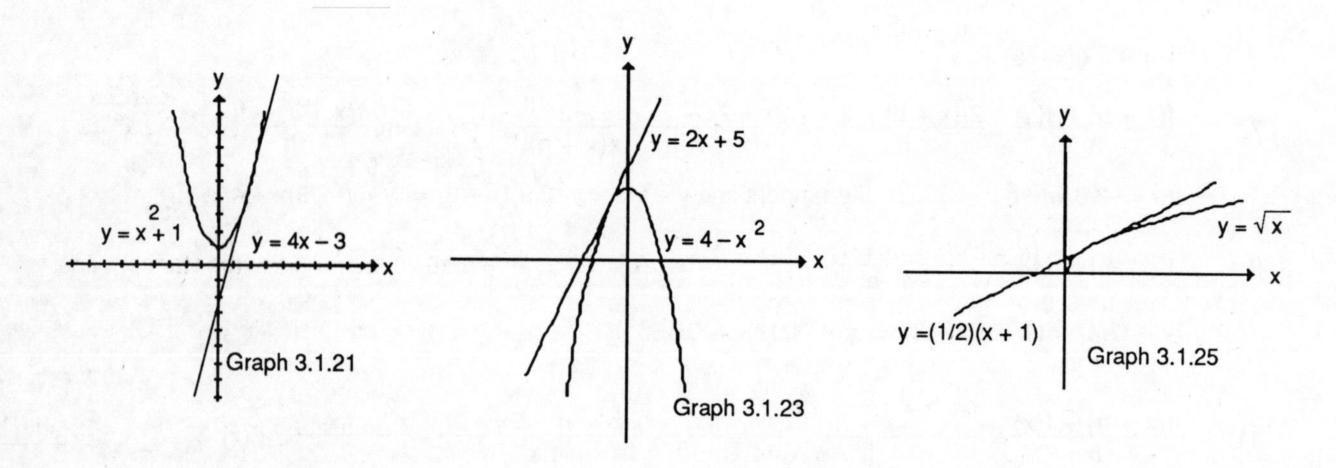


27. a) yes b) no c) no d) no This curve is continuous but appears to have vertical tangents at its extrema.

29. a) yes b) yes c) yes d) no This curve appears to be continuous with no vertical tangent lines at any point in its domain.

31. $\lim\limits_{x \to 0^+} f'(x) = 1$ while $\lim\limits_{x \to 0^-} f'(x) = 0$ implies $\lim\limits_{x \to 0} f'(x)$ does not exist. $f(x)$ is not differentiable at $x = 0$

33. $\lim_{x \to 1^+} f'(x) = 2$ while $\lim_{x \to 1^-} f'(x) = 1/2$ implies $\lim_{x \to 1} f'(x)$ does not exist. f(x) is not differentiable at x = 1

35. Since derivatives have the Intermediate Value Property, f(x) must assume every value between f(−1) = 0 and f(1) = 1 which it doesn't. Hence f(x) is not a derivative.

## 3.2 DIFFERENTIATION RULES

1. $\frac{dy}{dx} = 1, \frac{d^2y}{dx^2} = 0$

3. $\frac{dy}{dx} = 2x, \frac{d^2y}{dx^2} = 2$

5. $\frac{dy}{dx} = -2x, \frac{d^2y}{dx^2} = -2$

7. $\frac{dy}{dx} = 2, \frac{d^2y}{dx^2} = 0$

9. $\frac{dy}{dx} = x^2 + x + 1, \frac{d^2y}{dx^2} = 2x + 1$

11. $y' = 4x^3 - 21x^2 + 4x, y'' = 12x^2 - 42x + 4$

13. $y' = 8x - 8, y'' = 8$

15. $y' = 8x^3 - 8x, y'' = 24x^2 - 8$

17. $y' = 2x - 1, y'' = 2$

19. $y' = 2x^3 - 3x - 1, y'' = 6x^2 - 3, y''' = 12x, y^{(4)} = 12$

21. a) $y = (3x - 1)(2x + 5), y' = 3(2x + 5) + 2(3x - 1) = 12x + 13$

b) $y = 6x^2 + 13x - 5, y' = 12x + 13$

23. a) $y = (x + 1)(x^2 + 1), y' = 1(x^2 + 1) + 2x(x + 1) = 3x^2 + 2x + 1$

b) $y = x^3 + x^2 + x + 1, y' = 3x^2 + 2x + 1$

25. a) $y = x^2(x^3 - 1), y' = 2x(x^3 - 1) + x^2(3x^2) = 5x^4 - 2x$ b) $y = x^5 - x^2, y' = 5x^4 - 2x$

27. a) $y = (x - 1)(x^2 + x + 1), y' = 1(x^2 + x + 1) + (2x + 1)(x - 1) = 3x^2$ b) $y = x^3 - 1, y' = 3x^2$

29. $y = \frac{x - 1}{x + 7}, \frac{dy}{dx} = \frac{1(x + 7) - 1(x - 1)}{(x + 7)^2} = \frac{8}{(x + 7)^2}$

31. $y = \frac{2x + 1}{x^2 - 1}, \frac{dy}{dx} = \frac{2(x^2 - 1) - 2x(2x + 1)}{(x^2 - 1)^2} = \frac{-2(x^2 + x + 1)}{(x^2 - 1)^2}$

33. $y = (1 - x)(1 + x^2)^{-1} = \frac{1 - x}{1 + x^2}, \frac{dy}{dx} = \frac{-1(1 + x^2) - (2x)(1 - x)}{(1 + x^2)^2} = \frac{x^2 - 2x - 1}{(1 + x^2)^2}$

35. $y = \frac{x^2}{1 - x^3}, \frac{dy}{dx} = \frac{2x(1 - x^3) - x^2(-3x^2)}{(1 - x^3)^2} = \frac{x^4 + 2x}{(1 - x^3)^2}$

37. $y = \frac{3}{x^2} = 3x^{-2}, y' = -6x^{-3} = \frac{-6}{x^3}, y'' = 18x^{-4} = \frac{18}{x^4}$

39. $y = \frac{1}{2x^2} = \frac{1}{2}x^{-2}$, $y' = -x^{-3} = \frac{-1}{x^3}$, $y'' = 3x^{-4} = \frac{3}{x^4}$

41. $y = \frac{5}{x^4} = 5x^{-4}$, $y' = -20x^{-5} = \frac{-20}{x^5}$, $y'' = 100x^{-6} = \frac{100}{x^6}$

43. $y = x + 1 + \frac{1}{x} = x + 1 + x^{-1}$, $y' = 1 - x^{-2} = 1 - \frac{1}{x^2}$, $y'' = 2x^{-3} = \frac{2}{x^3}$

45. $y = \frac{x^3 + 7}{x} = x^2 + 7x^{-1}$, $\frac{dy}{dx} = 2x - 7x^{-2} = 2x - \frac{7}{x^2}$

47. $y = \frac{(x-1)(x^2+x+1)}{x^3} = \frac{x^3 - 1}{x^3} = 1 - x^{-3}$, $\frac{dy}{dx} = 3x^{-4} = \frac{3}{x^4}$

49. a) $\frac{d(u\,v)}{dx} = u\,v' + v\,u' = (5)(2) + (-1)(-3) = 13$ b) $\frac{d(u/v)}{dx} = \frac{u'v - u\,v'}{v^2} = \frac{(-3)(-1) - (5)(2)}{(-1)^2} = -7$

c) $\frac{d(v/u)}{dx} = \frac{v'u - u'v}{u^2} = \frac{(2)(5) - (-3)(-1)}{5^2} = \frac{7}{25}$ d) $\frac{d(7v - 2u)}{dx} = 7v' - 2u' = 7(2) - 2(-3) = 20$

51 $y = x^2 + 5x$, $m = 2x + 5 \therefore m = 11$ when $x = 3$

53. $y = x^3 - 3x + 1 \Rightarrow y' = 3x^2 - 3 \Rightarrow m = 3(2)^2 - 3 = 9$ the slope of the tangent line $\therefore \frac{-1}{9}$ is the slope of the normal line; the desired equation is $y - 3 = \frac{-1}{9}(x - 2) \Rightarrow y = \frac{-1}{9}x + \frac{29}{9}$

55. $y = 2x^3 - 3x^2 - 12x + 20 \Rightarrow y' = 6x^2 - 6x - 12 = 6(x^2 - x - 2) = 6(x-2)(x+1)$; since $m = y'$, solving $6(x-2)(x+1) = 0 \Rightarrow x = 2$ or $-1 \Rightarrow (2, 0)$ and $(-1, 27)$ are the desired points.

57. $y = \frac{4x}{x^2 + 1}$, $y' = \frac{4(x^2+1) - 2x(4x)}{(x^2+1)^2} = \frac{4(1 - x^2)}{(x^2+1)^2} \Rightarrow m = \frac{4(1-x^2)}{(x^2+1)^2}$; at $(0, 0)$ $m = 4$, $y - 0 = 4(x - 0) \Rightarrow$ $y = 4x$; at $(1, 2)$ $m = 0$, $y - 1 = 0(x - 2) \Rightarrow y = 2$

59. $P = \frac{nRT}{V - nb} - \frac{an^2}{V^2} \Rightarrow \frac{dP}{dV} = \frac{0(V - nb) - 1(nRT)}{(V - nb)^2} - \frac{0V^2 - 2V(an^2)}{(V^2)^2} = -\frac{nRT}{(V - nb)^2} + \frac{2an^2}{V^3}$

61. $R = M^2\left[\frac{C}{2} - \frac{M}{3}\right] = \frac{C}{2}M^2 - \frac{1}{3}M^3 \Rightarrow \frac{dR}{dM} = CM - M^2$

## 3.3 VELOCITY, SPEED AND OTHER RATES OF CHANGE

1. $s = .8t^2$ a) $\Delta s = s(2) - s(0) = 3.2$ m, $v_{av} = \frac{\Delta s}{\Delta t} = 1.6$ m/sec b) $v = 1.6t \Rightarrow v(2) = 3.2$m/sec, $a = 1.6 \Rightarrow a(2) = 1.6$ m/sec$^2$

3. $s = 2t^2 + 4t - 3$ a) $\Delta s = s(2) - s(0) = 18$ m, $v_{av} = \frac{\Delta s}{\Delta t} = 9$ m/sec b) $v = 4t + 5 \Rightarrow v(2) = 13$ m/sec, $a = 4 \Rightarrow a(2) = 4$ m/sec$^2$

5. $s = 4 - 2t - t^2$ a) $\Delta s = s(2) - s(0) = -8$ m, $v_{av} = \frac{\Delta s}{\Delta t} = -4$ m/sec b) $v = -2 - 2t \Rightarrow v(2) = -6$ m/sec, $a = -2 \Rightarrow a(2) = -2$ m/sec$^2$

7. $s = 4t + 3$ a) $\Delta s = s(2) - s(0) = 8$ m, $v_{av} = \frac{\Delta s}{\Delta t} = 4$ m/sec b) $v = 4 \Rightarrow v(2) = 4$ m/sec,

$a = 0 \Rightarrow a(2) = 0$ m/sec$^2$

9. $s_m = 1.86t^2 \Rightarrow v_m = 3.72t$, solving $3.72t = 16.6 \Rightarrow t = 4.46$ sec on Mars; $s_j = 11.44t^2 \Rightarrow v_j = 22.88t$, solving $22.88t = 16.6 \Rightarrow t = .726$ sec on Jupiter

11. $s = 24t - 4.9t^2 \Rightarrow v = s' = 24 - 9.8t$, solving $v = 0 \Rightarrow t = 24/9.8$ sec is the time it takes to reach its maximum height. The maximum is $s(24/9.8) = 1440/49$ m $\approx 29.39$ m.

13. $b(t) = 10^6 + 10^4 t - 10^3 t^2 \Rightarrow b'(t) = 10^4 - (2)(10^3 t) = 10^3[10 - 2t]$ a) $b'(0) = 10^4$ bacteria/hr

b) $b'(5) = 0$ bacteria/hr c) $b'(10) = -10^4$ bacteria/hr

15. $c(x) = 2000 + 100x - .1x^2 \Rightarrow c'(x) = 100 - .2x$ a) $c_{av} = \frac{c(100)}{100} = \$110$ per machine

b) Marginal cost = $c'(x) \Rightarrow$ the marginal cost of producing 100 machines is $c'(100) = \$80$.

c) The cost of producing the 101st machine is $c(101) - c(100) = 100 - \frac{201}{10} = \$79.90$.

17. $s = t^3 - 6t^2 + 9t \Rightarrow v = s' = 3t^2 - 12t + 9 \Rightarrow a = v' = s'' = 6t - 12$; $v = 0 \Rightarrow 3t^2 - 12t + 9 = 0 \Rightarrow$ $3(t^2 - 4t + 3) = 0 \Rightarrow 3(t-1)(t-3) = 0 \Rightarrow$ when $t = 1$ or 3, the velocity is zero; $a(1) = -6$ m/sec$^2$ and $a(3) = 6$ m/sec$^2$

19. a) 190 ft/sec b) 2 sec c) at 8 sec, 0 ft/sec d) at 10.8 sec, 90 ft/ sec

e) $10.8 - 8 = 2.8$ sec f) at 2 sec, between 2 and 10.8 sec

21. a) 0, 0 b) largest 1700, smallest about 1400

23. b 24. a 25. d 26. c

27. a)

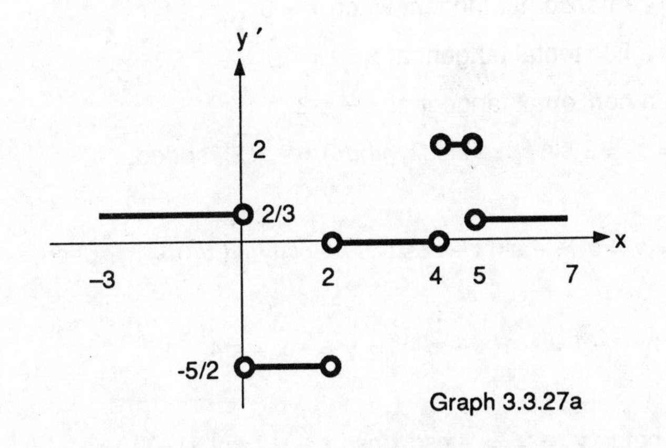

Graph 3.3.27a

b) 0, 2, 4, and 5

29. a)

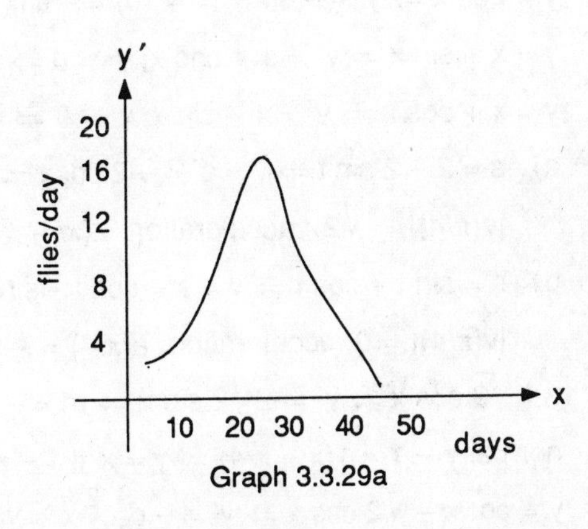

Graph 3.3.29a

b) The fastest is between the 20th and 30th days; slowest is between the 40th and 50th days.

## 3.4 DERIVATIVES OF TRIGONOMETRIC FUNCTIONS

1. $y = 1 + x - \cos x \Rightarrow \frac{dy}{dx} = 1 + \sin x$

3. $y = 1/x + 5 \sin x = x^{-1} + 5 \sin x \Rightarrow \frac{dy}{dx} = -x^{-2} + 5 \cos x = -\frac{1}{x^2} + 5 \cos x$

5. $y = \csc x - 5x + 7 \Rightarrow \frac{dy}{dx} = -\csc x \cot x - 5$

7. $y = x \sec x \Rightarrow \frac{dy}{dx} = \sec x + x \sec x \tan x$, product rule

9. $y = x^2 \cot x \Rightarrow \frac{dy}{dx} = 2x \cot x - x^2 \csc^2 x$, product rule

11. $y = 3x + x \tan x \Rightarrow \frac{dy}{dx} = 3 + \tan x + x \sec^2 x$, product rule

13. $y = \sin x \sec x = \tan x \Rightarrow \frac{dy}{dx} = \sec^2 x$

15. $y = \tan x \cot x = 1 \Rightarrow \frac{dy}{dx} = 0$

17. $y = \frac{4}{\cos x} = 4 \sec x \Rightarrow \frac{dy}{dx} = 4 \sec x \tan x$

19. $y = \frac{\cos x}{x} \Rightarrow \frac{dy}{dx} = \frac{(-\sin x)(x) - (1)(\cos x)}{x^2} = -\frac{x \sin x + \cos x}{x^2}$, quotient rule

21. $y = \frac{x}{1 + \cos x} \Rightarrow \frac{dy}{dx} = \frac{(1)(1 + \cos x) - (x)(-\sin x)}{(1 + \cos x)^2} = \frac{1 + \cos x + x \sin x}{(1 + \cos x)^2}$, quotient rule

23. $y = \frac{\cot x}{1 + \cot x} \Rightarrow \frac{dy}{dx} = \frac{(-\csc^2 x)(1 + \cot x) - (-\csc^2 x)(\cot x)}{(1 + \cot x)^2} = \frac{-\csc^2 x - \csc^2 x \cot x + \csc^2 x \cot x}{(1 + \cot x)^2} = -\frac{\csc^2 x}{(1 + \cot x)^2}$, quotient rule

25. $y = \csc x \Rightarrow y' = -\csc x \cot x \Rightarrow y'' = -[(-\csc x \cot x)(\cot x) + (\csc x)(-\csc^2 x)] = \csc x \cot^2 x + \csc^3 x = \csc^3 x + \csc x[\csc^2 x - 1] = \csc^3 x + \csc^3 x - \csc x = 2\csc^3 x - \csc x$

27. $y = \sin x \Rightarrow y' = \cos x \Rightarrow m = 1$, tangent $y - 0 = 1(x - 0) \Rightarrow y = x$, normal $y - 0 = -1(x - 0) \Rightarrow y = -x$

29. $y = \cos x \Rightarrow y' = -\sin x \Rightarrow m = 0$, tangent $y + 1 = 0(x - \pi) \Rightarrow y = -1$, normal $x = \pi$

31. $y = \sec x \Rightarrow y' = \sec x \tan x \Rightarrow y'(0) = \sec(0)\tan(x) = 0 \Rightarrow$ a horizontal tangent when $x = 0$; $y = \cos x \Rightarrow y' = -\sin x \Rightarrow y'(0) = -\sin(0) = 0 \Rightarrow$ a horizontal tangent when $x = 0$

33. $y = x + \sin x \Rightarrow y' = 1 + \cos x$; $y' = 0 \Rightarrow x = \pi \therefore$ a horizontal tangent at $x = \pi$

35. $y = x + \cos x \Rightarrow y' = 1 - \sin x$; $y' = 0 \Rightarrow x = 1 \therefore$ a horizontal tangent at $x = \pi/2$

37. a) $s = 2 - 2 \sin t \Rightarrow v = s' = -2 \cos t \Rightarrow a = v' = s'' = 2 \sin t \therefore$ velocity, $v(\pi/4) = -\sqrt{2}$; speed, $|v(\pi/4)| = \sqrt{2}$; acceleration, $a(\pi/4) = \sqrt{2}$

    b) $s = \sin t + \cos t \Rightarrow v = s' = \cos t - \sin t \Rightarrow a = v' = s'' = -\sin t - \cos t \therefore$ velocity, $v(\pi/4) = 0$; speed, $|v(\pi/4)| = 0$; acceleration, $a(\pi/4) = -\sqrt{2}$

39. $y = \sqrt{2} \cos x \Rightarrow y' = -\sqrt{2} \sin x \Rightarrow m = -1$, tangent $y - 1 = -1(x - \pi/4) \Rightarrow y = -x + \pi/4 + 1$, normal $y - 1 = 1(x - \pi/4) \Rightarrow y = x + 1 - \pi/4$

41. $y = \cot x - \sqrt{2} \csc x \Rightarrow y' = -\csc^2 x + \sqrt{2} \csc x \cot x$; $y' = 0 \Rightarrow -\csc x[\csc x - \sqrt{2} \cot x] = 0 \Rightarrow$ $\csc x \neq 0$ and $\csc x - \sqrt{2} \cot x = 0 \Rightarrow x = \pi/4$, now $y(\pi/4) = -1 \Rightarrow y = -1$ is the horizontal tangent

43.

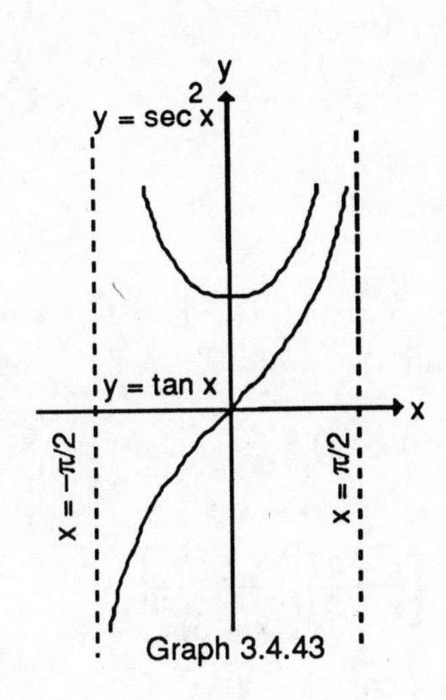

Graph 3.4.43

45.

| h | $\frac{1 - \cos h}{h^2}$ |
|---|---|
| 1 | .459697694 |
| .1 | .499583474 |
| .01 | .499958 |
| .001 | .5 |

The limit appears to be $\frac{1}{2}$.

47. $y = \sec x = \frac{1}{\cos x} \Rightarrow y' = \frac{(0)(\cos x) - (-\sin x)(1)}{\cos^2 x} = \left[\frac{1}{\cos x}\right]\left[\frac{\sin x}{\cos x}\right] = \sec x \tan x$

49. $y = \csc x = \frac{1}{\sin x} \Rightarrow y' = \frac{(0)(\sin x) - (\cos x)(1)}{\sin^2 x} = -\left[\frac{1}{\sin x}\right]\left[\frac{\cos x}{\sin x}\right] = -\csc x \cot x$

## 3.5 THE CHAIN RULE

1. $y = \sin(x + 1) \Rightarrow \frac{dy}{dx} = \cos(x + 1)$

3. $y = \cos(5x) \Rightarrow \frac{dy}{dx} = -5 \sin(5x)$

5. $y = \sin(2\pi x/5) \Rightarrow \frac{dy}{dx} = (2\pi/5)\cos(2\pi x/5)$

7. $y = \tan(2 - x) \Rightarrow \frac{dy}{dx} = -\sec^2(2 - x)$

9. $y = \sec(2x - 1) \Rightarrow \frac{dy}{dx} = 2 \sec(2x - 1) \tan(2x - 1)$

11. $y = \csc(x^2 + 7x) \Rightarrow \frac{dy}{dx} = [-\csc(x^2 + 7x) \cot(x^2 + 7x)](2x + 7) = -(2x + 7)(\csc(x^2 + 7x) \cot(x^2 + 7x)$

13. $y = \cot(3x + \pi) \Rightarrow \frac{dy}{dx} = [-\csc^2(3x + \pi)](3) = -3\csc^2(3x + \pi)$

15. $y = \sec^2 x - \tan^2 x = \sec^2 x - (\sec^2 x - 1) = 1 \Rightarrow \frac{dy}{dx} = 0$

17. $y = \sin^3 x \Rightarrow \frac{dy}{dx} = 3(\sin^2 x)(\cos x)$

19. $y = (2x + 1)^5 \Rightarrow \frac{dy}{dx} = 5(2x + 1)^4(2) = 10(2x + 1)^4$

21. $y = (x + 1)^{-3} \Rightarrow \frac{dy}{dx} = (-3)(x + 1)^{-4}(1) = -3(x + 1)^{-4}$

23. $y = \left[1 - \frac{x}{7}\right]^{-7} \Rightarrow \frac{dy}{dx} = [-7]\left[1 - \frac{x}{7}\right]^{-8}\left[-\frac{1}{7}\right] = \left[1 - \frac{x}{7}\right]^{-8}$

25. $y = [1 + x - 1/x]^3 = \left[1 + x - x^{-1}\right]^3 \Rightarrow \frac{dy}{dx} = 3\left[1 + x - x^{-1}\right]^2\left[1 - (-1)(x^{-2})\right] =$
$3\left[1 + x - x^{-1}\right]^2\left[1 + x^{-2}\right] = 3[1 + x - 1/x]^2\left[1 + 1/x^2\right]$

27. $y = (x^2 + 2x + 3)^3 \Rightarrow \frac{dy}{dx} = 3(x^2 + 2x + 3)^2(2x + 2) = 6(x^2 + 2x + 3)(x + 1)$

29. $y = \cos(\sin x) \Rightarrow \frac{dy}{dx} = -[\sin(\sin x)][\cos x]$

31. $y = 1 + x + \sec^2 x \Rightarrow \frac{dy}{dx} = 1 + 2(\sec x)(\sec x \tan x) = 1 + 2(\sec^2 x)(\tan x)$

33. $y = (\csc x + \cot x)^{-1} \Rightarrow \frac{dy}{dx} = (-1)(\csc x + \cot x)^{-2}(-\csc x \cot x - \csc^2 x) =$
$(\csc x)(\cot x + \csc x)(\csc x + \cot x)^{-2} = (\csc x)(\cot x + \csc x)^{-1}$

35. $y = \sin\left[\frac{x-2}{x+3}\right] \Rightarrow \frac{dy}{dx} = \cos\left[\frac{x-2}{x+3}\right]\left[\frac{(1)(x+3) - (1)(x-2)}{(x+3)^2}\right] = \cos\left[\frac{x-2}{x+3}\right]\left[\frac{5}{(x+3)^2}\right]$

37. $y = \sin^2(3x - 2) \Rightarrow \frac{dy}{dx} = [2\sin(3x - 2)][\cos(3x - 2)][3] = 6[\sin(3x - 2)]\cos(3x - 2)$

39. $y = (1 + \cos 2x)^2 \Rightarrow \frac{dy}{dx} = 2(1 + \cos 2x)(-\sin 2x)(2) = -4(1 + \cos 2x)(\sin 2x)$

41. $y = \sin(\cos(2x - 5)) \Rightarrow \frac{dy}{dx} = [\cos(\cos(2x - 5))](-\sin(2x - 5))(2) = -2[\cos(\cos(2x - 5))][\sin(2x - 5)]$

43. $y = \tan x \Rightarrow y' = \sec^2 x \Rightarrow y'' = 2(\sec x)(\sec x \tan x) = 2(\sec^2 x)\tan x$

45. $y = \cot x \Rightarrow y' = -\csc^2 x \Rightarrow y'' = -2(\csc x)(-\csc x \cot x) = 2(\csc^2 x)(\cot x)$

47. $f(u) = u^5 + 1 \Rightarrow f'(u) = 5u^4$, $u = g(x) = \sqrt{x} \Rightarrow g'(x) = (1/2)\,x^{-1/2}$, $x = 1 \Rightarrow u = 1 \Rightarrow (f\circ g)' =$
$\left[5u^4 \Big|_{u=1}\right]\left[(1/2)x^{-1/2}\Big|_{x=1}\right] = (5)(1/2) = 5/2$

49. $f(u) = \cot\frac{\pi u}{10} \Rightarrow f'(u) = -\left[\csc^2\frac{\pi u}{10}\right]\left[\frac{\pi}{10}\right]$, $u = g(x) = 5\sqrt{x} \Rightarrow g'(x) = \frac{5}{2\sqrt{x}}$, $x = 1 \Rightarrow u = 5$, $(f\circ g)' =$
$\left[-\left[\csc^2\frac{\pi u}{10}\right]\left[\frac{\pi}{10}\right]\Big|_{u=5}\right]\left[\frac{5}{2\sqrt{x}}\Big|_{x=1}\right] = \left[-\frac{\pi}{10}\right]\left[\frac{5}{2}\right] = -\frac{\pi}{4}$

51. $f(u) = \frac{2u}{u^2 + 1} \Rightarrow f'(u) = \frac{2(u^2 + 1) - 2u(2u)}{(u^2 + 1)^2} = \frac{2 - 2u^2}{(u^2 + 1)^2}$, $u = g(x) = 10x^2 + x + 1 \Rightarrow g'(x) = 20x + 1$,
$x = 0 \Rightarrow u = 1$, $(f\circ g)' = \left[\frac{2 - 2u^2}{(u^2 + 1)^2}\Big|_{u=1}\right]\left[(20x + 1)\Big|_{x=0}\right] = (0)(1) = 0$

53. a) $y = \cos u \Rightarrow \frac{dy}{du} = -\sin u$, $u = 6x + 2 \Rightarrow \frac{du}{dx} = 6$, $\frac{dy}{dx} = \frac{dy}{du}\frac{du}{dx} = \left[(-\sin u)\Big|_{u=6x+2}\right][6] =$
$(-\sin(6x + 2))(6) = -6\sin(6x + 2)$

b) $y = \cos 2u \Rightarrow \frac{dy}{du} = -2\sin 2u$, $u = 3x + 1 \Rightarrow \frac{du}{dx} = 3$, $\frac{dy}{dx} = \frac{dy}{du}\frac{du}{dx} = \left[(-2\sin 2u)\Big|_{u=3x+1}\right][3] =$
$[-2\sin(6x + 2)][3] = -6\sin(6x + 2)$

55. a) $y = u/5 + 7 \Rightarrow \frac{dy}{du} = 1/5$, $u = 5x - 35 \Rightarrow \frac{du}{dx} = 5$, $\frac{dy}{dx} = \frac{dy}{du}\frac{du}{dx} = \left[(1/5)\Big|_{u=5x-35}\right][5] = (1/5)(5) = 1$

b) $y = 1 + 1/u = 1 + u^{-1} \Rightarrow \frac{dy}{du} = -u^{-2}$, $u = 1/(x - 1) = (x - 1)^{-1} \Rightarrow \frac{du}{dx} = -(x - 1)^{-2}$,
$\frac{dy}{dx} = \frac{dy}{du}\frac{du}{dx} = \left[-u^{-2}\Big|_{u=1/(x-1)}\right]\left[-(x - 1)^{-2}\right] = \left[-(x - 1)^2\right]\left[-(x - 1)^{-2}\right] = 1$

57. $s = \cos\theta \Rightarrow \frac{ds}{d\theta} = -\sin\theta,\ \frac{d\theta}{dt} = 5 \Rightarrow \left.\frac{ds}{dt}\right|_{\theta = 3\pi/2} = \left.\frac{ds}{d\theta}\frac{d\theta}{dt}\right|_{\theta = 3\pi/2} = (-\sin 3\pi/2)(5) = -(-1)(5) = 5$

59. $y = \sin(x/2) \Rightarrow m = y' = (1/2)\cos(x/2)$ which has a maximum value of 1/2

61. $y = 2\tan(\pi x/4) \Rightarrow m\,|_{x=1} = y'\,|_{x=1} = (\pi/2)\sec^2(\pi x/4)\,|_{x=1} = \pi,\ y(1) = 2\ \therefore$ the tangent line is $y - 2 = \pi(x-1) \Rightarrow y = \pi x + 2 - \pi$, while the normal line is $y - 2 = -\frac{1}{\pi}(x-1) \Rightarrow y = -\frac{1}{\pi}x + 2 + \frac{1}{\pi}$

63. a) $\left.\frac{d}{dx}\{2\,f(x)\}\right|_{x=2} = 2\,f'(2) = (2)(1/3) = 2/3$

b) $\left.\frac{d}{dx}\{f(x) + g(x)\}\right|_{x=3} = f'(3) + g'(3) = 2\pi + 5$

c) $\left.\frac{d}{dx}\{f(x)\cdot g(x)\}\right|_{x=3} = f'(3)\cdot g(3) + f(3)\cdot g'(3) = (2\pi)(-4) + (3)(5) = -8\pi + 15$

d) $\left.\frac{d}{dx}\left\{\frac{f(x)}{g(x)}\right\}\right|_{x=2} = \frac{f'(2)\,g(2) - g'(2)\,f(2)}{(g(2))^2} = \frac{(1/3)(2) - (-3)(8)}{(2)^2} = \frac{2/3 + 24}{4} = \frac{37}{6}$

e) $\left.\frac{d}{dx}\{f(g(x))\}\right|_{x=2} = f'(g(2))\,g'(2) = f'(2)\cdot(-3) = (1/3)(-3) = -1$

f) $\left.\frac{d}{dx}\left\{\sqrt{f(x)}\right\}\right|_{x=2} = \frac{1}{2\sqrt{f(2)}}\,f'(2) = \frac{1/3}{2\sqrt{8}} = \frac{1}{12\sqrt{2}} = \frac{\sqrt{2}}{24}$

g) $\left.\frac{d}{dx}\left\{\frac{1}{g(x)^2}\right\}\right|_{x=3} = \frac{-2}{g(3)^3}\,g'(3) = \frac{-10}{-64} = \frac{5}{32}$

h) $\left.\frac{d}{dx}\left\{\sqrt{f(2)^2 + g(2)^2}\right\}\right|_{x=2} = \frac{1}{2\sqrt{f(x)^2 + g(x)^2}}\,[2\,f(2)\,f'(2) + 2\,g(2)\,g'(2)] =$

$\frac{(8)(1/3) + (2)(-3)}{\sqrt{8^2 + 2^2}} = \frac{-10}{3\sqrt{68}} = \frac{-5}{3\sqrt{17}}$

65. $V = \frac{4}{3}\pi r^3,\ r = 10$ cm, $\frac{dr}{dt} = 1/2$ cm/sec $\Rightarrow \left.\frac{dV}{dt}\right|_{r=10} = \left.\frac{4}{3}\pi\,3r^2\frac{dr}{dt}\right|_{r=10} = \frac{4}{3}\pi\,3\cdot 10^2\cdot\frac{1}{2} =$

$200\pi$ cm$^3$/sec

## 3.6 IMPLICIT DIFFERENTIATION AND FRACTIONAL POWERS

1. $y = x^{9/4} \Rightarrow \frac{dy}{dx} = (9/4)x^{5/4}$

3. $y = \sqrt[3]{x} = x^{1/3} = (1/3)x^{-2/3} = \frac{1}{3\sqrt[3]{x^2}}$

5. $y = (2x + 5)^{-1/2} \Rightarrow y' = (-1/2)(2x + 5)^{-3/2}(2) = -(2x + 5)^{-3/2}$

7. $y = x\sqrt{x^2 + 1} = x(x^2 + 1)^{1/2} \Rightarrow y' = (x^2 + )^{1/2} + (x/2)(x^2 + 1)^{-1/2}(2x) = \frac{2x^2 + 1}{\sqrt{x^2 + 1}}$

9. $x^2y + xy^2 = 6 \Rightarrow 2xy + x^2\frac{dy}{dx} + y^2 + 2xy\frac{dy}{dx} = 0 \Rightarrow (x^2 + 2xy)\frac{dy}{dx} = -2xy - y^2 \therefore \frac{dy}{dx} = \frac{-2xy - y^2}{x^2 + 2xy}$

11. $2xy + y^2 = x + y \Rightarrow 2y + 2x\frac{dy}{dx} = 1 + \frac{dy}{dx} \Rightarrow (2x + 2y - 1)\frac{dy}{dx} = 1 - 2y \therefore \frac{dy}{dx} = \frac{1 - 2y}{2x + 2y - 1}$

13. $x^2y^2 = x^2 + y^2 \Rightarrow 2xy^2 + 2x^2y\frac{dy}{dx} = 2x + 2y\frac{dy}{dx} \Rightarrow (2x^2y - 2y)\frac{dy}{dx} = 2x - 2xy^2 \therefore \frac{dy}{dx} = \frac{x(1 - y^2)}{y(x^2 - 1)}$

15. $y^2 = \frac{x - 1}{x + 1} \Rightarrow 2y\frac{dy}{dx} = \frac{(1)(x + 1) - (1)(x - 1)}{(x + 1)^2} \Rightarrow 2y\frac{dy}{dx} = \frac{2}{(x + 1)^2} \therefore \frac{dy}{dx} = \frac{1}{y(x + 1)^2}$

17. $y = \sqrt{1 - \sqrt{x}} = (1 - x^{1/2})^{1/2} \Rightarrow \frac{dy}{dx} = (1/2)(1 - x^{1/2})^{-1/2}(-1/2)(x^{-1/2}) = \frac{-1}{4\sqrt{x}\sqrt{1 - \sqrt{x}}}$

19. $y = \sqrt{1 + \cos 2x} = (1 + \cos 2x)^{1/2} \Rightarrow \frac{dy}{dx} = (1/2)(1 + \cos 2x)^{-1/2}(-\sin 2x)(2) = \frac{-\sin 2x}{\sqrt{1 + \cos 2x}}$

21. $y = 3(\csc x)^{3/2} \Rightarrow \frac{dy}{dx} = (3)\left[(3/2)(\csc x)^{1/2}(-\csc x)(\cot x)\right] = (-9/2)(\csc^{3/2}x)(\cot x)$

23. $x = \tan y \Rightarrow 1 = \sec^2 y\frac{dy}{dx} \Rightarrow \frac{dy}{dx} = \frac{1}{\sec^2 y} = \cos^2 y$

25. $x + \tan(xy) = 0 \Rightarrow 1 + \left[\sec^2(xy)\right]\left[y + x\frac{dy}{dx}\right] = 0 \Rightarrow x\sec^2(xy)\frac{dy}{dx} = -1 - y\sec^2(xy) \Rightarrow$

$$\frac{dy}{dx} = \frac{-1 - y\sec^2(xy)}{x\sec^2(xy)} = \frac{-1}{x\sec^2(xy)} - \frac{y}{x} = \frac{-\cos^2(xy)}{x} - \frac{y}{x} = \frac{-\cos^2(xy) - y}{x}$$

27. $x^2 + y^2 = 1 \Rightarrow 2x + 2yy' = 0 \Rightarrow y' = -\frac{x}{y} \therefore \frac{dy}{dx} = -\frac{x}{y}$, also $2x + 2yy' = 0 \Rightarrow x + yy' = 0 \Rightarrow$

$$1 + (y')^2 + yy'' = 0 \Rightarrow yy'' = -1 - (y')^2 \Rightarrow y'' = \frac{-1 - (y')^2}{y} \Rightarrow y'' = \frac{-1 - \left(-\frac{x}{y}\right)^2}{y} = -\frac{1}{y} - \frac{x^2}{y^3} = \frac{-y^2 - x^2}{y^3}$$

$$\therefore \frac{d^2y}{dx^2} = \frac{-y^2 - x^2}{y^3}$$

29. $y^2 = x^2 + 2x \Rightarrow 2yy' = 2x + 2 \Rightarrow yy' = x + 1 \Rightarrow y' = \frac{x + 1}{y} \therefore \frac{dy}{dx} = \frac{x + 1}{y}$, also $yy' = x + 1 \Rightarrow$

$$(y')^2 + yy'' = 1 \Rightarrow y'' = \frac{1}{y} - \frac{(y')^2}{y} \Rightarrow y'' = \frac{1}{y} - \left(\frac{1}{y}\right)\left(\frac{x + 1}{y}\right)^2 \Rightarrow y'' = \frac{1}{y} - \frac{(x + 1)^2}{y^3} \Rightarrow$$

$$y'' = \frac{y^2 - (x + 1)^2}{y^3} \therefore \frac{d^2y}{dx^2} = \frac{y^2 - (x + 1)^2}{y^3}$$

31. $y + 2\sqrt{y} = x \Rightarrow y + 2y^{1/2} = x \Rightarrow y' + (2)(1/2)y^{-1/2}y' = 1 \Rightarrow (1 + y^{-1/2})y' = 1 \Rightarrow y' = \dfrac{1}{1 + y^{-1/2}} \Rightarrow$

$y' = \dfrac{1}{1 + \dfrac{1}{\sqrt{y}}} \Rightarrow y' = \dfrac{\sqrt{y}}{\sqrt{y} + 1} \therefore \dfrac{dy}{dx} = \dfrac{\sqrt{y}}{\sqrt{y} + 1}$, also $y' + (2)(1/2)y^{-1/2}y' = 1 \Rightarrow y' + y^{-1/2}y' = 1 \Rightarrow$

$y'' + (-1/2)y^{-3/2}(y')^2 + y^{-1/2}y'' = 0 \Rightarrow (1 + y^{-1/2})y'' = \dfrac{(y')^2}{2\sqrt{y^3}} \Rightarrow y'' = \dfrac{(y')^2}{2\sqrt{y^3}(1 + y^{-1/2})} \Rightarrow$

$y'' = \dfrac{\left(\dfrac{1}{1 + y^{-1/2}}\right)^2}{2\sqrt{y^3}(1 + y^{-1/2})} \Rightarrow y'' = \dfrac{1}{2\sqrt{y^3}(1 + y^{-1/2})^3} \Rightarrow y'' = \dfrac{1}{2\sqrt{y^3}\left(1 + \dfrac{1}{\sqrt{y}}\right)^3} = \dfrac{1}{\dfrac{2\sqrt{y^3}(\sqrt{y} + 1)^3}{\sqrt{y^3}}} \Rightarrow$

$y'' = \dfrac{1}{2(\sqrt{y} + 1)^3} \therefore \dfrac{d^2y}{dx^2} = \dfrac{1}{2(\sqrt{y} + 1)^3}$

33. $x^2 + xy - y^2 = 1 \Rightarrow 2x + y + xy' - 2yy' = 0 \Rightarrow (x - 2y)y' = -2x - y \Rightarrow y' = \dfrac{2x + y}{2y - x}$

a) the slope of the tangent line $m = y'|_{(2,3)} = \dfrac{2x + y}{2y - x}\Big|_{(2,3)} = \dfrac{7}{4}$, the tangent line is

$y - 3 = \frac{7}{4}(x - 2) \Rightarrow y = \frac{7}{4}x - \frac{1}{2}$

b) the normal line is $y - 3 = -\frac{4}{7}(x - 2) \Rightarrow y = -\frac{4}{7}x + \frac{29}{7}$

35. $x^2y^2 = 9 \Rightarrow 2xy^2 + 2x^2yy' = 0 \Rightarrow x^2yy' = -xy^2 \Rightarrow y' = -\dfrac{y}{x}$

a) the slope of the tangent line , $= y'|_{(-1,3)} = -\dfrac{y}{x}\Big|_{(-1,3)} = 3$, the tangent line is $y - 3 = 3(x + 1) \Rightarrow$

$y = 3x + 6$

b) the normal line is $y - 3 = -\frac{1}{3}(x + 1) \Rightarrow y = -\frac{1}{3}x + \frac{8}{3}$

37. Solving $x^2 + xy + y^2 = 7$ and $y = 0 \Rightarrow x^2 = 7 \Rightarrow x = \pm\sqrt{7} \Rightarrow (-\sqrt{7}, 0)$ and $(\sqrt{7}, 0)$ are the points where the curve crosses the x–axis. Now $x^2 + xy + y^2 = 7 \Rightarrow 2x + y + xy' + 2yy' = 0 \Rightarrow$

$(x + 2y)y' = -2x - y \Rightarrow y' = -\dfrac{2x + y}{x + 2y} \Rightarrow m = -\dfrac{2x + y}{x + 2y} \Rightarrow$ the slope at $(-\sqrt{7}, 0)$ is $m = -\dfrac{-2\sqrt{7}}{-\sqrt{7}} = -2$ and

the slope at $(\sqrt{7}, 0)$ is $m = -\dfrac{2\sqrt{7}}{\sqrt{7}} = -2$ Since the slope is $-2$ in each case, the corresponding

tangents must be parallel.

39. $2xy + \pi \sin y = 2\pi \Rightarrow 2y + 2xy' + (\pi \cos y)y' = 0 \Rightarrow (2x + \pi \cos y)y' = -2y \Rightarrow y' = \dfrac{-2y}{2x + \pi \cos y} \Rightarrow$

$\dfrac{dy}{dx} = \dfrac{-2y}{2x + \pi \cos y} \Rightarrow \dfrac{dy}{dx}\Big|_{(1,\pi/2)} = \dfrac{-2y}{2x + \pi \cos y}\Big|_{(1,\pi/2)} = \dfrac{-\pi}{2 + \pi \cos(\pi/2)} = -\dfrac{\pi}{2}$

41. $y^4 = y^2 - x^2 \Rightarrow 4y^3y' = 2yy' - 2x \Rightarrow (2)(2y^3 - y)y' = -2x \Rightarrow y' = \dfrac{x}{y - 2y^3}$; the slope of the tangent line

at $(\frac{\sqrt{3}}{4}, \frac{\sqrt{3}}{2})$ is $\dfrac{x}{y - 2y^3}\Big|_{(\frac{\sqrt{3}}{4}, \frac{\sqrt{3}}{2})} = \dfrac{\frac{\sqrt{3}}{4}}{\frac{\sqrt{3}}{2} - \frac{6\sqrt{3}}{8}} = \dfrac{\frac{1}{4}}{\frac{1}{2} - \frac{3}{4}} = \dfrac{1}{2 - 3} = -1$; the slope of the tangent line

at $(\frac{\sqrt{3}}{4}, \frac{1}{2})$ is $\dfrac{x}{y - 2y^3}\Big|_{(\frac{\sqrt{3}}{4}, \frac{1}{2})} = \dfrac{\frac{\sqrt{3}}{4}}{\frac{1}{2} - \frac{2}{8}} = \dfrac{2\sqrt{3}}{4 - 2} = \sqrt{3}$

43. a) $f(x) = \frac{3}{2}x^{2/3} - 3 \Rightarrow f'(x) = x^{-1/3} \Rightarrow f''(x) = -\frac{1}{3}x^{-4/3}$ $\therefore$ a is false

b) $f(x) = \frac{9}{10}x^{5/3} - 7 \Rightarrow f'(x) = \frac{3}{2}x^{2/3} \Rightarrow f''(x) = x^{-1/3}$ $\therefore$ b is true

c) $f''(x) = x^{-1/3} \Rightarrow f'''(x) = -\frac{1}{3}x^{-4/3}$ $\therefore$ c is true

d) $f'(x) = \frac{3}{2}x^{2/3} + 6 \Rightarrow f''(x) + x^{-1/3}$ $\therefore$ d is true

45. $s = \sqrt{1 + 4t} = (1 + 4t)^{1/2} \Rightarrow v = s' = (1/2)(1 + 4t)^{-1/2}(4) = 2(1 + 4t)^{-1/2} = \frac{2}{\sqrt{1 + 4t}} \Rightarrow$

$a = v' = s'' = (2)(-1/2)(1 + 4t)^{-3/2}(4) = (-4)(1 + 4t)^{-3/2} = \frac{-4}{\sqrt{(1 + 4t)^3}}$

$\therefore v|_{t=6} = \frac{2}{5}$ m/sec and $a|_{t=6} = -\frac{4}{125}$ m/sec$^2$

## 3.7 LINEAR APPROXIMATIONS AND DIFFERENTIALS

1. $f(x) = x^4 \Rightarrow f'(x) = 4x^3 \Rightarrow L(x) = f'(1)(x - 1) + f(1) = 4(x - 1) + 1 \Rightarrow L(x) = 4x - 3$

3. $f(x) = x^3 - x \Rightarrow f'(x) = 3x^2 - 1 \Rightarrow L(x) = f'(1)(x - 1) + f(1) = 2(x - 1) + 0 \Rightarrow L(x) = 2x - 2$

5. $f(x) = \sqrt{x} = x^{1/2} \Rightarrow f'(x) = (1/2)x^{-1/2} = \frac{1}{2\sqrt{x}} \Rightarrow L(x) = f'(4)(x - 4) + f(4) = \frac{1}{4}(x - 4) + 2 \Rightarrow$

$L(x) = \frac{1}{4}x + 1$

7. $f(x) = x^2 + 2x \Rightarrow f'(x) = 2x + 2 \Rightarrow L(x) = f'(0)(x - 0) + f(0) = 2(x - 0) + 0 \Rightarrow L(x) = 2x$

9. $f(x) = 2x^2 + 4x - 3 \Rightarrow f'(x) = 4x + 4 \Rightarrow L(x) = f'(-1)(x + 1) + f(-1) = 0(x + 1) + (-5) \Rightarrow L(x) = -5$

11. $f(x) = \sqrt[3]{x} = x^{1/3} \Rightarrow f'(x) = (1/3)x^{-2/3} \Rightarrow L(x) = f'(8)(x - 8) + f(8) = \frac{1}{12}(x - 8) + 2 \Rightarrow L(x) = \frac{1}{12}x + \frac{4}{3}$

13. $f(x) = \sin x \Rightarrow f'(x) = \cos x \Rightarrow L(x) = f'(0)(x - 0) + f(0) =$

$1(x - 0) + 0 \Rightarrow L(x) = x$

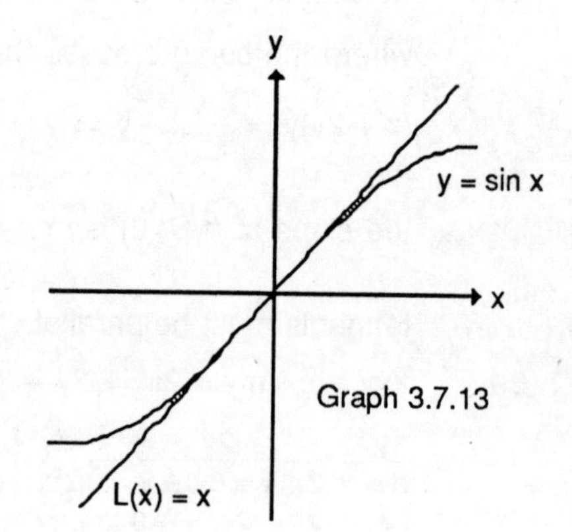

Graph 3.7.13

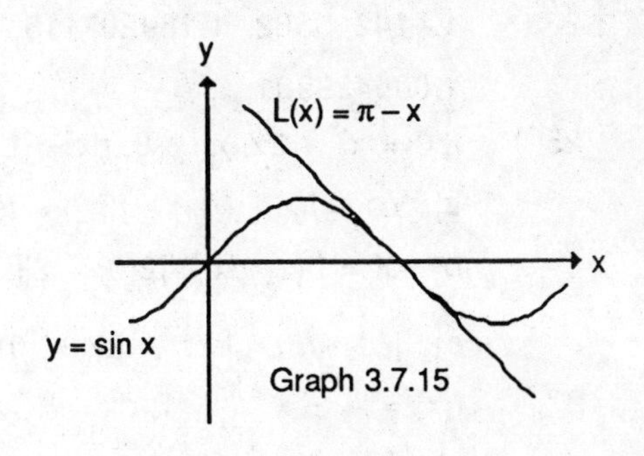

Graph 3.7.15

15. $f(x) = \sin x \Rightarrow f'(x) = \cos x \Rightarrow L(x) = f'(\pi)(x - \pi) + f(\pi) =$
$(-1)(x - \pi) + 0 \Rightarrow L(x) = -x + \pi$

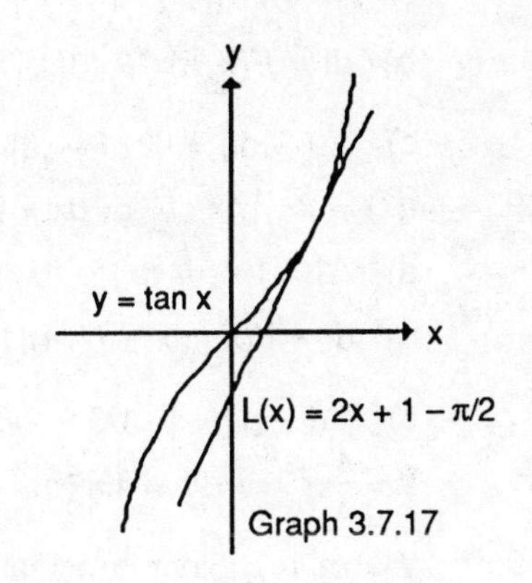

Graph 3.7.17

17. $f(x) = \tan x \Rightarrow f'(x) = \sec^2 x \Rightarrow$
$L(x) = f'(\pi/4)(x - \pi/4) + f(\pi/4) = (2)(x - \pi/4) + 1 \Rightarrow$
$L(x) = 2x + 1 - \pi/2$

| 19. | $f(x)$ | $(1+x)^k \approx 1+kx$ |
|---|---|---|
| a) | $(1+x)^2$ | $1 + 2x$ |
| b) | $\frac{1}{(1+x)^5} = (1+x)^{-5}$ | $1 + (-5)x = 1 - 5x$ |
| c) | $\frac{2}{1-x} = 2(1+(-x))^{-1}$ | $2[1 + (-1)(-x)] = 2 + 2x$ |
| d) | $(1-x)^6 = (1+(-x))^6$ | $1 + (6)(-x) = 1 - 6x$ |
| e) | $3(1+x)^{1/3}$ | $3[1 + (1/3)(x)] = 3 + x$ |
| f) | $\frac{1}{\sqrt{1+x}} = (1+x)^{-1/2}$ | $1 + (-1/2)(x) = 1 - \frac{x}{2}$ |

21. $f(x) = \sqrt{x+1} + \sin x = (x+1)^{1/2} + \sin x \Rightarrow f'(x) = (1/2)(x+1)^{-1/2} + \cos x \Rightarrow$
$L_f(x) = f'(0)(x - 0) + f(0) = \frac{3}{2}(x - 0) + 1 \Rightarrow L_f(x) = \frac{3}{2}x + 1$, the linearization of $f(x)$; $g(x) = \sqrt{x+1} =$
$(x+1)^{1/2} \Rightarrow g'(x) = (1/2)(x+1)^{-1/2} \Rightarrow L_g(x) = g'(0)(x - 0) + g(0) = \frac{1}{2}(x - 0) + 1 \Rightarrow L_g(x) = \frac{1}{2}x + 1$,
the linearization of $g(x)$; $h(x) = \sin x \Rightarrow h'(x) = \cos x \Rightarrow L_h(x) = h'(0)(x - 0) + h(0) =$
$(1)(x - 0) + 0 \Rightarrow L_h(x) = x$, the linearization of $h(x)$. $L_f(x) = L_g(x) + L_h(x)$ implies that the
linearization of a sum is equal to the sum of the linearizations.

23. 1.414213562, 1.189207115, 1.090507733,1.044273782, 1.021897149, 1.010889286, 1.005429901, ...

25. $f(x) = x^2 + 2x,\ x_o = 0,\ dx = .1 \Rightarrow f'(x) = 2x + 2$

a) $\Delta f = f(x_o + dx) - f(x_o) = f(.1) - f(0) = .01 + .2 = .21$

b) $df = f'(x_o)dx = [2(0) + 2][.1] = .2$

c) $|\Delta f - df| = |.21 - .22| = .01$

27. $f(x) = x^3 - x,\ x_o = 1,\ dx = .1 \Rightarrow f'(x) = 3x^2 - 1$

a) $\Delta f = f(x_o + dx) - f(x_o) = f(1.1) - f(1) = .231$

b) $df = f'(x_o)dx = (3(1)^2 - 1)(.1) = .2$

c) $|\Delta f - df| = |,231 - .2| = .031$

29. $f(x) = x^{-1},\ x_o = .5,\ dx = .1 \Rightarrow f'(x) = -x^{-2}$

a) $\Delta f = f(x_o + dx) - f(x_o) = f(.6) - f(.5) = -1/3$

b) $df = f'(x_o)dx = (-4)(1/10) = -2/5$

c) $|\Delta f - df| = |-1/3 - 2/5| = 1/15$

31. $V = \frac{4}{3}\pi r^3 \Rightarrow dV = 4\pi r_o^2 dr$

33. $V = x^3 \Rightarrow dV = 3x_o^2 dx$

35. $V = \pi r^2 h$, height constant $\Rightarrow dV = 2\pi r_o h dr$

37. Given r = 2 m, dr = .02 m a) $A = \pi r^2 \Rightarrow dA = 2\pi r dr = 2\pi(2)(.02) = .08\pi\ m^2$

b) $\left[\frac{.08\pi}{4\pi}\right][100\%] = 2\%$

39. The error in measurement $dx = (1\%)(10) = .1$ cm, $V = x^3 \Rightarrow dV = 3x^2dx = 3(10)^2(.1) = 30\ cm^2 \Rightarrow$ the percentage error in the volume calculation is $\left[\frac{30}{1000}\right][100\%] = 3\%$

41. Given $d = 100$ cm, $dd = 1$ cm, $V = \frac{4}{3}\pi\left(\frac{d}{2}\right)^3 = \frac{\pi d^3}{6} \Rightarrow dV = \frac{\pi}{2}d^2(dd) = \frac{\pi}{2}(100)^2(1) = \frac{10^4\pi}{2}$,

now $\frac{dV}{V}(10^2\%) = \left[\frac{\frac{10^4\pi}{2}}{\frac{10^6\pi}{6}}\right][10^2\%] = \left[\frac{\frac{10^6\pi}{2}}{\frac{10^6\pi}{6}}\right]\% = 3\%$

43. $V = \pi h^3 \Rightarrow dV = 3\pi h^2 dh$; recall that $\Delta V \approx dV$. $|\Delta V| \le (1\%)(V) = \left(\frac{(1)(\pi h^3)}{100}\right) \Rightarrow |dV| \le \left(\frac{(1)(\pi h^3)}{100}\right) \Rightarrow$ $|3\pi h^2 dh| \le \left(\frac{(1)(\pi h^3)}{100}\right) \Rightarrow |dh| \le \left(\frac{1}{300}\right)(h) = \left(\frac{1}{3}\%\right)(h)$. $\therefore$ the greatest tolerated error in the measurement of h is $\frac{1}{3}$ %.

45. $V = \pi r^2 h$, h is constant $\Rightarrow dV = 2\pi rh dr$. Recall that $\Delta V \approx dV$. We want $|\Delta V| \le \frac{1}{1000}V \Rightarrow$ $|dV| \le \frac{\pi r^2 h}{1000} \Rightarrow |2\pi rh dr| \le \frac{\pi r^2 h}{1000} \Rightarrow |dr| \le \frac{r}{2000} = (.05\%)(r)$. A .05% variation in the radius can be tolerated.

47. A 5% error in measuring t $\Rightarrow dt = 5\%t = \frac{t}{20}$. If $s = 16t^2$, then $ds = 32tdt = 32t\left(\frac{t}{20}\right) = \frac{32t^2}{20} = \frac{16t^2}{10} = \left(\frac{1}{10}\right)s = (10\%)(s) \Rightarrow$ a 10% error in the calculation of s.

49. a) Clearly, $3 < 4 \Rightarrow \sqrt{3} < 2 \Rightarrow \sqrt{3} - 2 < 0 \Rightarrow \sqrt{3} + 2 - 4 < 0 \Rightarrow \sqrt{3} + \sqrt{4} - 4 < 0 \Rightarrow g(3) < 0$; $5 > 4 \Rightarrow \sqrt{5} > 2 \Rightarrow \sqrt{5} - 2 > 0 \Rightarrow \sqrt{4} + \sqrt{5} - 4 > 0 \Rightarrow g(4) > 0$. By the Intermediate Value Theorem, there exists an x between 3 and 4 such that $g(x) = 0$.

b) $g(x) = \sqrt{x} + \sqrt{x+1} - 4 = x^{1/2} + (x+1)^{1/2} - 4 \Rightarrow g(3) = \sqrt{3} - 2$, $g'(x) = (1/2)x^{-1/2} + (1/2)(x+1)^{-1/2}$ and $g'(3) = \frac{1}{2\sqrt{3}} + \frac{1}{4}$. The linearization of g(x) is

$L(x) = g'(3)(x-3) + g(3) = \left(\frac{1}{2\sqrt{3}} + \frac{1}{4}\right)(x-3) + \sqrt{3} - 2 = \left(\frac{2\sqrt{3}+3}{12}\right)(x-3) + \sqrt{3} - 2 =$

$\left(\frac{2\sqrt{3}+3}{12}\right)x - \frac{2\sqrt{3}}{4} - \frac{3}{4} + \frac{4\sqrt{3}}{4} - \frac{8}{4} \Rightarrow L(x) = \left(\frac{2\sqrt{3}+3}{12}\right)x + \frac{\sqrt{3}}{2} - \frac{11}{4}$. Solving $L(x) = 0 \Rightarrow$

$x = \left(\frac{11 - 2\sqrt{3}}{4}\right)\left(\frac{12}{2\sqrt{3}+3}\right) = 28\sqrt{3} - 45 \approx 3.497422612$

c) $g(3.497422612) = -0.009147509$

51. $y = x^3 - 3x \Rightarrow dy = (3x^2 - 3)dx$

53. $y = \frac{2x}{1+x^2} \Rightarrow dy = \left(\frac{(2)(1+x^2) - (2x)(2x)}{(1+x^2)^2}\right)dx = \left(\frac{2-2x^2}{(1+x^2)^2}\right)dx$

55. $y + xy - x = 0 \Rightarrow dy + ydx + xdy - dx = 0 \Rightarrow (1+x)dy = dx - ydx \Rightarrow dy = \frac{(1-y)}{1+x}dx$

57. $y = \sin(5x) \Rightarrow dy = [\cos(5x)][5]dx = 5\cos(5x)dx$

59. $y = 4\tan(x/2) \Rightarrow dy = 4[\sec^2(x/2)][1/2]dx = 2\sec^2(x/2)dx$

61. $y = 3\csc(1 - x/3) \Rightarrow dy = 3[-\csc(1 - x/3)\cot(1 - x/3)][-1/3]dx = \csc(1 - x/3)\cot(1 - x/3)dx$

## 3.8 NEWTON'S METHOD FOR APPROXIMATING SOLUTIONS OF EQUATIONS

1. $y = x^2 + x - 1 \Rightarrow y' = x^2 + 1$, $x_{n+1} = x_n - \frac{x_n^2 + x_n - 1}{2x_n + 1}$ $\therefore x_o = 1 \Rightarrow x_2 = .619047619$ and $x_o = -1 \Rightarrow x_2 = -1.66666667$ The roots are 0.618033989 and $-1.61803399$, used the Calculus Tool Kit.

3. $y = x^4 + x - 3 \Rightarrow y' = 4x^3 + 1$, $x_{n+1} = x_n - \frac{x_n^4 + x_n - 3}{4x_n^3 + 1}$ $\therefore x_o = 1 \Rightarrow x_2 = 1.16541962$ and $x_o = -1 \Rightarrow x_2 = -1.64516129$. The roots are 1.16403514 and $-1.45262688$, used the Calculus Tool Kit.

5. $y = x^4 - 2 \Rightarrow y' = 4x^3$, $x_{n+1} = x_n - \frac{x_n^4 - 2}{4x_n^3}$ $\therefore x_o = 1 \Rightarrow x_2 = 1.1935$. The root is 1.18920711.

7. If $f'(x_o) \neq 0$ then $x_o = x_1 = x_2 = x_3 = \ldots$.

9.

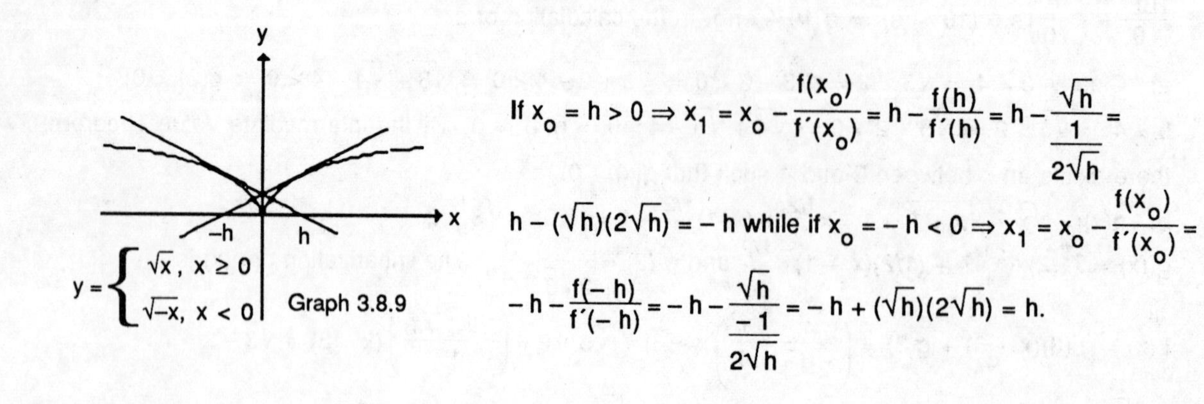


Graph 3.8.9

If $x_o = h > 0 \Rightarrow x_1 = x_o - \dfrac{f(x_o)}{f'(x_o)} = h - \dfrac{f(h)}{f'(h)} = h - \dfrac{\sqrt{h}}{\frac{1}{2\sqrt{h}}} =$ $h - (\sqrt{h})(2\sqrt{h}) = -h$ while if $x_o = -h < 0 \Rightarrow x_1 = x_o - \dfrac{f(x_o)}{f'(x_o)} =$ $-h - \dfrac{f(-h)}{f'(-h)} = -h - \dfrac{\sqrt{h}}{\frac{-1}{2\sqrt{h}}} = -h + (\sqrt{h})(2\sqrt{h}) = h.$

11. If $f(x) = x^3 + 2x - 4$, then $f(1) = -1 < 0$, $f(2) = 8 > 0$; and by the Intermediate Value Theorem, the equation $x^3 + 2x - 4 = 0$ has a solution between 1 and 2. Consequently, $f'(x) = 3x^2 + 2$ and $x_{n+1} = x_n - \dfrac{x_n^3 + 2x_n - 4}{3x_n^2 + 2}$. The root is 1.17951.

13. $f(x) = \tan x \Rightarrow f'(x) = \sec^2 x$ $\therefore x_{n+1} = x_n - \dfrac{\tan(x_n)}{\sec^2(x_n)}$. With $x_o = 3$ we approximate $\pi$ to be 3.14159.

15. Let $f(x) = \dfrac{1}{x^2 + 1} - x = (x^2 + 1)^{-1} - x \Rightarrow f'(x) = -(x^2 + 1)^{-2}(2x) - 1 = \dfrac{-2x}{(x^2 + 1)^2} - 1.$

$\therefore x_{n+1} = x_n - \dfrac{\dfrac{1}{x_n^2 + 1} - x_n}{\dfrac{-2x_n}{(x_n^2 + 1)^2} - 1}$. If $x_o = 1$, then $x_4 = 0.682327804$ which is 0.68233 to five decimal places.

17. Let $f(x) = \tan(x) - 2x \Rightarrow f'(x) = \sec^2(x) - 2$. $\therefore x_{n+1} = x_n - \dfrac{\tan(x_n) - 2x_n}{\sec^2(x_n) - 2}$. If $x_o = 1$, then $x_6 = x_7 = x_8 = 1.16556119$.

19. $f(x) = 2x^4 - 4x^2 + 1 \Rightarrow f'(x) = 8x^3 - 8x$ $\therefore x_{n+1} = x_n - \dfrac{2x_n^4 - 4x_n^2 + 1}{8x_n^3 - 8x_n}$. If $x_o = -2$, then $x_6 = -1.30656296$; if $x_o = -.5$, then $x_3 = -0.5411961$. The positive roots are 0.5411961 and 1.30656296 because f(x) is an even function.

## PRACTICE EXERCISES

1. $y = x^5 - \frac{1}{8}x^2 + \frac{1}{4}x \Rightarrow \frac{dy}{dx} = 5x^4 - \frac{1}{4}x + \frac{1}{4}$

3. $y = (x+1)^2(x^2+2x) \Rightarrow \frac{dy}{dx} = 2(x+1)(1)(x^2+2x) + (x+1)^2(2x+2) =$
$2(x+1)\left[(x^2+2x) + 2(x+1)^2\right] = 2(x+1)(2x^2+4x+1)$

5. $y = 2\sin x\cos x = \sin(2x) \Rightarrow \frac{dy}{dx} = [\cos(2x)](2) = 2\cos 2x$

7. $y = \frac{x}{x+1} \Rightarrow \frac{dy}{dx} = \frac{1(x+1) - x(1)}{(x+1)^2} = \frac{1}{(x+1)^2}$

9. $y = (x^3+1)^{-4/3} \Rightarrow \frac{dy}{dx} = (-4/3)(x^3+1)^{-7/3}(3x^2) = -4x^2(x^3+1)^{-7/3}$

11. $y = \cos(1-2x) \Rightarrow \frac{dy}{dx} = [-\sin(1-2x)][-2] = 2\sin(1-2x)$

13. $y = (x^2+x+1)^3 \Rightarrow \frac{dy}{dx} = 3(x^2+x+1)^2(2x+1)$

15. $y = \sqrt{2u+u^2} = (2u+u^2)^{1/2},\ u = 2x+3 \Rightarrow$
$\frac{dy}{dx} = \left[\frac{dy}{du}\Big|_{u=2x+3}\right]\left[\frac{du}{dx}\right] = \left[(1/2)(2u+u^2)^{-1/2}(2+2u)\Big|_{u=2x+2}\right][2] = \frac{4(x+2)}{\sqrt{4x^2+16x+15}}$

17. $xy + y^2 = 1 \Rightarrow y + xy' + 2yy' = 0 \Rightarrow (x+2y)y' = -y \Rightarrow \frac{dy}{dx} = y' = \frac{-y}{x+2y}$

19. $x^2 + xy + y^2 - 5x = 2 \Rightarrow 2x + y + xy' + 2yy' - 5 = 0 \Rightarrow (x+2y)y' = 5 - 2x - y \Rightarrow \frac{dy}{dx} = = y' = \frac{5-2x-y}{x+2y}$

21. $5x^{4/5} + 10y^{6/5} = 15 \Rightarrow 4x^{-1/5} + 12y^{1/5}y' = 0 \Rightarrow 3y^{1/5}y' = -x^{-1/5} \Rightarrow \frac{dy}{dx} = y' = -\frac{x^{-1/5}}{3y^{1/5}} = -\frac{1}{3(xy)^{1/5}}$

23. $y^2 = \frac{x}{x+1} \Rightarrow 2yy' = \frac{1(x+1) - 1(x)}{(x+1)^2} = \frac{1}{(x+1)^2} \Rightarrow \frac{dy}{dx} = = y' = \frac{1}{2y(x+1)^2}$

25. $y^2 = \frac{(5x^2+2x)^{3/2}}{3} \Rightarrow 2yy' = \left(\frac{1}{3}\right)\left(\frac{3}{2}\right)\left(5x^2+2x\right)^{1/2}(10x+2) \Rightarrow \frac{dy}{dx} = y' = \frac{(5x^2+2x)^{1/2}(5x+1)}{2y}$

27. $y = \sqrt{x} + 1 + \frac{1}{\sqrt{x}} = x^{1/2} + 1 + x^{-1/2} \Rightarrow y' = (1/2)x^{-1/2} + (-1/2)x^{-3/2} = \frac{x-1}{2\sqrt{x^3}}$

29. $y = \sec(1+3x) \Rightarrow y' = [\sec(1+3x)\tan(1+3x)][3] \Rightarrow \frac{dy}{dx} = y' = 3\sec(1+3x)\tan(1+3x)$

31. $y = \cot(x^2) \Rightarrow y' = [-\csc^2(x^2)][2x] \Rightarrow \frac{dy}{dx} = y' = -2x\csc^2(x^2)$

33. $y = \sqrt{\frac{1-x}{1+x^2}} = \left(\frac{1-x}{1+x^2}\right)^{1/2} \Rightarrow y' = \left(\frac{1}{2}\right)\left(\frac{1-x}{1+x^2}\right)^{-1/2}\left(\frac{(-1)(1+x^2) - (2x)(1-x)}{(1+x^2)^2}\right) \Rightarrow$
$\frac{dy}{dx} = y' = \frac{x^2-2x-1}{2\sqrt{(1+x^2)(1-x)}}$

35. a)



Graph 3.P.35

$y = \begin{cases} x, & 0 \le x \le 1 \\ 2-x, & 1 < x \le 2 \end{cases}$

b) yes

c) No, $\lim_{x \to 1^-} f'(x) = 1 \ne -1 = \lim_{x \to 1^+} f'(x)$

37. $y = 2x^3 - 3x^2 - 12x + 20 \Rightarrow y' = 6x^2 - 6x - 12 = 6(x - 2)(x + 1)$, $y' = 0 \Rightarrow x = 2$ or $x = -1$, the points are (2, 0) and (– 1, 27)

39. $s(t) = 10\cos(t + \pi/4) \Rightarrow v(t) = s'(t) = -10\sin(t + \pi/4) \Rightarrow a(t) = v'(t) = s''(t) = -1o\cos(t + \pi/4)$.

a) $s(0) = 10\cos(\pi/4) = \frac{10}{\sqrt{2}}$

b) Solving $10\cos(t + \pi/4) = -10 \Rightarrow \cos(t + \pi/4) = -1 \Rightarrow t = \frac{3\pi}{4}$ when the particle is farthest to the left. Solving $10\cos(t + \pi/4) = 10 \Rightarrow \cos(t + \pi/4) = 1 \Rightarrow t = -\frac{\pi}{4}$, but $t \geq 0 \Rightarrow t = 2\pi + \frac{-\pi}{4} = \frac{7\pi}{4}$ when the particle is farthest to the right.

c) From parts a and b we have: $v\left(\frac{3\pi}{4}\right) = 0$, $v\left(\frac{7\pi}{4}\right) = 0$, $a\left(\frac{3\pi}{4}\right) = 10$, and $a\left(\frac{7\pi}{4}\right) = -10$.

d) Solving $10\cos(t+\pi/4) = 0 \Rightarrow t = \frac{\pi}{4} \Rightarrow v\left(\frac{\pi}{4}\right) = -10$, $\left|v\left(\frac{\pi}{4}\right)\right| = 10$ and $a\left(\frac{\pi}{4}\right) = 0$.

41. $s = 490t^2 \Rightarrow v = 980t \Rightarrow a = 980$. a) Solving $160 = 490t^2 \Rightarrow t = \frac{4}{7}$ sec. The average velocity was $\frac{s(4/7) - s(0)}{4/7} = 280$ cm/sec. b) At the 160 cm mark the balls are falling at $v(4/7) = 560$ cm/sec. The acceleration at the 160 cm mark was 980 cm/sec$^2$. c) The light was flashing at a rate of $\frac{17}{4/7} = 29.75$ flashes per second.

43. a) iii b) i c) ii

45.

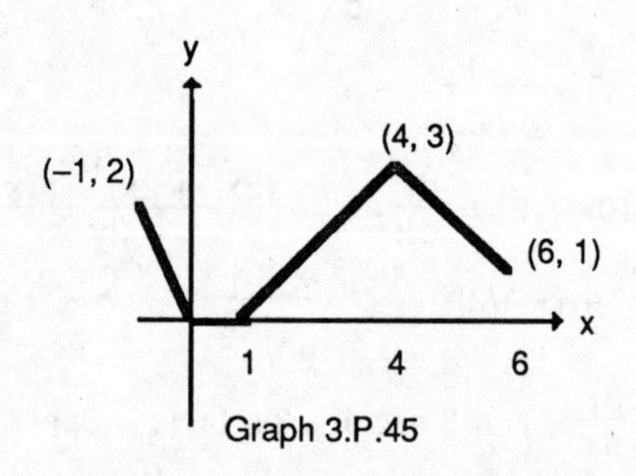

Graph 3.P.45

47. $V = \pi\left[10 - \frac{x}{3}\right]x^2 \Rightarrow \frac{dV}{dt} = \pi\left[\left(-\frac{1}{3}\right)x^2 + \left(10 - \frac{x}{3}\right)(2x)\right]\left[\frac{dx}{dt}\right] = \pi\left[20x - x^2\right]\left[\frac{dx}{dt}\right] =$ $\pi\left[20x - x^2\right]$ in$^3$/ inch increase of depth.

49. a) When $y = 4 + \cot x - 2\csc x \Rightarrow y' = -\left[\frac{1}{\sin x}\right]\left[\frac{1 - 2\cos x}{\sin x}\right]$. To find the location of the horizontal tangent set $y' = 0 \Rightarrow x = \frac{\pi}{3}$ radians. When $x = \frac{\pi}{3}$, then $y = 4 - \sqrt{3}$, the horizontal tangent.

b) When $x = \frac{\pi}{2}$, then $y' = -1$. The tangent line is $y = -x + \frac{\pi + 4}{2}$.

51. If $y = \sin(x - \sin x)$ then $y' = [\cos(x - \sin x)][1 - \cos x]$. When $y' = 0 \Rightarrow x = 0 + n(2\pi)$ where n is an integer. The curve has an infinite number of horizontal tangents as indicated above.

53. a) $5f'(1) - g'(1) = 1$  b) $[f'(0)][g(0)]^3 + 3[f(0)][g(0)]^2[g'(0)] = 6$

c) $\dfrac{f'(1)[g(1)+1] - f(1)g'(1)}{(g(1)+1)^2} = 1$  d) $f'(g(0))g'(0) = -\dfrac{1}{9}$

e) $g'(f(0))f'(0) = -\dfrac{40}{3}$  f) $[3/2][1+f(1)]^{1/2}[1 + f'(1)] = 2$

g) $f'(0 + g(0))(1 + g'(0)) = -\dfrac{4}{9}$

55. The derivative of $\sin(x + a) = \sin x \cos a + \cos x \sin a$ is $\cos(x + a) = \cos x \cos a - \sin x \sin a$, an identity. This principle does not apply to the equation $x^2 - 2x - = 0$ because it is not an identity.

57. When $s = t^2 + 5t$ and $t = (u^2 + 2u)^{1/3}$, then $\dfrac{ds}{dt} = 2t + 5$ and $\dfrac{dt}{du} = (1/3)(u^2 + 2u)^{-2/3}(2u + 2)$. $\therefore \dfrac{ds}{du} = \left[\dfrac{ds}{dt}\Big|_{t = (u^2+2u)^{1/3}}\right]\left[\dfrac{dt}{du}\right] = \left[2(u^2 + 2u)^{1/3} + 5\right]\left[\left(\dfrac{2}{3}\right)(u^2 + 2u)^{-2/3}(u + 1)\right]$. Accordingly, $\dfrac{ds}{du}\Big|_{u=2} = \dfrac{9}{2}$.

59. If $y = \sqrt{x}$, then $y' = \dfrac{1}{2\sqrt{x}}$. The equation of the tangent line is $y - 2 = \dfrac{1}{4}(x - 4) \Rightarrow \dfrac{x}{-4} + \dfrac{y}{1} = 1$. The intercepts are $(0, 1)$ and $(-4, 0)$.

61. a) If $x^2 + 2y^2 = 9$, then $y' = -\dfrac{x}{2y}$. The tangent line is $y - 2 = \left(-\dfrac{1}{4}\right)(x - 1) \Rightarrow y = -\dfrac{1}{4}x + \dfrac{9}{4}$. The normal line is $y - 2 = 4(x - 1) \Rightarrow y = 4x - 2$.

b) If $x^3 + y^2 = 2$, then $y' = -\dfrac{3x^2}{2y}$. The tangent line is $y - 1 = \left(-\dfrac{3}{2}\right)(x - 1) \Rightarrow y = -\dfrac{3}{2}x + \dfrac{5}{2}$. The normal line is $y - 1 = \left(\dfrac{2}{3}\right)(x - 1) \Rightarrow y = \dfrac{2}{3}x + \dfrac{1}{3}$.

c) If $xy + 2x - 5y = 2$, then $y' = \dfrac{y + 2}{5 - x}$. The tangent line is $y - 2 = 2(x - 3) \Rightarrow y = 2x - 4$. The normal line is $y - 2 = \left(-\dfrac{1}{2}\right)(x - 3) \Rightarrow y = -\dfrac{1}{2}x + \dfrac{7}{2}$.

63.

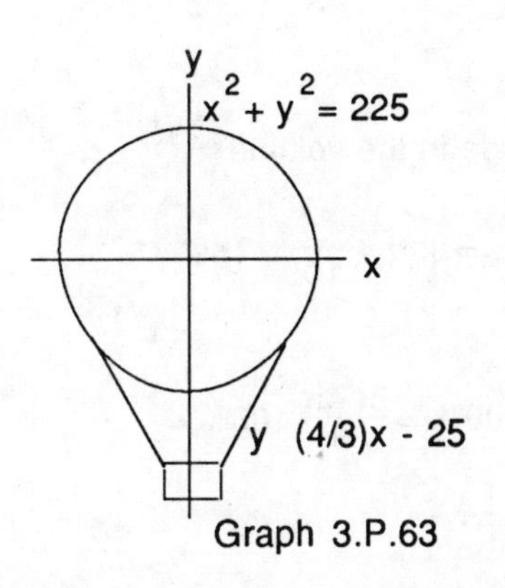

Graph 3.P.63

When $x^2 + y^2 = 225$, then $y' = -\dfrac{x}{y}$. The tangent line to the balloon at $(12, -9)$ is $y + 9 = \dfrac{4}{3}(x - 12) \Rightarrow y = \dfrac{4}{3}x - 25$. The top of the gondola is 23 ft below the center of the balloon. The intersection of $y = -23$ and $y = \dfrac{4}{3}x - 25$ is at the far right edge of the gondola. The gondola is 3 ft wide.

65. a) $x^3 + y^3 = 1 \Rightarrow 3x^2 + 3y^2y' = 0 \Rightarrow y' = -\dfrac{x^2}{y^2} \Rightarrow y'' = \dfrac{-2xy^2 + 2x^2yy'}{y^4} \Rightarrow \dfrac{d^2y}{dx^2} = \dfrac{-2xy^3 - 2x^4}{y^5}$

b) $y^2 = 1 - \dfrac{2}{x} = 1 - 2x^{-1} \Rightarrow y' = \dfrac{x^{-2}}{y} \Rightarrow y'' = \dfrac{-2x^{-3}y - x^{-2}y'}{y^2} \Rightarrow \dfrac{d^2y}{dx^2} = \dfrac{-2xy^2 - 1}{x^4y^3}$

67. a) $y = \sqrt{2x+7} = (2x+7)^{1/2} \Rightarrow y' = (1/2)(2x+7)^{-1/2}(2) = (2x+7)^{-1/2} \Rightarrow$
$\frac{d^2y}{dx^2} = (-1/2)(2x+7)^{-3/2}(2) = \frac{-1}{\sqrt{(2x+7)^3}}$ b) $x^2 + y^2 = 1 \Rightarrow x + yy' = 0 \Rightarrow$
$y' = -\frac{x}{y} \Rightarrow y'' = -\frac{y - xy'}{y^2} = -\left[-\frac{1}{y^3}\right] = \frac{1}{y^3}$; see exercise 66 for missing details.

69. a) If $f(x) = \tan x$ and $x = -\frac{\pi}{4}$, then $f'(x) = \sec^2 x$, $f(-\frac{\pi}{4}) = -1$ and $f'(-\frac{\pi}{4}) = 2$. The linearization of f(x) is $L(x) = 2(x + \frac{\pi}{4}) + (-1) = 2x + \frac{\pi - 2}{2}$.

b) If $f(x) = \sec x$ and $x = -\frac{\pi}{4}$, then $f'(x) = \sec x \tan x$, $f(-\frac{\pi}{4}) = \sqrt{2}$ and $f'(-\frac{\pi}{4}) = -\sqrt{2}$. The linearization of f(x) is $L(x) = -\sqrt{2}(x + \frac{\pi}{4}) + \sqrt{2} = -\sqrt{2}x + \frac{\sqrt{2}(4 - \pi)}{4}$.

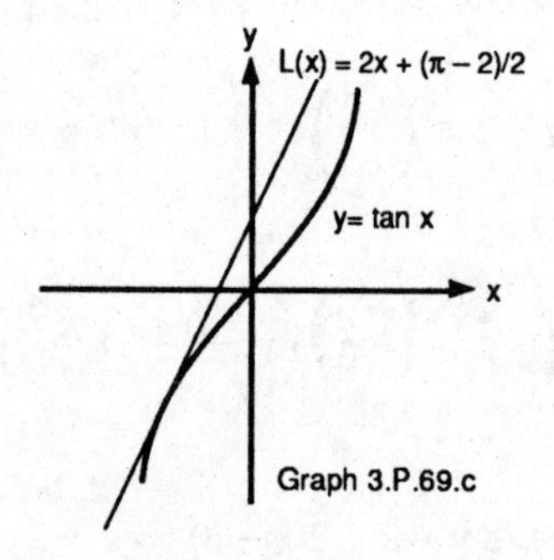

Graph 3.P.69.c

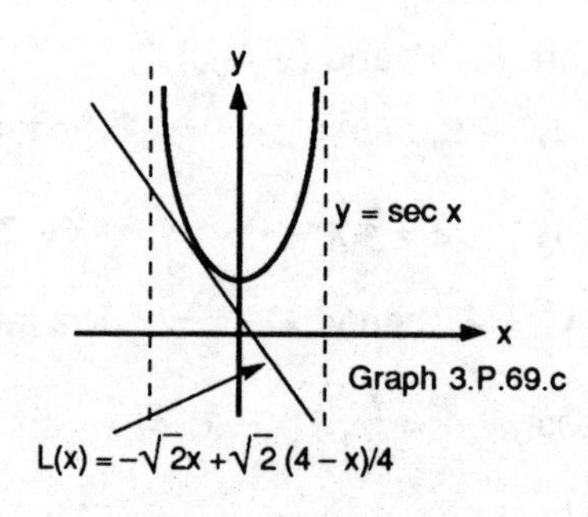

Graph 3.P.69.c

71. a) $f(x) = \sqrt{1+x} + \sin x - .5 \Rightarrow f(-\frac{\pi}{4}) \approx -0.74$ and $f(0) = .5$.

b) From exercise 21 in section 3.7, we have that the linearization of $\sqrt{1+x} + \sin x$ is $\frac{3}{2}x + 1$. Accordingly, the linearization of f(x) is $L(x) = \frac{3}{2}x + \frac{1}{2}$. Solving $L(x) = 0 \Rightarrow x = -\frac{1}{3}$.

c) $f(-\frac{1}{3}) \approx -0.0107$

73. When the volume is $V = \frac{1}{3}\pi r^2 h$, then $dV = \frac{2}{3}\pi r_o h dr$ estimates the change in the volume.

75. a) $S = 6r^2 \Rightarrow dS = 12r dr$. We want $|dS| \le (2\%)(S) \Rightarrow |12r dr| \le \frac{12r^2}{100} \Rightarrow |dr| \le \frac{r}{100}$. The measurement must have an error less than 1%.

b) When $V = r^3$ then $dV = 3r^2 dr$. The accuracy of the volume is $\frac{dV}{V} 100\% = \frac{3r^2 dr}{r^3} 100\% =$
$\left(\frac{3}{r}\right) dr\ 100\% = \left(\frac{3}{r}\right)\left(\frac{r}{100}\right) 100\% = 3\%$

77. With h representing the height of the tree, we have $h = 100 \tan\theta$ with $\theta$ being the angle of elevation of the tree top. Recall that $1° = \frac{\pi}{180}$ radians. Now $dh = 100 \sec^2\theta\, d\theta \Rightarrow$
$dh = 100\sec^2\left(\frac{\pi}{6}\right)\left(\frac{\pi}{180}\right) = \frac{20\pi}{27}$. The error is $\pm \frac{20\pi}{27}$ ft $\approx \pm 0.7757$ ft.

79. Condition 1, $E(a) = 0 \Rightarrow f(a) - g(a) = 0 \Rightarrow f(a) - [m(a - a) + c] = 0 \Rightarrow c = f(a)$ and $g(x) = m(x - a) + f(a)$. Condition 2, $\lim_{x \to a} \frac{E(x)}{x - a} = 0 \Rightarrow \lim_{x \to a} \frac{f(x) - g(x)}{x - a} \Rightarrow$

$$\lim_{x \to a} \frac{f(x) - [m(x - a) + f(a)]}{x - a} = 0 \Rightarrow \lim_{x \to a} \frac{f(x) - f(a)}{x - a} - \lim_{x \to a} m = 0 \Rightarrow f'(a) - m = 0 \Rightarrow m = f'(a).$$

$\therefore$ $g(x) = f'(a)(x - a) + f(a)$ which is $L(x)$.

81. We wish to solve $x^3 - 4 = 0$. Let $y = x^3 - 4 \Rightarrow y' = 3x^2$ $\therefore$ $x_{n+1} = x_n - \frac{x_n^3 - 4}{3x_n^2}$. If $x_o = 1.2$, then $x_7 = 1.58740105$, an approximation for the root.

83. We wish to solve $\sec x - 4 = 0$ when $0 \le x \le \frac{\pi}{2}$. Let $f(x) = \sec x - 4 \Rightarrow f'(x) = \sec x \tan x$ $\therefore$ $x_{n+1} = x_n - \frac{\sec x_n - 4}{\sec x_n \tan x_n}$. If $x_o = 1.4$, then $x_{10} = 1.313811607$, an approximation for the root.

85. We wish to solve $2 \cos x - \sqrt{1 + x} = 0$. Let $f(x) = 2 \cos x - \sqrt{1 + x}$ then $f'(x) = -2 \sin x - \frac{1}{2\sqrt{1 + x}}$

$\therefore$ $x_{n+1} = x_n - \dfrac{2 \cos x_n - \sqrt{1 + x_n}}{-2 \sin x_n - \frac{1}{2\sqrt{1 + x_n}}}$. If $x_o = .8$, then $x_{10} = 0.828360808$, an approximation of the root.

# CHAPTER 4

# APPLICATIONS OF DERIVATIVES

## 4.1 RELATED RATES OF CHANGE

1. $A = \pi r^2 \Rightarrow \frac{dA}{dt} = 2\pi r \frac{dr}{dt}$

3. $V = x^3 \Rightarrow \frac{dV}{dt} = 3x^2 \frac{dx}{dt}$

5. $V = (1/3)\pi r^2 h \Rightarrow \frac{dV}{dt} = (1/3)\pi r^2 \frac{dh}{dt}$

7. Given $A = \pi r^2$, $\frac{dr}{dt} = 0.01$ cm/sec, and $r = 50$ cm. Since $\frac{dA}{dt} = 2\pi r \frac{dr}{dt}$, then $\left.\frac{dA}{dt}\right|_{r=50} = 2\pi(50)\left(\frac{1}{100}\right) = \pi$ cm$^2$/min.

9. Given $\frac{dl}{dt} = -2$ cm/sec, $\frac{dw}{dt} = 2$ cm/sec, $l = 12$ cm and $w = 5$ cm.

   a) $A = lw \Rightarrow \frac{dA}{dt} = l\frac{dw}{dt} + w\frac{dl}{dt} \Rightarrow \frac{dA}{dt} = 12(2) + 5(-2) = 14$ cm$^2$/sec, increasing

   b) $P = 2l + 2w \Rightarrow \frac{dP}{dt} = 2\frac{dl}{dt} + 2\frac{dw}{dt} = 2(-2) + 2(2) = 0$ cm/sec, constant

   c) $D = \sqrt{w^2 + l^2} = (w^2 + l^2)^{1/2} \Rightarrow \frac{dD}{dt} = (1/2)(w^2 + l^2)^{-1/2}\left[2w\frac{dw}{dt} + 2l\frac{dl}{dt}\right] \Rightarrow$

   $\frac{dD}{dt} = \frac{w\frac{dw}{dt} + l\frac{dl}{dt}}{\sqrt{w^2 + l^2}} = -\frac{14}{13}$ cm/sec, decreasing

11. Given: $\frac{dx}{dt} = 5$ ft/sec and the ladder is 13 ft. Since $x^2 + y^2 = 169 \Rightarrow \frac{dy}{dt} = -\frac{x}{y}\frac{dx}{dt}$, the ladder is sliding down the wall at $-12$ ft/sec. The area of the triangle formed by the ladder and walls is $A = \frac{1}{2}xy \Rightarrow \frac{dA}{dt} = \left(\frac{1}{2}\right)\left[y\frac{dy}{dt} + x\frac{dx}{dt}\right]$. The area is changing at $-\frac{119}{2}$ ft$^2$/sec.

13. The radius = 1.9 in and $\frac{dr}{dt} = \frac{1}{3000}$ in/min. $V = 6\pi r^2 \Rightarrow \frac{dV}{dt} = 12\pi r\frac{dr}{dt} \Rightarrow \frac{dh}{dt} = \frac{4}{\pi h^2}\frac{dV}{dt}$. The volume is changing at $\approx 0.0239$ in$^3$/min.

15. If $V = \frac{4}{3}\pi r^3$, $S = 4\pi r^2$, and $\frac{dV}{dt} = kS = 4k\pi r^2$, then $\frac{V}{dt} = 4\pi r^2\frac{dr}{dt} = 4k\pi r^2 \Rightarrow \frac{dr}{dt} = k$, a constant.

    Therefore, the radius is increasing at a constant rate.

17. Let x represent the length of the rope and y the horizontal distance the boat is from the dock. We have $x^2 = y^2 + 36 \Rightarrow \frac{dy}{dt} = \frac{x}{y}\frac{dx}{dy}$ $\therefore$ the boat is approaching the dock at 2.5 ft/sec.

19. The volume of the coffee in the pot is $V = 9\pi h \Rightarrow \frac{dV}{dt} = 9\pi\frac{dh}{dt} \Rightarrow$ the rate the coffee is rising is $\frac{dh}{dt} = \frac{1}{9\pi}\frac{dV}{dt} = \frac{10}{9\pi}$ in/min. From the figure we have that $r = h/2 \Rightarrow V = \frac{1}{3}\pi r^2 h = \frac{\pi h^3}{12}$, the volume of the filter. The rate the coffee is falling is $\frac{dh}{dt} = \frac{4}{\pi h^2}\frac{dV}{dt} = \frac{4}{25\pi}(-10) = -\frac{8}{5\pi}$ in/min.

21. When $y = QD^{-1} \Rightarrow \frac{dy}{dt} = D^{-1}\frac{dQ}{dt} - QD^{-2}\frac{dD}{dt} = \frac{1}{41}(0) - \frac{233}{(41)^2}(-2) = \frac{466}{1681} \approx 0.2772$ L/min.

23. Let P(x, y) represent a point on the curve $y = x^2$ and $\theta$ the angle of inclination of a line containing P and the origin. Consequently, $\tan\theta = \frac{y}{x} \Rightarrow \tan\theta = \frac{x^2}{x} = x \Rightarrow \sec^2\theta\frac{d\theta}{dt} = \frac{dx}{dt} \Rightarrow \frac{d\theta}{dt} = \cos^2\theta\frac{dx}{dt}$. Recall that $\frac{dx}{dt} = 10$ m/sec $\Rightarrow \left.\frac{d\theta}{dt}\right|_{x=3} = 1$ rad/sec. $\lim\limits_{x\to\infty} \frac{x}{\sqrt{x^4+x^2}}(10) = (10)\lim\limits_{x\to\infty}\frac{1}{\sqrt{x^2+1}} = 0$ rad/sec

25. When s represents the length of the shadow and x the distance the man is from the streetlight, then $s = \frac{3}{5}x \Rightarrow \frac{ds}{dt} = \frac{3}{5}\frac{dx}{dt} \Rightarrow$ the shadow is decreasing at a rate of 3 ft/sec since $\frac{dx}{dt} = -5$ ft/sec. If l represents the distance the tip of the shadow is from the streetlight $\Rightarrow l = s + x \Rightarrow \frac{dl}{dt} = \frac{ds}{dt} + \frac{dx}{dt} =$ $3 + 5 = 8$ ft/sec, the speed the tip of the shadow is moving.

27. Let y represent the distance between the girl and kite and x represents the horizontal distance between the girl and kite $\Rightarrow y^2 = (300)^2 + x^2 \Rightarrow \frac{dy}{dx} = \frac{x}{y}\frac{dx}{dt} = \frac{400(25)}{500} = 20$ ft/sec.

29. Let s represent the horizontal distance between the car and plane while r is the distance between the car and plane $\Rightarrow 9 + s^2 = r^2 \Rightarrow \frac{ds}{dt} = \frac{r}{\sqrt{r^2-9}}\frac{dr}{dt} \Rightarrow \left.\frac{ds}{dt}\right|_{r=5} = \frac{5}{\sqrt{16}}(-160) \Rightarrow$ 200 MPH = speed of plane + speed of car $\Rightarrow$ the speed of the car is 80 MPH.

31. Let a represent the distance between point O and ship A , b the distance between point O and ship B, and D the distance between the ships. By the Law of Cosines, $D^2 =$ $a^2 + b^2 - 2ab\cos 120° \Rightarrow \frac{dD}{dt} = \frac{1}{2D}\left[2a\frac{da}{dt} + 2b\frac{db}{dt} + a\frac{db}{dt} + b\frac{da}{dt}\right]$. When $a = 5$, $\frac{da}{dt} = 14$, $b = 3$, and $\frac{db}{dt} = 21$, then $\frac{dD}{dt} = \frac{413}{2D}$ where $D = 7$. The ships are moving $\frac{dD}{dt} \approx 29.5$ knots apart.

## 4.2 MAXIMA, MINIMA, AND THE MEAN VALUE THEOREM

1. With $f(-2) = 11 > 0$, $f(-1) = -1 < 0$, and the Intermediate Value Theorem, we may conclude that $f(x) = 0$ has at least one root between $-2$ and $-1$ when $f(x) = x^4 + 3x + 1$. When $-2 < x < -1 \Rightarrow$ $-8 < x^3 < -1 \Rightarrow -32 < 4x^3 < -4 \Rightarrow -29 < 4x^3 + 3 < -1 \Rightarrow f'(x) < 0$ for $-2 < x < -1 \Rightarrow f(x)$ is decreasing for all $-2 < x < -1 \Rightarrow f(x) = 0$ has exactly one solution when $-2 < x < -1$.

3. With $f(1) = -1 < 0$, $f(3) = 7/3 > 0$, and the Intermediate Value Theorem, we may conclude that $f(x) = 0$ has at least one root between 1 and 3 when $f(x) = x - 2x^{-1}$. Since $f'(x) = 1 + 2/x^2 \Rightarrow$ $f'(x) > 0$ for $1 < x < 3 \Rightarrow f(x)$ is increasing for all $1 < x < 3 \Rightarrow f(x) = 0$ has exactly one solution when $1 < x < 3$.

5. a)

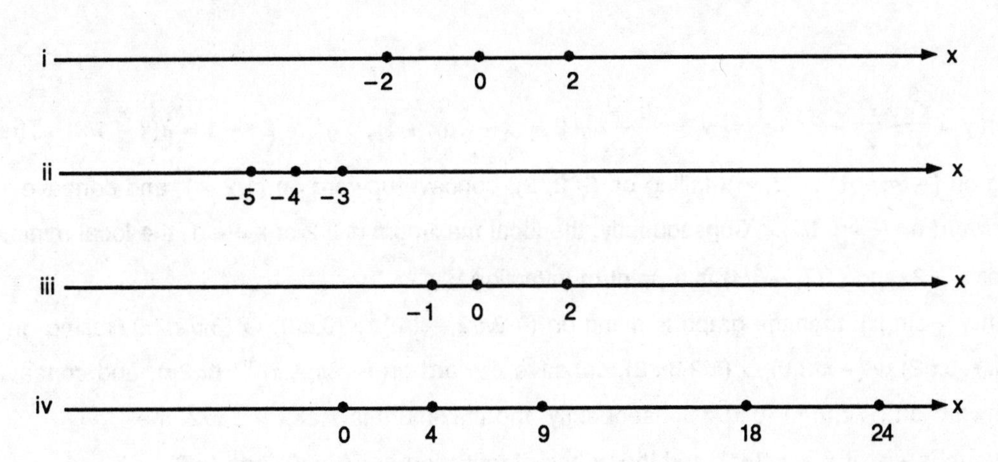


b) Let $r_1$ and $r_2$ be zeros of the polynomial $P(x) = x^n + a_{n-1}x^{n-1} + \ldots + \alpha_1 x + a_0$. Then $P(r_1) = P(r_2) = 0$. Since polynomials are everywhere continuous and differentiable by Rolle's Theorem, $P'(r) = 0$ for some $r_1 < r < r_2$. $P'(x) = nx^{n-1} + (n-1)a_{n-1}x^{n-2} + \ldots + a_1$.

7. When $f(x) = x^2 + 2x - 1$ for $0 \le x \le 1$, then $\frac{f(1) - f(0)}{1 - 0} = f'(c) \Rightarrow 3 = 2c + 2 \Rightarrow c = 1/2$.

9. When $f(x) = x + 1/x$ for $1/2 \le x \le 2$, then $\frac{f(2) - f(1/2)}{2 - 1/2} = f'(c) \Rightarrow 0 = 1 - 1/c^2 \Rightarrow c = 1$.

11. If s(t) = distance traveled in time t, then $s(0) = 0$ and $s(2) = 159$. From the Mean Value Theorem there exists a c between 0 and 2 such that $\frac{s(2) - s(0)}{2 - 0} = s'(c) = v(c) \Rightarrow v(c) = 79.5$ MPH. The speed of the truck at time c was 79.5 MPH while the speed limit was 65 MPH.

13. If s(t) = the distance traveled in time t, then $s(0) = 0$ and $s(24) = 184$. From the Mean Value Theorem there exists a c between 0 and 24 such that $\frac{s(24) - s(0)}{24 - 0} = s'(c) = v(c) \Rightarrow v(c) \approx 7.67$ knots.

15. Let $f(x) = \sin x$ for $a \le x \le b$. From the Mean Value Theorem there exists a c between a and b such that $\frac{\sin b - \sin a}{b - a} = \cos c \Rightarrow -1 \le \frac{\sin b - \sin a}{b - a} \le 1 \Rightarrow \left|\frac{\sin b - \sin a}{b - a}\right| \le 1 \Rightarrow |\sin b - \sin a| \le |b - a|$.

17. When $f(x) = 1/x \Rightarrow f'(x) = -1/x^2 \Rightarrow f'(x) < 0$ for all non-zero numbers $\Rightarrow f(x) = 1/x$ is always decreasing on any interval on which it is defined.

19. $f'(x) = \left(1 + x^4 \cos x\right)^{-1} \Rightarrow f''(x) = -\left(1 + x^4 \cos x\right)^{-2}\left(4x^3 \cos x - x^3 \sin x\right) = -x^3\left(1 + x^4 \cos x\right)^{-2}(4 \cos x - \sin x) < 0$ for $0 < x < 0.1 \Rightarrow f'(x)$ is decreasing when $0 < x < 0.1 \Rightarrow$ $\min f' \approx 0.999900509$ and $\max f' = 1$. Now we have $0.999900509 \le \frac{f(0.1) - 1}{0.1} \le 1 \Rightarrow 0.09999005 \le f(0.1) - 1 \le 0.1 \Rightarrow 1.09999005 \le f(0.1) \le 1.1 \Rightarrow f(0.1) \approx 1.099995025$.

21. If $f(0) = 3$ and $f'(x) = 0$ for all x. From the Mean Value Theorem we have that $\frac{f(x) - f(0)}{x - 0} = f'(c)$ where c is between x and 0. Now $\frac{f(x) - 3}{x} = 0 \Rightarrow f(x) - 3 = 0 \Rightarrow f(x) = 3$.

## 4.3 CURVE SKETCHING WITH Y′ AND Y″

1. When $y = \frac{x^3}{3} - \frac{x^2}{2} - 2x + \frac{1}{3} \Rightarrow y' = x^2 - x - 2 = (x - 2)(x + 1) \Rightarrow y'' = 2x - 1 = 2(x - 1/2)$. The graph is rising on $(-\infty, -1) \cup (2, \infty)$, falling on $(-2, 2)$, concave upward on $(1/2, \infty)$, and concave downward on $(-\infty, 1/2)$. Consequently, the local maximum is 3/2 at $x = -1$, the local minimum is $-3$ at $x = 2$, and $(1/2, -3/4)$ is a point of inflection.

3. When $y = \sin|x|$, then the graph is rising on $(-3\pi/2, -\pi/2) \cup (0, \pi/2) \cup (3\pi/2, 2\pi)$, falling on $(-2\pi, -3\pi/2) \cup (-\pi/2, 0) \cup (\pi/2, 3\pi/2)$, concave upward on $(-2\pi, -\pi) \cup (\pi, 2\pi)$, and concave downward on $(-\pi, 0) \cup (0, \pi)$. Consequently, the maximum is 1 at $x = \pm\pi/2$, the minimum is $-1$ at $x = \pm 3\pi/2$, and the points of inflection are $(-\pi, 0)$ and $(\pi, 0)$.

5. When $y = x^2 - 4x + 3$, then $y' = 2x - 4 = 2(x - 2)$ and $y'' = 2$. The curve rises on $(2, \infty)$ and falls on $(-\infty, 2)$. At $x = 2$ there is a minimum. Since $y'' > 0$, the curve is concave upward for all x.

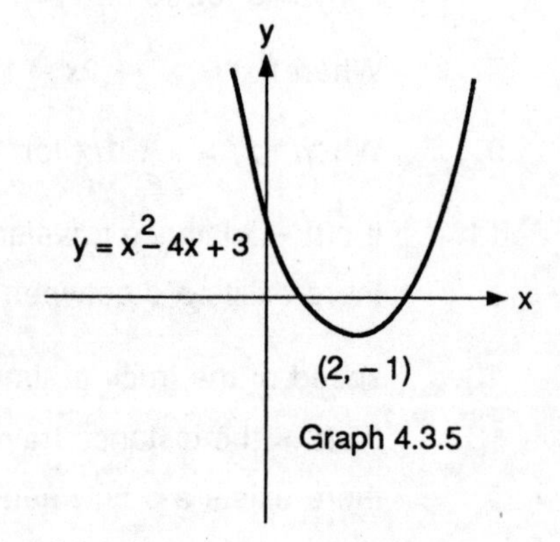

Graph 4.3.5

7. When $y = 2x - x^2$, then $y' = 2 - 2x = 2(1 - x)$ and $y'' = -2$. The curve rises on $(-\infty, 1)$ and falls on $(1, \infty)$. At $x = 1$ there is a maximum. Since $y'' < 0$, the curve is concave downward for all x.

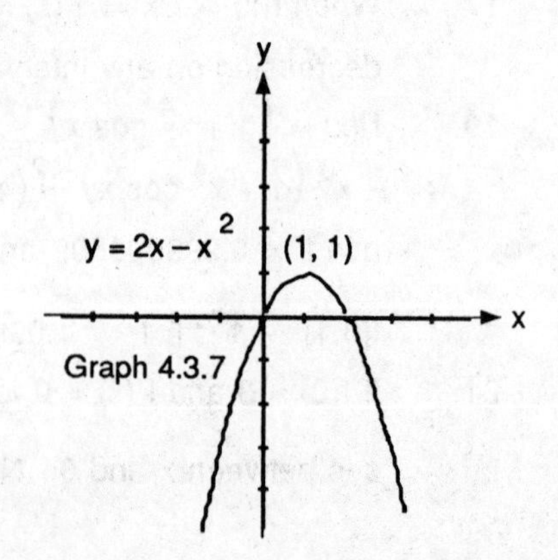

Graph 4.3.7

9. When $y = x^3 - 3x + 3$, then $y' = 3x^2 - 3 = 3(x-1)(x+1)$ and $y'' = 6x$. The curve rises on $(-\infty, -1) \cup (1, \infty)$ and falls on $(-1, 1)$. At $x = -1$ there is local maximum and at $x = 1$ a local minimum. The curve is concave downward on $(-\infty, 0)$ and concave upward on $(0, \infty)$. There is a point of inflection at $x = 0$.

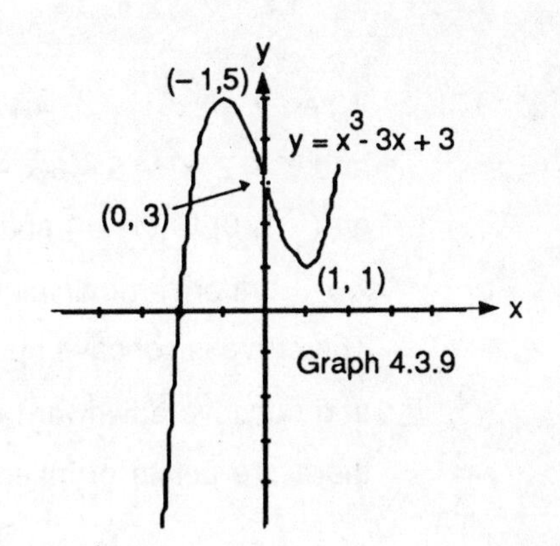

Graph 4.3.9

11. When $y = 4 + 3x - x^3$, then $y' = 3 - 3x^2 = 3(1-x)(1+x)$ and $y'' = -6x$. The curve rises on $(-1, 1)$ and falls on $(-\infty, -1) \cup (1, \infty)$. At $x = -1$ there is a local minimum and at $x = 1$ a local maximum. The curve is concave upward on $(-\infty, 0)$ and concave downward on $(0, \infty)$. At $x = 0$ there is a point of inflection.

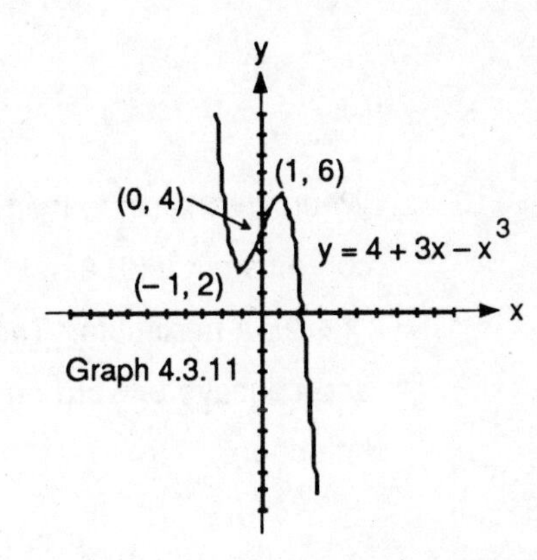

Graph 4.3.11

13. When $y = x^3 - 6x^2 + 9x + 1$, then $y' = 3x^2 - 12x + 9 = 3(x-3)(x-1)$ and $y'' = 6x - 12 = 6(x-2)$. The curve rises on $(-\infty, 1) \cup (3, \infty)$ and falls on $(1, 3)$. At $x = 1$ there is a local maximum and at $x = 3$ a local minimum. The curve is concave downward on $(-\infty, 2)$ and concave upward on $(2, \infty)$. At $x = 2$ there is a point of inflection.

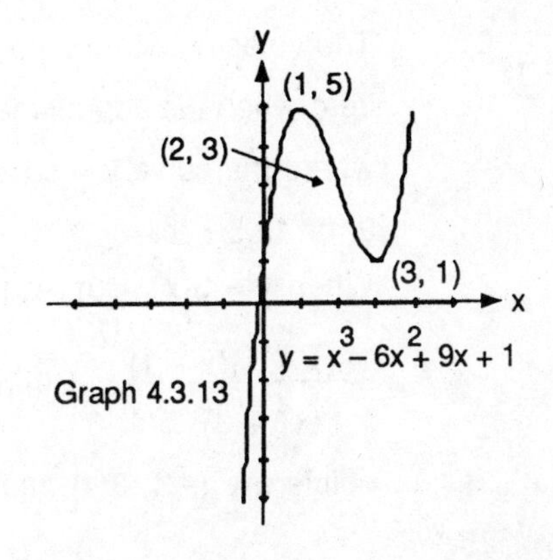

Graph 4.3.13

15. When $y = \frac{x^3}{9} - 3x + 4$, then $y' = \frac{x^2}{3} - 3 = \frac{1}{3}(x-3)(x+3)$ and $y'' = \frac{2x}{3}$. The curve rises on $(-\infty, -3) \cup (3, \infty)$ and falls on $(-3, 3)$. At $x = -3$ there is a local maximum and at $x = 3$ a local minimum. The curve is concave downward on $(-\infty, 0)$ and concave upward on $(0, \infty)$. At $x = 0$ there is a point of inflection.

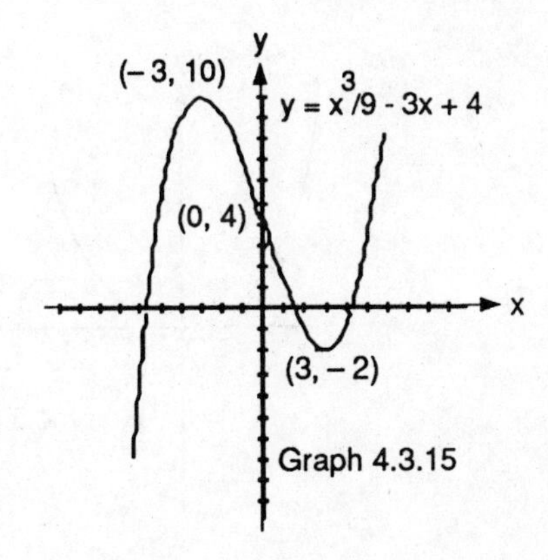

Graph 4.3.15

17. When $y = 2x^4 - 4x^2 + 1$, then $y' = 8x^3 - 8x = 8x(x + 1)(x - 1)$ and $y'' = 24x^2 - 8 = 8(x - 1/\sqrt{3})(x + 1/\sqrt{3})$. The curve rises on $(-1, 0) \cup (1, \infty)$ and falls on $(-\infty, -1) \cup (0, 1)$. At $x = \pm 1$ there are minimums and at $x = 0$ a local maximum. The curve is concave upward on $(-\infty, -\sqrt{3}/3) \cup (\sqrt{3}/3, \infty)$ and concave downward on $(-\sqrt{3}/3, \sqrt{3}/3)$. At $x = \pm\sqrt{3}/3$ there are points of inflection.

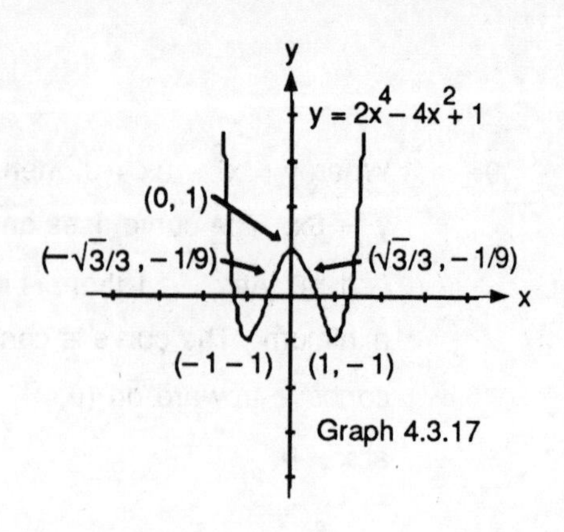

Graph 4.3.17

19. When $y = x + \sin x$, then $y' = 1 + \cos x$ and $y'' = -\sin x$. The curve rises on $(0, 2\pi)$. At $x = 0$ there is a minimum and at $x = 2\pi$ a maximum. The curve is concave downward on $(0, \pi)$ and concave upward on $(\pi, 2\pi)$. At $x = \pi$ there is a point of inflection.

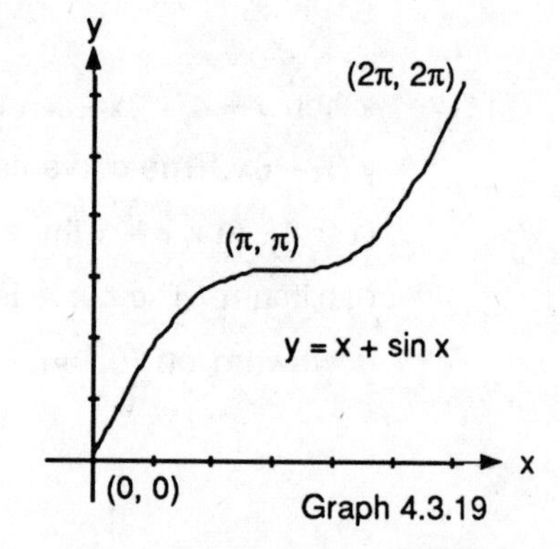

Graph 4.3.19

21. The velocity will be zero when the slope of the tangent line for $y = s(t)$ is horizontal. The velocity is zero when t is approximately 2, 6 ,or 9.5 sec. The acceleration will be zero at those values of t where the curve $y = s(t)$ has points of inflection. The acceleration is zero when t is approximately 4, 8, or 12.5 sec.

23. When $y = 3(x^2 + 3)^{-1}$, then $y' = -3(x^2 + 3)^{-2}(2x)$ and $y'' = -6(x^2 + 3)^{-2} + 24x^2(x^2 + 3)^{-3} = \dfrac{18(x + 1)(x - 1)}{(x^2 + 3)^3}$. The points of inflection are located at $x = \pm 1$. The coordinates of the inflection points are $(-1, 3/4)$ and $(1, 3/4)$.

25.

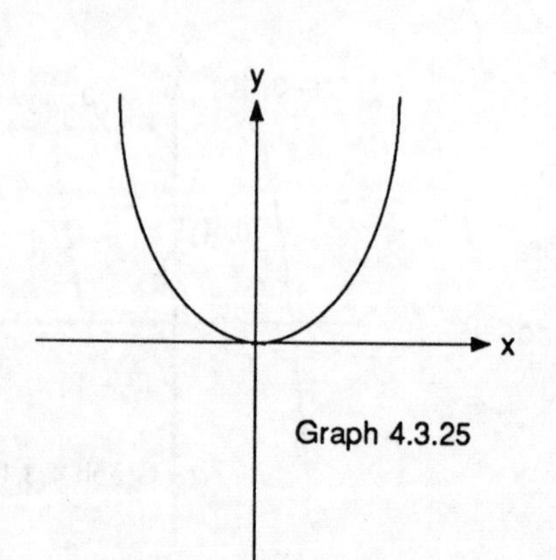

Graph 4.3.25

27.

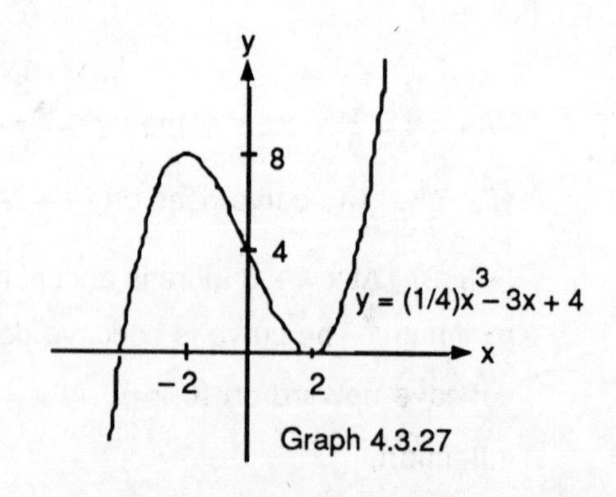

Graph 4.3.27

29. When $y' = (x-1)^2(x-2)$, then $y'' = 2(x-1)(x-2) + (x-1)^2$. The curve falls on $(-\infty, 2)$ and rises on $(2, \infty)$. At $x = 2$ there is a local minimum. The curve is concave upward on $(-\infty, 1) \cup (5/3, \infty)$ and concave downward on $(1, 5/3)$. At $x = 1$ or $x = 5/3$ there are inflection points.

31. When $y = x + 1/x$, then $y' = 1 - x^{-2} = \dfrac{(x+1)(x-1)}{x^2}$. The curve rises on $(-\infty, -1) \cup (1, \infty)$ and falls on $(-1, 1)$. At $x = -1$ the local maximum is $-2$ and at $x = 1$ the local minimum is 2.

33. No. When $f(x) = x^3$, then $f'(x) = 3x^2$ and $f'(0) = 0$ but $x = 0$ is neither a local minimum nor maximum.

35. True. If $y = ax^2 + bx + c$ where $a \neq 0$, then $y' = 2ax + b$ and $y'' = 2a$. Since $2a$ is a constant, it is not possible for $y''$ to change signs.

37. a)

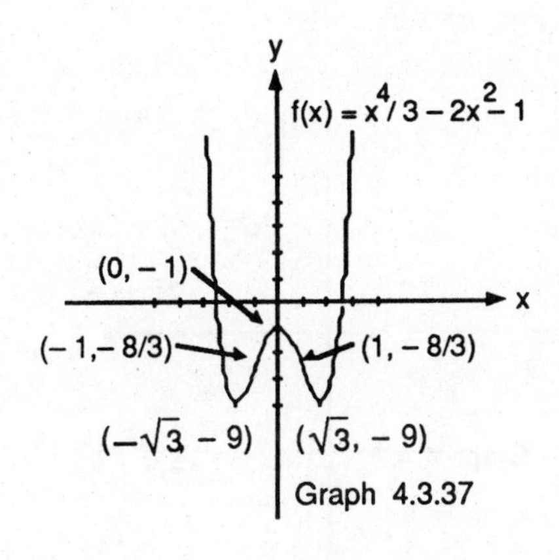

Graph 4.3.37

b) Recall that even functions are symmetric to the y–axis. Consequently, we have to find only one of the roots. Newton's method indicates that the positive root is approximately 2.54245976. The negative root must be approximately – 2.54245976.

## 4.4 GRAPHING RATIONAL FUNCTIONS --ASYMPTOTES AND DOMINANT TERMS

1.

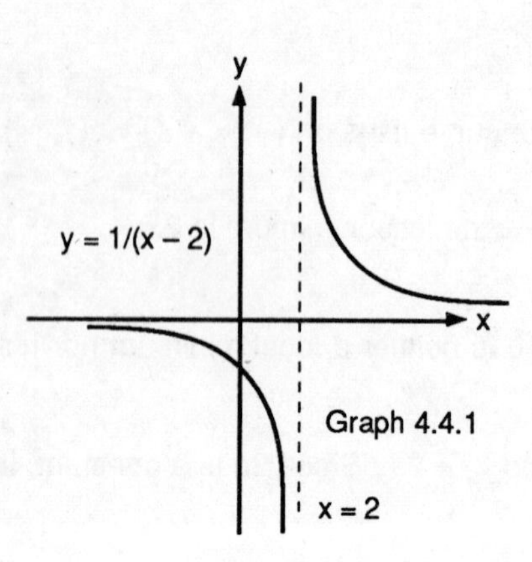


Graph 4.4.1

3.

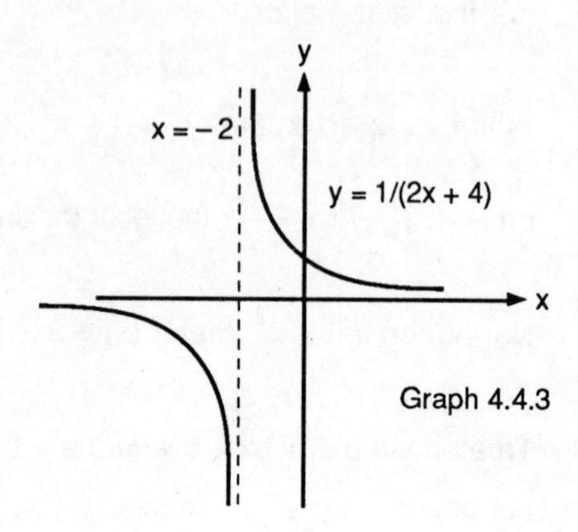


Graph 4.4.3

5,

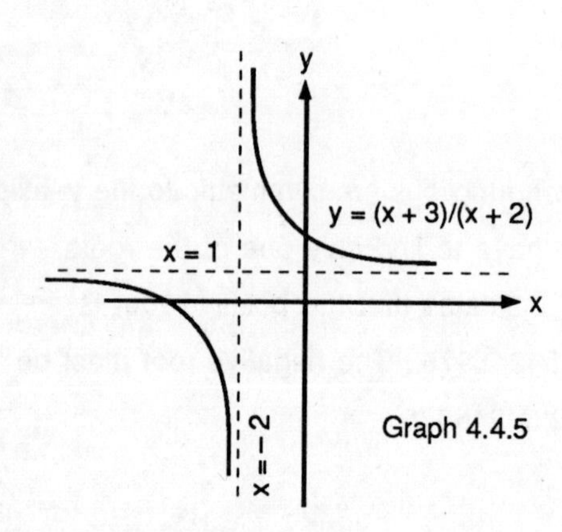


Graph 4.4.5

7.

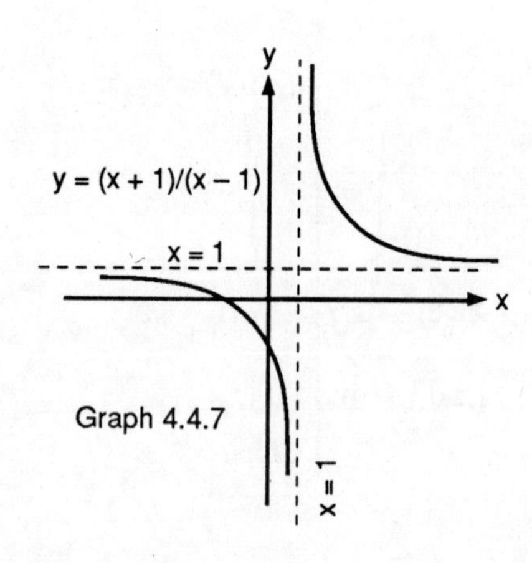


Graph 4.4.7

9.

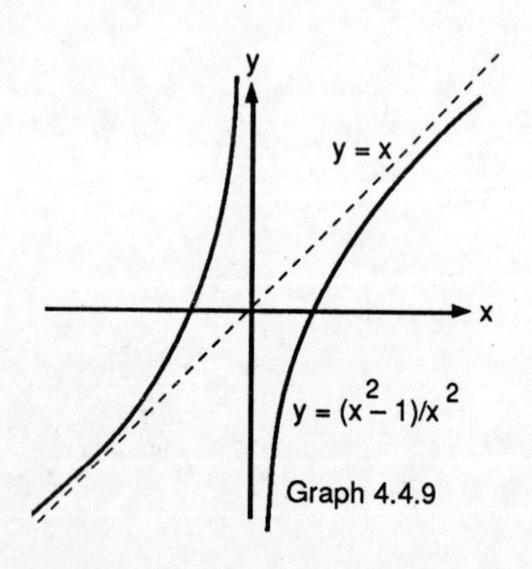


Graph 4.4.9

11.

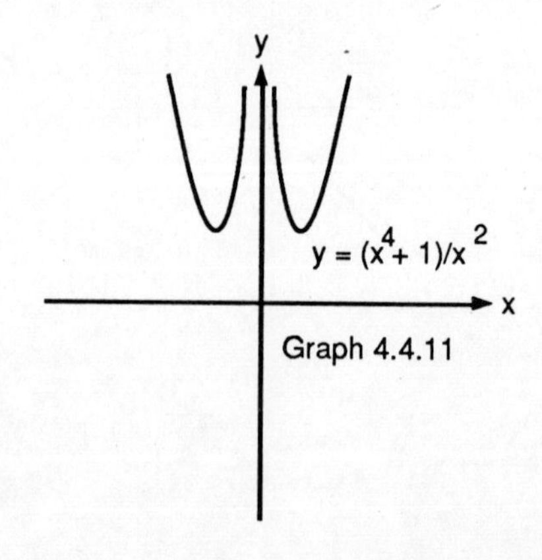


Graph 4.4.11

13.

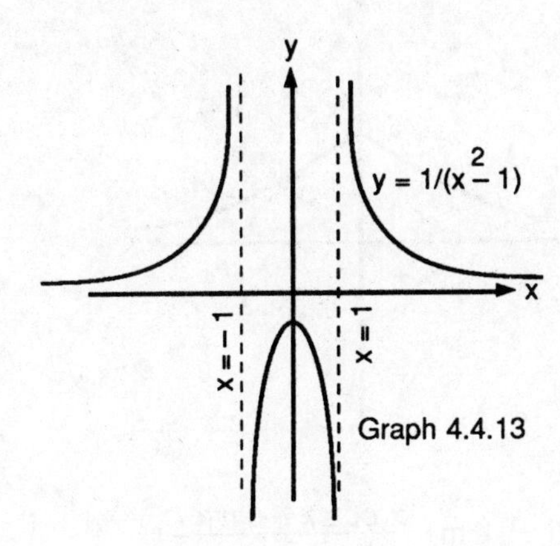


15.

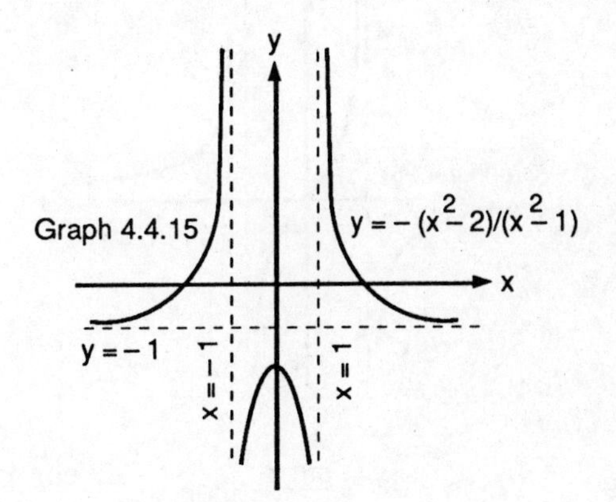


17.

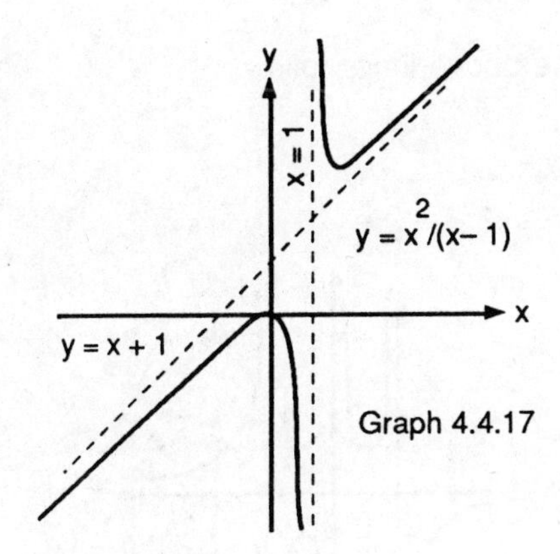


19.

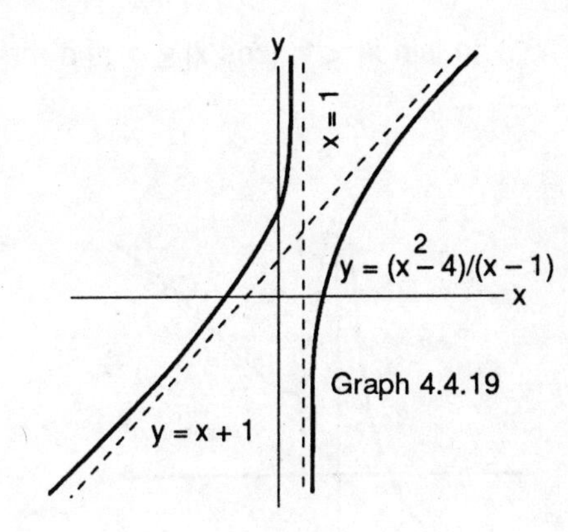


21.

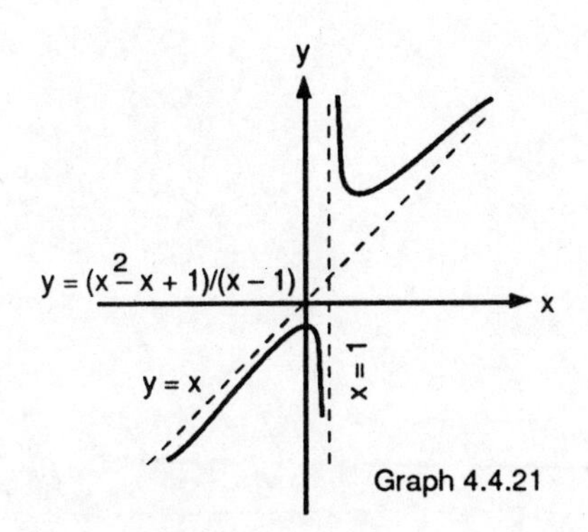


23.

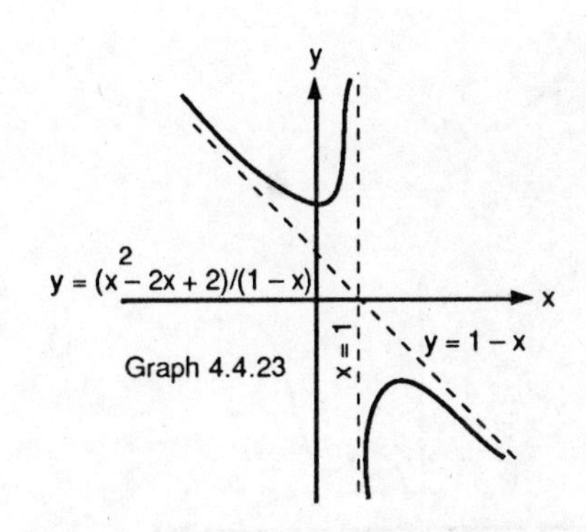

25.

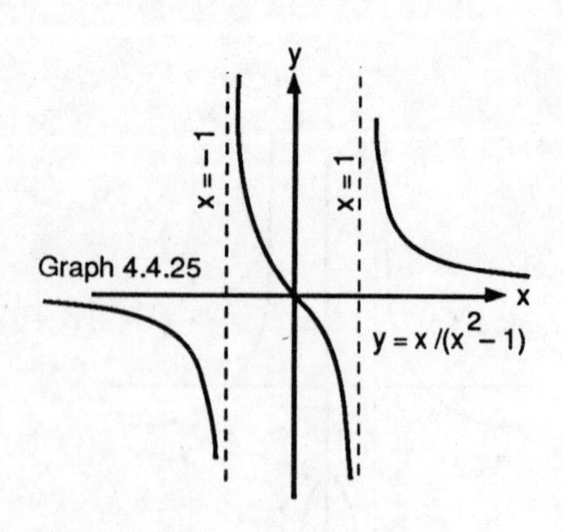

Graph 4.4.25

27.

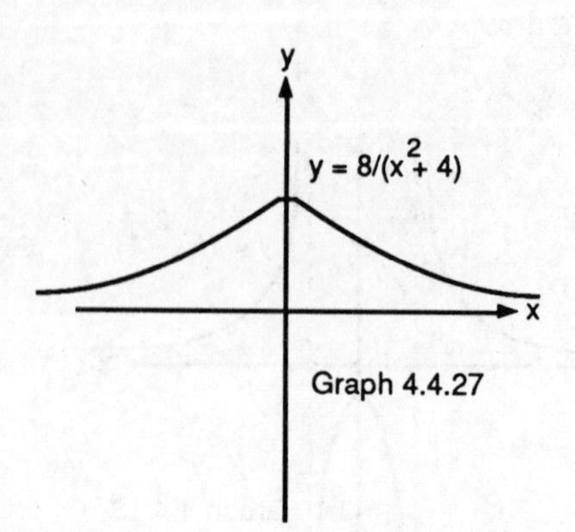

Graph 4.4.27

29. When $y = 2 + \frac{\sin x}{x}$, then $y' = \frac{x \cos x - \sin x}{x^2}$. Consequently, $\lim_{x \to \infty} \frac{x \cos x - \sin x}{x^2} =$

$\lim_{x \to \infty} \frac{\cos x}{x} - \lim_{x \to \infty} \frac{\sin x}{x^2} = 0 - 0 = 0$ which represents the slope of the tangent line as $x \to \infty$.

With $|\sin x| < 1$, $|\cos x| < 1$ and the Sandwich Theorem, the above limits follow.

31.

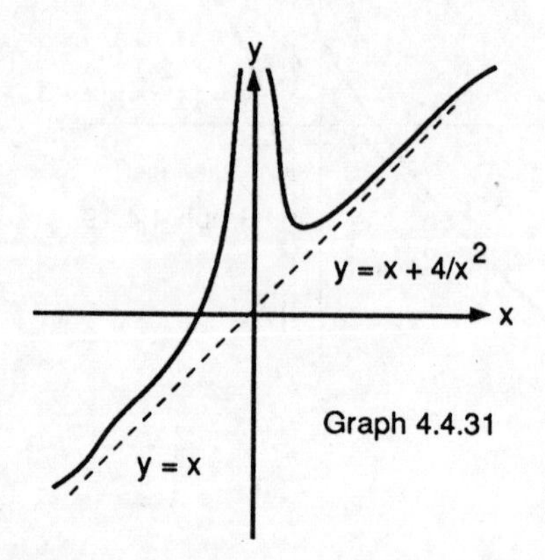

Graph 4.4.31

33.

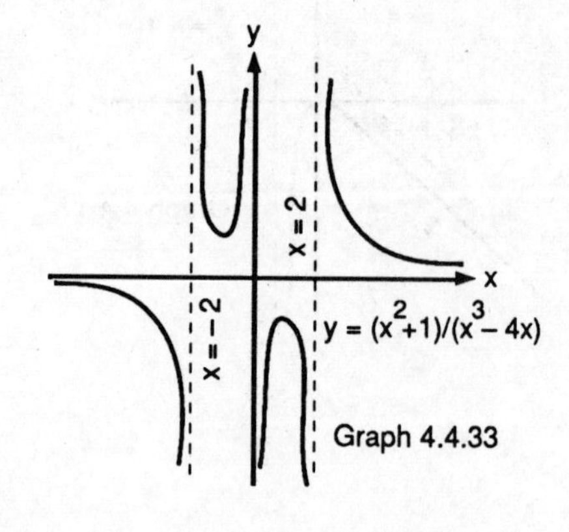

Graph 4.4.33

35.

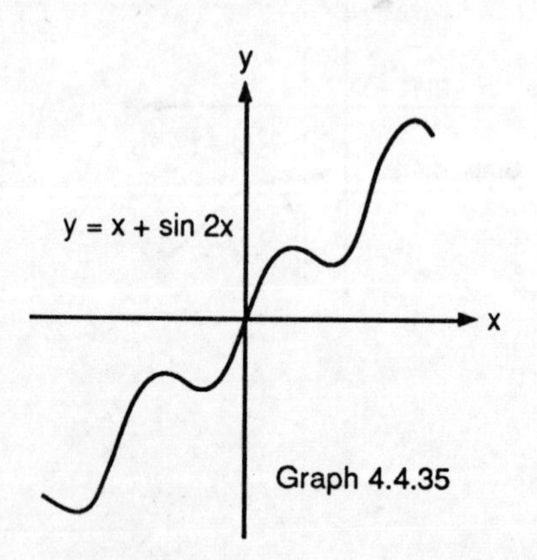

Graph 4.4.35

## 4.5 OPTIMIZATION

1. Let x and 20 – x represent the numbers where $0 \le x \le 20$.

   a) Let $s(x) = x^2 + (20 - x)^2$, then $s'(x) = 4(x - 10)$. The solution of $s'(x) = 0$ is 10. The maximum value of s(x) is the largest of s(0) = 400, s(10) = 200, or s(20) = 400. ∴ the numbers are 0 and 20.

   b) Let $s(x) = \sqrt{x} + (20 - x)$, then $s'(x) = \frac{1 - 2\sqrt{x}}{2\sqrt{x}}$. The critical numbers are 0 or 1/4. The maximum value of s(x) is the largest of s(0) = 20, $s(\frac{1}{4}) = \frac{1}{2} + 19\frac{3}{4} = 20\frac{1}{4}$, and $s(20) = \sqrt{20} \approx 4.47$. The numbers are $\frac{1}{4}$ and $19\frac{3}{4}$.

3. Let l and w represent the length and width of the rectangle respectively. With an area of 16 $in^2$, we have that $(l)(w) = 16 \Rightarrow w = 16l^{-1}$ and the perimeter, $P = 2l + 2w = 2l + 32l^{-1}$. Solving $P'(l) = 0 \Rightarrow \frac{2(l + 4)(l - 4)}{l^2} = 0 \Rightarrow l = -4, 4$, or 0. Due to the fact that – 4 and 0 are not possible lengths of a rectangle, l must be 4, w = 4, and the perimeter is 16 in.

5. a) The line containing point P also contains the points (0, 1) and (1, 0). ∴ the line containing P is y = 1 – x and a general point on that line is (x, 1 – x).

   b) The area A(x) is 2x(1 – x), where $0 \le x \le 5$.

   c) When $A(x) = 2x - 2x^2$, then $A'(x) = 0 \Rightarrow x = 1/2$. Due to A(0) = 0 and A(1) = 0, we may conclude that A(1/2) = 1/2 sq units, the largest area.

7. The volume of the box is V(x) = x(15 – 2x)(8 – 2x), where $0 \le x \le 4$. Solving $V'(x) = 0 \Rightarrow x = 5/3$ or 6, but 6 is not in the domain. Since V(0) = V(4) = 0, V(5/3) must be the maximum volume. The dimensions are: 14/3 x 35/3 x 5/3 inches.

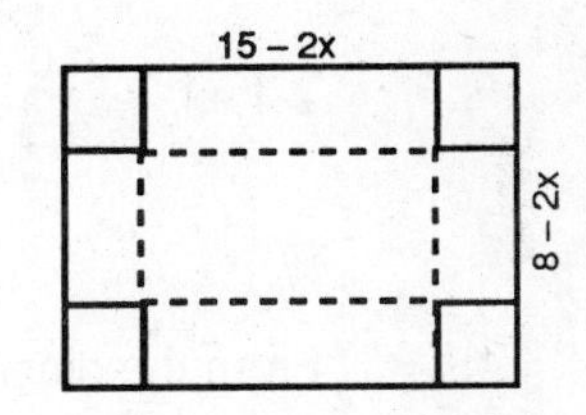


9. The area is A(x) = x(800 – 2x), where $0 \le x \le 400$. Solving $A'(x) = 0 \Rightarrow$ x = 200. With A(0) = A(400) = 0, the maximum area is A(200). The largest area that an be enclosed is A(200) = 80000 $m^2$.

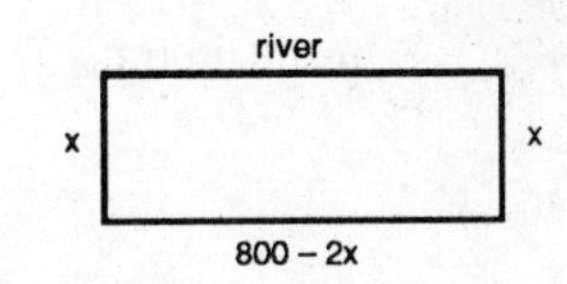


11. The surface area is $S(x) = x^2 + 4x\left(\frac{500}{x^2}\right)$, where 0 < x. The critical points are 0 or 10, but 0 is not in the domain. $S''(10) > 0 \Rightarrow$ at x = 10 there is a minimum. ∴ the dimensions should be 10 ft on the base edge and 5 ft for the height.

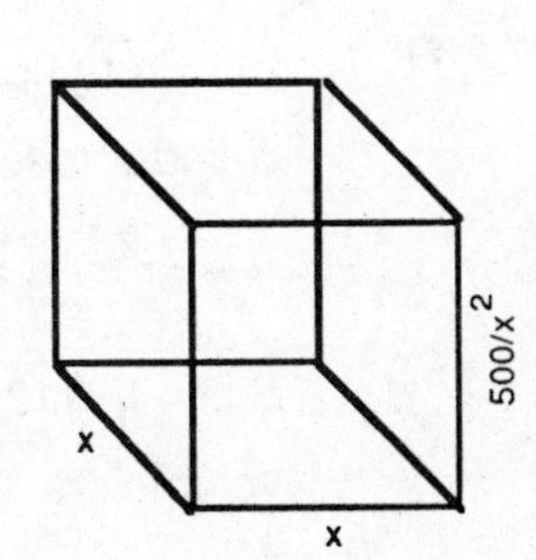

13. The area of the printing is $(y - 4)(x - 8) = 50$. Consequently, $y = (15/(x - 8)) + 4$. The area of the paper is $A(x) = x\left(\frac{50}{x-8} + 4\right)$, where $8 < x$. The critical points are $-2$, or 18, but $-2$ is not in the domain. Now $A''(18) > 0 \Rightarrow$ that at $x = 18$ we have a minimum. $\therefore$ the dimensions which will minimize the amount of paper are 18 by 9 inches.

15. The area of the triangle is $A(\theta) = \frac{ab \sin \theta}{2}$, where $0 < \theta < \pi$. Solving $A'(\theta) = 0 \Rightarrow \theta = \frac{\pi}{2}$. $A''\left(\frac{\pi}{2}\right) < 0 \Rightarrow$ there is a maximum at $\theta = \frac{\pi}{2}$.

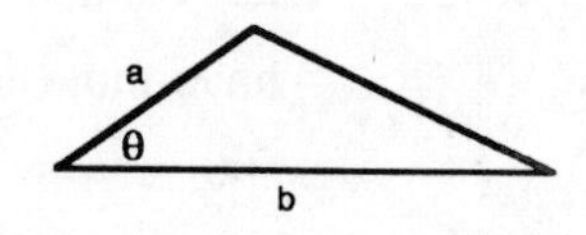


17. With a volume of 1000 cm and $V = \pi r^2 h$, then $h = \frac{1000}{\pi r^2}$. The surface area is $S = 2\pi h + \pi r^2 = 2000r^{-1} + \pi r^2$, where $0 < r$. The critical points are 0 and $\frac{10}{\sqrt[3]{\pi}}$, but 0 is not in the domain. At $r = \frac{10}{\sqrt[3]{\pi}}$ we have a minimum because $s''\left(\frac{10}{\sqrt[3]{\pi}}\right) < 0$.

$\therefore r = h = \frac{10}{\sqrt[3]{\pi}}$

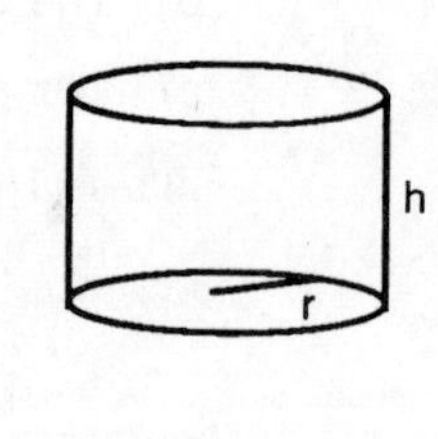


19. From the diagram we have $4x + l = 108$ and $V = x^2 l$. The volume of the box is $V(x) = x^2(108 - 4x)$, where $0 \le x \le 27$. The critical points are 0 and 18, but $x = 0$ would result in no box. At $x = 18$ we have $V''(x) < 0 \Rightarrow$ a maximum. The dimensions of the box are 18 x 18 x 36 in.

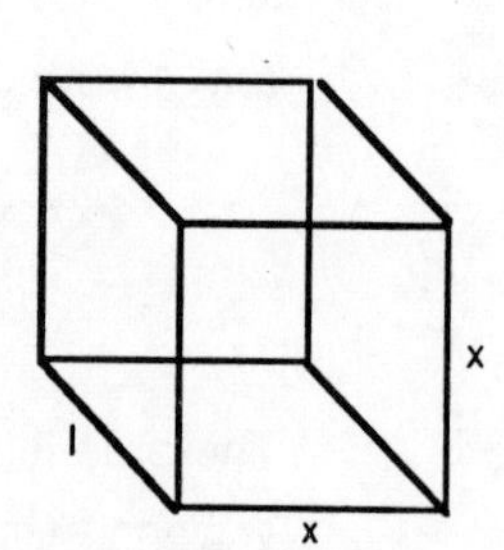


21. a) From figure 4.39 we have $P = 2x + 2y \Rightarrow y = P/2 - x$. If $P = 36$, then $y = 18 - x$. When the cylinder is formed, $x = 2\pi r \Rightarrow r = x/2\pi$ and $h = y \Rightarrow h = 18 - x$. The volume of the cylinder is $V = \pi r^2 h \Rightarrow V(x) = \frac{18x^2 - x^3}{4\pi}$. Solving $V'(x) = 0 \Rightarrow x = 0$ or 12; but when $x = 0$, there is no cylinder. Since $V''(12) < 0$ at $x = 12$, there is a maximum. The values of $x = 12$ and $y = 6$ will give the largest volume.

21. b) From part a) we have $V(x) = \pi x^2(18 - x)$. Solving $V'(x) = 0 \Rightarrow x = 0$ or 12, but $x = 0$ would result in no cylinder. Since $V''(12) < 0$ at $x = 12$, there is a maximum. The values of $x = 12$ and $y = 6$ will give the largest volume.

23. a) $f(x) = x^2 + \frac{a}{x} \Rightarrow f'(x) = x^{-2}\left(2x^3 - a\right)$. Solving $f'(x) = 0$ at $x = 2$ implies that $a = 16$.

b) $f(x) = x^2 + \frac{a}{x} \Rightarrow f''(x) = 2x^{-3}\left[x^3 + a\right]$. Solving $f''(x) = 0$ at $x = 2$ implies that $a = -1$.

25. If $f(x) = x^3 + ax^2 + bx$, then $f'(x) = 3x^2 + 2ax + b$ and $f''(x) = 6x + 2a$.

a) A local maximum at $x = -1$ and local minimum at $x = 3 \Rightarrow f'(-1) = 0$ and $f'(3) = 0$. $\therefore 3 - 2a + b = 0$ and $27 + 6a + b = 0 \Rightarrow a = -3$ and $b = -9$.

b) A local maximum at $x = -1$ and a point of inflection at $x = 1 \Rightarrow f'(-1) = 0$ and $f''(1) = 0$. $\therefore 48 + 8a + b = 0$ and $6 + 2a = 0 \Rightarrow a = -3$ and $b = -24$.

27. From the diagram we have $d^2 = 4r^2 - w^2$. The strength of the beam is $S = kwd^2 = kw(4r^2 - w^2)$. When $r = 6$, then $S = 144kw - kw^3$. $S'(w) = 144k - 3kw^2 = 3k(48 - w^2)$. $S'(w) = 0 \Rightarrow w = \pm 4\sqrt{3}$. $S''(4\sqrt{3}) < 0$ and $-4\sqrt{3}$ is not acceptable. $\therefore S(4\sqrt{3})$ is the maximum strength. The dimensions of the strongest beam are $4\sqrt{3}$ by $4\sqrt{6}$ inches.

29. If $y = \cot x - \sqrt{2} \csc x$ where $0 < x < \pi$, then $y' = [\csc x][\sqrt{2} \cot x - \csc x]$. Solving $y' = 0 \Rightarrow \cos x = \frac{1}{\sqrt{2}} \Rightarrow x = \frac{\pi}{4}$. For $0 < x < \frac{\pi}{4}$ we have $y' > 0$ and $y' < 0$ when $\frac{\pi}{4} < x < \pi$ $\therefore$ at $x = \frac{\pi}{4}$ there is a maximum. The maximum value of y is $-1$.

31. The distance between a point on the curve $y = \sqrt{x}$ and $(3/2, 0)$ is represented by $D(x) = \sqrt{(3/2 - x)^2 + (0 - \sqrt{x})^2}$. $D(x)$ is minimized when $I(x) = (3/2 - x)^2 + (0 - \sqrt{x})^2 = (3/2 - x)^2 + x$ is minimized. $I'(x) = 2(3/2 - x)(-1) + 1 = 2(x - 1)$. The solution of $I'(x) = 0$ is $x = 1$ and $I''(1) = 2 > 0$. $\therefore$ the minimum value of $I(x)$ is $I(1) = \frac{5}{4} \Rightarrow$ the minimum distance is $D(1) = \sqrt{\frac{5}{4}} = \frac{\sqrt{5}}{2}$.

33. If $f(x) = x^2 - x + 1$, then $f'(x) = 2x - 1$ and $f''(x) = 2$. At $x = \frac{1}{2}$ there is a minimum. $f(x)$ is never negative because $f\left(\frac{1}{2}\right) = \frac{3}{4} > 0$.

35. From the diagram the area of the cross section is $A(\theta) = \cos\theta + \sin\theta\cos\theta$, where $0 < \theta < \pi/2$. $A'(\theta) = -\sin\theta + \cos^2\theta - \sin^2\theta = -\left(2\sin^2\theta + \sin\theta - 1\right) = -(2\sin\theta - 1)(\sin\theta + 1)$. $A'(\theta) = 0 \Rightarrow \sin\theta = 1/2$ or $\sin\theta = -1 \Rightarrow \theta = \pi/6$ because $\sin\theta \neq -1$ when $0 < \theta < \pi/2$. $A'(\theta) > 0$ for $0 < \theta < \pi/6$ and $A'(\theta) < 0$ for $\pi/6 < \theta < \pi/2$ $\therefore$ at $\theta = \pi/6$ there is a maximum. The volume of the trough is maximized when the area of the cross section is maximized.

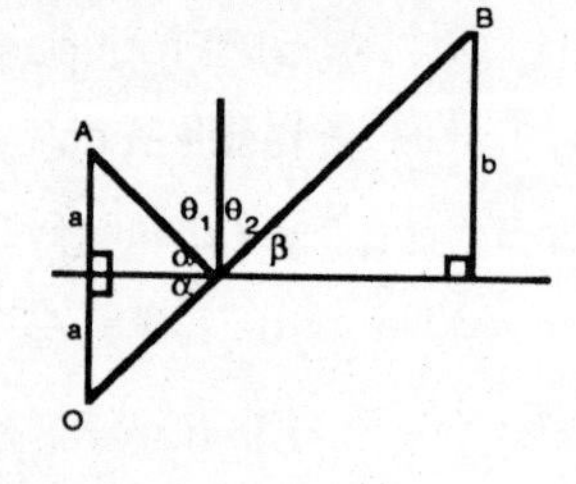


37. The distance $\overline{OB}$ is minimized when $\overline{OB}$ is a straight line. Hence, $\angle\alpha = \angle\beta \Rightarrow \theta_1 = \theta_2$.

39. If $v = kax - kx^2$, then $v' = ka - 2kx$ and $v'' = -2k$. $v' = 0 \Rightarrow x = a/2$. At $x = a/2$ there is a maximum since $v''(a/2) = -2k < 0$. The maximum value of v is $ka^2/4$.

41. If $v = cr_o r^2 - cr^3$, then $v' = 2cr_o r - 3cr^2 = cr[2r_o - 3r]$ and $v'' = 2cr_o - 6cr = 2c[r_o - 3r]$. The solution of $v' = 0$ is $r = 0$ or $2r_o/3$, but 0 is not in the domain. $v' > 0$ for $r < 2r_o/3$ and $v' < 0$ for $r > 2r_o/3 \Rightarrow$ at $r = 2r_o/3$ there is a maximum.

43. The profit is $p = nx - nc = n(x - c) = \left[a(x-c)^{-1} + b(100 - x)\right](x - c) = a + b(100 - x)(x - c) =$ $a + (bc + 100b)x - 100bc - bx^2$. $p'(x) = bc + 100b - 2bx \Rightarrow p''(x) = -2b$. Solving $p'(x) = 0 \Rightarrow$ $x = \frac{c}{2} + 50$. At $x = \frac{c}{2} + 50$ there is a maximum since $p''\left(\frac{c}{2} + 50\right) < 0$.

45. $A(q) = kmq^{-1} + cm + \frac{h}{2}q$, where $q > 0 \Rightarrow A'(q) = -kmq^{-2} + \frac{h}{2} = \frac{hq^2 - 2km}{2q^2}$ and $A''(q) = 2kmq^{-3}$.

The critical points are $-\sqrt{\frac{2km}{h}}$, 0, and $\sqrt{\frac{2km}{h}}$, but only $\sqrt{\frac{2km}{h}}$ is in the domain.

$A''\left(\sqrt{\frac{2km}{h}}\right) > 0 \Rightarrow q = \sqrt{\frac{2km}{h}}$ is a minimum.

47. The profit $p(x) = r(x) - c(x) = (6x) - (x^3 - 6x^2 + 15x) = -x^3 + 6x^2 - 9x$, where $x \ge 0$. $p'(x) = -3x^2 + 12x - 9 = -3(x - 3)(x - 1)$ and $p''(x) = -6x + 12$. The critical points are 1 and 3. $p''(1) = 6 > 0 \Rightarrow$ at $x = 1$ there is a local minimum and $P''(3) = -6 < 0 \Rightarrow$ at $x = 3$ there is a maximum. $P(3) = 0 \Rightarrow$ the best you can do is break even.

## 4.6 ANTIDERIVATIVES AND DIFFERENTIAL EQUATIONS

1. a) $x^2 + C$ b) $3x + C$ c) $x^2 + 3x + C$

3. a) $x^3 + C$ b) $\frac{x^3}{3} + C$ c) $\frac{x^3}{3} + x^2 + x + C$

5. a) $x^{-3} + C$ b) $-\frac{1}{3}x^{-3} + C$ c) $-\frac{1}{3}x^{-3} + x^2 + 3x + C$

7. a) $x^{3/2} + C$ b) $\frac{8}{3}x^{3/2} + C$ c) $\frac{x^3}{3} - \frac{8}{3}x^{3/2} + C$

9. a) $x^{2/3} + C$ b) $x^{1/3} + C$ c) $x^{-1/3} + C$

11. a) $\frac{\cos 3x}{3} + C$ b) $-3\cos(x) + C$ c) $-3\cos x + \frac{\cos 3x}{3} + C$

13. a) $\tan(x) + C$ b) $\tan(5x) + C$ c) $\frac{\tan 5x}{5} + C$

15. a) $\sec(x) + C$ b) $\sec(2x) + C$ c) $2\sec(2x) + C$

17. $(\sin x - \cos x)^2 = \sin^2 x - 2\sin x \cos x + \cos^2 x = \sin^2 x + \cos^2 x - \sin 2x = 1 - \sin 2x$. The antiderivative is $x + \frac{\cos 2x}{2} + C$.

19. a) $-\sqrt{x} + C$ b) $x + C$ c) $\sqrt{x} + C$
d) $-x + C$ e) $-\sqrt{x} + x + C$ f) $-3\sqrt{x} - 2x + C$
g) $\frac{x^2}{2} - \sqrt{x} + C$ h) $-3x + C$

21. $\frac{dy}{dx} = 2x - 7$, $y = 0$ when $x = 2 \Rightarrow y = x^2 - 7x + C$ at $(2, 0) \Rightarrow C = 10 \Rightarrow y = x^2 - 7x + 10$

23. $\frac{dy}{dx} = x^2 + 1$, $y = 1$ when $x = 0 \Rightarrow y = \frac{x^3}{3} + x + C$ at $(0, 1) \Rightarrow C = 1 \Rightarrow y = \frac{x^3}{3} + x + 1$

25. $\frac{dy}{dx} = -\frac{5}{x^2} = -5x^{-2}$, $x > 0$; $y = 3$ when $x = 5 \Rightarrow y = 5x^{-1} + C$ at $(5, 3) \Rightarrow C = 2 \Rightarrow$
$y = 5x^{-1} + 2$ or $y = \frac{5}{x} + 2$.

27. $\frac{dy}{dx} = 3x^2 + 2x + 1$, $y = 0$ when $x = 1 \Rightarrow y = x^3 + x^2 + x + C$ at $(1, 0) \Rightarrow C = -3 \Rightarrow y = x^3 + x^2 + x - 3$.

29. $\frac{dy}{dx} = 1 + \cos x$, $y = 4$ when $x = 0 \Rightarrow y = x + \sin x + C$ at $(0,4) \Rightarrow C = 4 \Rightarrow y = x + \sin x + 4$.

31. $\frac{d^2y}{dx^2} = 2 - 6x$, $y = 1$ and $\frac{dy}{dx} = 4$ when $x = 0 \Rightarrow \frac{dy}{dx} = 2x - 3x^2 + C_1$ at $(0, 4) \Rightarrow C_1 = 4 \Rightarrow$
$\frac{dy}{dx} = 2x - 3x^2 + 4$. $y = x^2 - x^3 + 4x + C_2$ at $(0, 1) \Rightarrow C_2 = 1 \Rightarrow y = x^2 - x^3 + 4x + 1$.

33. $v = s'(t) = 9.8t$, $s = 10$ when $t = 0 \Rightarrow s = 4.9t^2 + C$ at $(0, 10) \Rightarrow C = 10 \Rightarrow s = 4.9t^2 + 10$.

35. $a = s''(t) = 32$, $v = 20$ and $s = 0$ when $t = 0 \Rightarrow v = 32t + C_1 \Rightarrow C_1 = 20 \Rightarrow v = 32t + 20$.
$s = 16t^2 + 20t + C_2$ at $(0, 0) \Rightarrow C_2 = 0 \Rightarrow s = 16t^2 + 20t$.

37. $m = y' = 3\sqrt{x} = 3x^{1/2} \Rightarrow y = 2x^{3/2} + C$ at $(9, 4) \Rightarrow C = -50 \Rightarrow y = 2x^{3/2} - 50$ or $y = 2\sqrt{x^3} - 50$.

39. $r'(x) = 3x^2 - 6x + 12$, $r = 0$ when $x = 0 \Rightarrow r(x) = x^3 - 3x^2 + 12x + C$ at $(0,0) \Rightarrow C = 0 \Rightarrow$
$r(x) = x^3 - 3x^2 + 12x$.

41. $a(t) = v'(t) = 1.6 \Rightarrow v(t) = 1.6t + C$ at $(0,0) \Rightarrow C = 0 \Rightarrow v(t) = 1.6t$. When $t = 30$, then
$v(30) = 48$ m/sec.

43. $a(t) = v'(t) = 9.8 \Rightarrow v(t) = 9.8t + C_1$ at $(0,0) \Rightarrow C_1 = 0 \Rightarrow v(t) = s'(t) = 9.8t \Rightarrow s(t) = 4.9t^2 + C_2$ at
$(0, 0) \Rightarrow C_2 = 0 \Rightarrow s(t) = 4.9t^2$. Solving $s(t) = 10 \Rightarrow t^2 = \frac{10}{4.9} \Rightarrow t = \sqrt{\frac{10}{4.9}}$. $v\left(\sqrt{\frac{10}{4.9}}\right) = 9.8\sqrt{\frac{10}{4.9}} =$
$\frac{(2)(4.9)\sqrt{10}}{\sqrt{4.9}} = (2)\sqrt{4.9}\sqrt{10} = 14$ m/sec.

45. a) $y^{-1/2}dy = -kdt \Rightarrow 2y^{1/2} + C_1 = -kt + C_2$.

b) $2y^{1/2} = -kt + C_2 - C_1 = -kt + C \Rightarrow y^{1/2} = \frac{C - kt}{2} \Rightarrow y = \frac{(C - kt)^2}{4}$.

47. Part a) $\Rightarrow y = x^2 - 1$, which does not contain the point $(1,1)$.
Part b) $\Rightarrow y = x^3$, which is symmetric to the origin.
Part c) $\Rightarrow y = x^2 + 2x - 2$, which has $(0, -2)$ as a y-intercept.
Part d) $\Rightarrow y = x^2$, which satisfies all the conditions. $\therefore$ d)

## PRACTICE EXERCISES

1. $A = \pi r^2 \Rightarrow \frac{dA}{dt} = 2\pi r \frac{dr}{dt}$, when r = 10 and $\frac{dr}{dt} = -\frac{2}{\pi} \Rightarrow \frac{dA}{dt} = 2\pi(10)\left(-\frac{2}{\pi}\right) =$
$-40\ m^2$/sec.

3. $V = x^3 \Rightarrow \frac{dV}{dt} = 3x^2 \frac{dx}{dt}$, when $\frac{dV}{dt} = 1200$ and x = 20. Consequently, $\frac{dx}{dt} = \frac{1}{3x^2}\frac{dV}{dt} = 1$ cm/min.

5. a) From Fig. 4.48 we have $\frac{10}{h} = \frac{4}{r} \Rightarrow r = .4h \Rightarrow r = \frac{2}{5}h$.

   b) $V = \frac{1}{3}\pi r^2 h = \frac{1}{3}\pi\left(\frac{2}{5}h\right)^2 h = \frac{4\pi h^3}{75} \Rightarrow \frac{dV}{dt} = \frac{4\pi h^2}{25}\frac{dh}{dt}$, when $\frac{dV}{dt} = -5$ and h = 6.
   $\therefore \frac{dh}{dt} = -\frac{125}{144\pi}$ ft/min.

7. a) From the diagram in the text, we have $\tan\theta = \frac{s}{132} \Rightarrow \sec^2\theta \frac{d\theta}{dt} = \frac{1}{132}\frac{ds}{dt} \Rightarrow \frac{d\theta}{st} = \frac{\cos^2\theta}{132}\frac{ds}{dt}$
   when $\frac{ds}{dt} = 264$ and $\theta = 0$. $\therefore \frac{d\theta}{dt} = -2$ rad/sec.

   b) The $\left|\frac{d\theta}{dt}\right|$ is the same at 1/2 sec before $\theta = 0$ or after, the only difference is the sign. Hence,
   $\left.\frac{d\theta}{dt}\right|_{after} = -\left.\frac{d\theta}{dt}\right|_{before} = -\frac{\cos^2(\pi/4)}{132}\left(\frac{264}{-1}\right) = 1$ rad/sec. 1/2 sec before or after $\Rightarrow |\theta| = \pi/4$.

9. $f(x) = x^4 + 2x^2 - 2 \Rightarrow f'(x) = 4x^3 + 4x$. Since $f(0) = -2 < 0$, $f(1) = 1 > 0$ and $f'(x) \ge 0$ for $0 \le x \le 1$, we may conclude that f(x) has exactly one solution when $0 \le x \le 1$.

11. $y = \frac{x}{x+1} \Rightarrow y' = \frac{1}{(x+1)^2} > 0$, for all x in the domain of $\frac{x}{x+1}$. $\therefore y = \frac{x}{x+1}$ is increasing.

13.

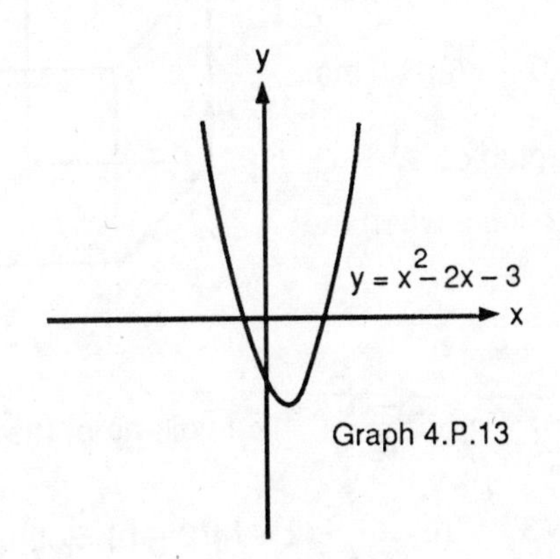

Graph 4.P.13

15.

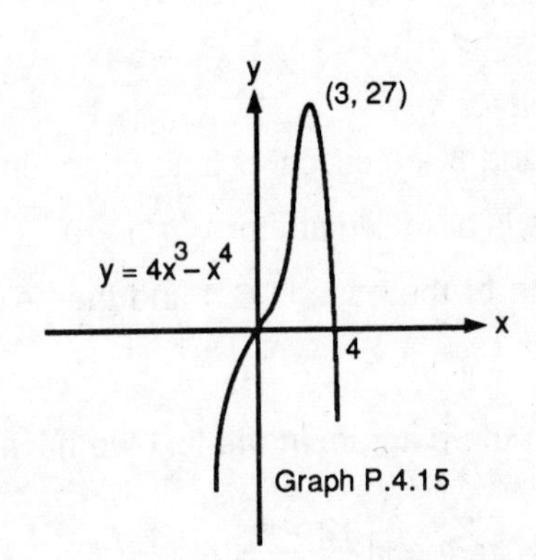

Graph P.4.15

17.

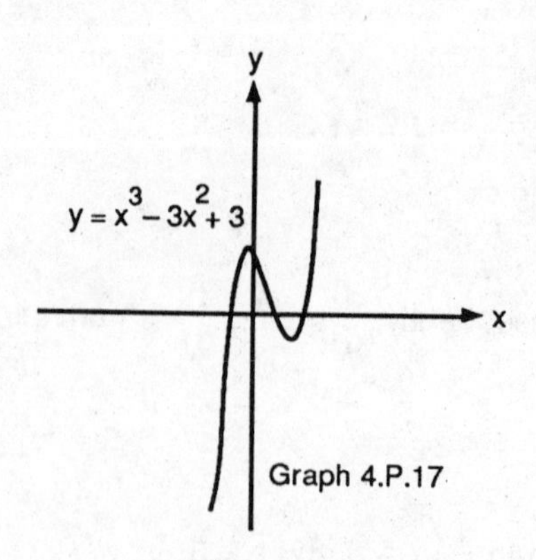

Graph 4.P.17

19.

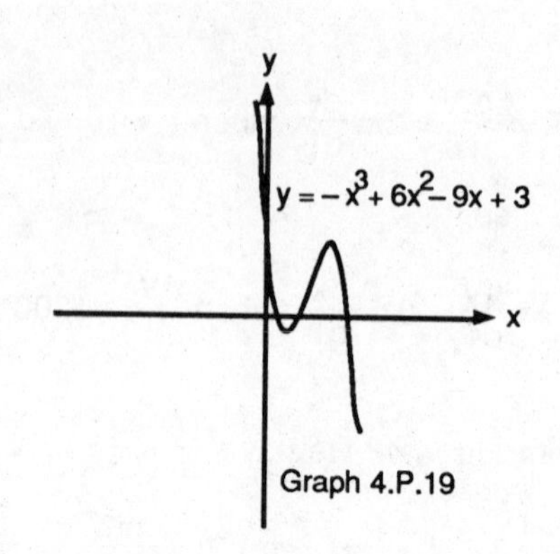

Graph 4.P.19

21. If $y' = 6(x + 1)(x - 2)^2$, then $y' < 0$ for $x < -1$, $y' > 0$ for $x > -1$. f has a local minimum at $x = -1$. $y'' = 6(x - 2)(3x) \Rightarrow y'' > 0$ for $x < 0$ or $x > 2$, while $y'' < 0$ for $0 < x < 2$. f has points of inflection at $x = 0$ and 2.

23. $y' = 3x^2 + 2bx - 5 \Rightarrow y'' = 6x + 2b$. Solving $y'' = 0 \Rightarrow x = -\frac{b}{3}$ and $y''(1) = 0 \Rightarrow b = -3$.

25. a) At point T the curve is decreasing $\Rightarrow y' < 0$, and concave downward $\Rightarrow y'' < 0$.
b) At point P the curve is decreasing $\Rightarrow y' < 0$, and concave upward $\Rightarrow y'' > 0$.

27. $f(x) = 4x^3 - 12x^2 + 9x \Rightarrow f'(x) = 12x^2 - 24x + 9 = 3(2x - 1)(2x - 3)$. $f'(x) < 0$ for $1/2 < x < 3/2$ and $f'(x) > 0$ for $x < 1/2$ or $x > 3/2$. f has a local maximum at $x = 1/2$ and a local minimum at $x = 3/2$. $f(0) = 0$, $f(1/2) = 2$, $f(3/2) = 0$, and $f(2) = 2$. The minimum value is 0 and it occurs at $x = 0$ or $3/2$. The maximum value is 2; and this occurs when $x = 1/2$ or 2.

29. If $A = \frac{1}{2}rs$ and $P = s + 2r = 100$, then $A(r) = \frac{1}{2}r(100 - 2r) = 50r - r^2$. $A'(r) = 50 - 2r \Rightarrow 25$ is a critical point. At $r = 25$ there is a maximum because $A''(r) < 0$ for all possible r. The values of $r = 25$ ft and $s = 50$ ft will give the sector its greatest area.

31. The surface area is $S = x^2 + 4xh = 108 \Rightarrow h = 27x^{-1} - \frac{1}{4}x$. The volume is $V(x) = x^2\left(27x^{-1} - \frac{1}{4}x\right) = 27x - \frac{1}{4}x^3$. $V'(x) = (3 + x/2)(3 - x/2) \Rightarrow$ that $-6$ and 6 are critical points, but a length of $-6$ is not acceptable. At $x = 6$ there is a maximum for $V''(6) < 0$. The maximum volume occurs when the length of the base is 6 ft and the height is 3 ft.

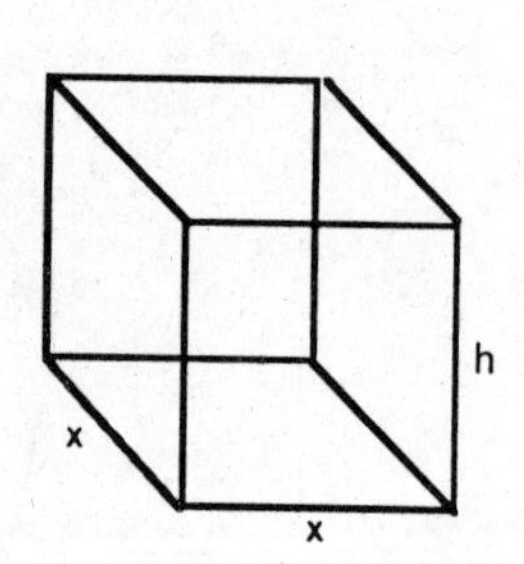


33. From the diagram in the text we have $\left(\frac{h}{2}\right)^2 + r^2 = (\sqrt{3})^2 \Rightarrow r^2 = \frac{12 - h^2}{4}$. The volume of the cylinder is $V = \pi r^2 h = \pi\left(\frac{12 - h^2}{4}\right)h = \frac{\pi}{4}(12h - h^3)$, where $0 \le h \le 2\sqrt{3}$. $V'(h) = \frac{3\pi}{4}(2 + h)(2 - h) \Rightarrow$ the critical points $-2$ or 2, but $-2$ is not in the domain. At $h = 2$ there is a maximum for $V''(2) = -3\pi < 0$. The dimensions of the largest cylinder are radius $= \sqrt{2}$ and height $= 2$.

35. From the diagram in the text we have the cost $c = (50000)x + (30000)(20 - y)$ and $x = \sqrt{y^2 + 144}$, where $0 \le y \le 20$. $c(y) = 50000\sqrt{y^2 + 144} + 30000(20 - y) \Rightarrow c'(y) = \dfrac{10000\left(5y - 3\sqrt{y^2 + 144}\right)}{\sqrt{y^2 + 144}}$.

The critical points are $-9$ or $9$, but $-9$ is not in the domain. $c(0) = \$1,200,000$, $c(9) = \$1,080,000$, and $c(20) = \$1,166,190$. The mimimum cost occurs when $x =$ 15 miles and $y = 9$ miles.

37. The profit $P = 2px + py = 2px + p\left[\dfrac{40 - 10x}{5 - x}\right]$, where p is the profit on grade B tires and $0 \le x \le 4$. $P'(x) = \dfrac{2p}{(5 - x)^2}\left(x^2 - 10x + 20\right) \Rightarrow$ the critical points $(5 - \sqrt{5})$, 5 or $(5 + \sqrt{5})$, but only $(5 - \sqrt{5})$ is in the domain. $P'(x) > 0$ for $0 < x < (5 - \sqrt{5})$ and $P'(x) < 0$ for $(5 - \sqrt{5}) < x < 4 \Rightarrow$ at $x = (5 - \sqrt{5})$ there is a local maximum. $P(0) = 8p$, $P(5 - \sqrt{5}) = 4p(5 - \sqrt{5}) \approx 11p$, and $P(4) = 8p \Rightarrow$ at $x = (5 - \sqrt{5})$ there is absolute maximum. The maximum occurs when $x = (5 - \sqrt{5})$ and $y = 2(5 - \sqrt{5})$ i.e. $x \approx 276$ tires and $y \approx 553$ tires.

39. Yes, all differentiable functions having 3 as a derivative differ by only a constant. Consequently, the difference $3x -$ g(x) is a constant K because $g'(x) = 3$. $\therefore g(x) = 3x + K$, the same form as F(x).

41. a) $C$ b) $x + C$ c) $\dfrac{x^2}{2} + C$ d) $\dfrac{x^3}{3} + C$

e) $\dfrac{x^{11}}{11} + C$ f) $-x^{-1} + C$ g) $-\frac{1}{4}x^{-4} + C$ h) $\frac{2}{7}x^{7/2} + C$

i) $\frac{3}{7}x^{7/3} + C$ j) $\frac{4}{7}x^{7/4} + C$ k) $\frac{2}{3}x^{3/2} + C$ l) $2x^{1/2} + C$

m) $\frac{7}{4}x^{4/7} + C$ n) $-\frac{3}{4}x^{-4/3} + C$

43. $x^3 + \frac{5}{2}x^2 - 7x + C$ 45. $\frac{2}{3}\sqrt{x^3} + 2\sqrt{x} + C$ 47. $\frac{3}{5}\sin(5x) + C$

49. $3\tan(3x) + C$ 51. $\frac{x}{2} - \sin(x) + C$ 53. $3\sec(x/3) + 5x + C$

55. $\tan(x) - x + C$ 57. $x - \frac{1}{2}\sin(2x) + C$

59. $\dfrac{dy}{dx} = 1 + x + \dfrac{x^2}{2} \Rightarrow y = x + \dfrac{x^2}{2} + \dfrac{x^3}{6} + C$ at $(0,1) \Rightarrow C = 1$ $\therefore y = x + \dfrac{x^2}{2} + \dfrac{x^3}{6} + 1$.

61. $\dfrac{dy}{dx} = \dfrac{x^2 + 1}{x^2} = 1 + x^{-2} \Rightarrow y = x - x^{-1} + C$ at$(1,-1) \Rightarrow C = -1 \Rightarrow y = x - x^{-1} - 1 = x - \dfrac{1}{x} - 1$.

63. $\dfrac{d^2y}{dx^2} = -\sin x \Rightarrow \dfrac{dy}{dx} = \cos(x) + C_1$ at $(0,1) \Rightarrow C_1 = 0 \Rightarrow \dfrac{dy}{dx} = \cos(x)$. $y = \sin(x) + C_2$ at $(0,0) \Rightarrow$ $C_2 = 0 \Rightarrow y = \sin x$.

65. $y = x$

67. Given: $a(t) = -32$ ft/sec$^2$ and $v(0) = 32$ ft/sec. $v(t) = -32t + C_1$ at $(0,32) \Rightarrow C_1 = 32 \Rightarrow$ $v(t) = -32t + 32$. $s(t) = -16t^2 + 32t + C_2$ at $(0,0) \Rightarrow C_2 = 0 \Rightarrow s(t) = -16t^2 + 32t$.

$V(t) = 0 \Rightarrow t = 1$ sec and $s(1) = 16$ ft. The dirt will reach a height of 16 ft from the bottom of the hole and fall back into the hole. You better duck!

69. $f(x) = x^3 + x - 1 \Rightarrow f'(x) = 3x^2 + 1 > 0$ for all $x \Rightarrow$ the curve is always increasing. Since, $f(0) = -1$, $f(1) = 1$ and continuously increasing, there must be only one solution for the equation $x^3 + x - 1 = 0$. From Newton's method we find that 0.682 is a root.

71. a) b)

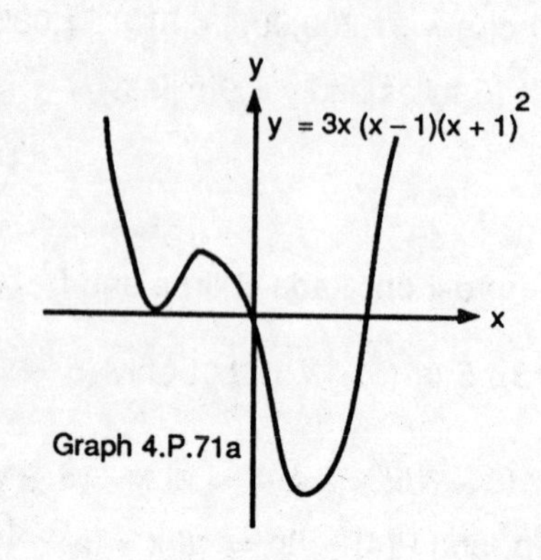


Graph 4.P.71a

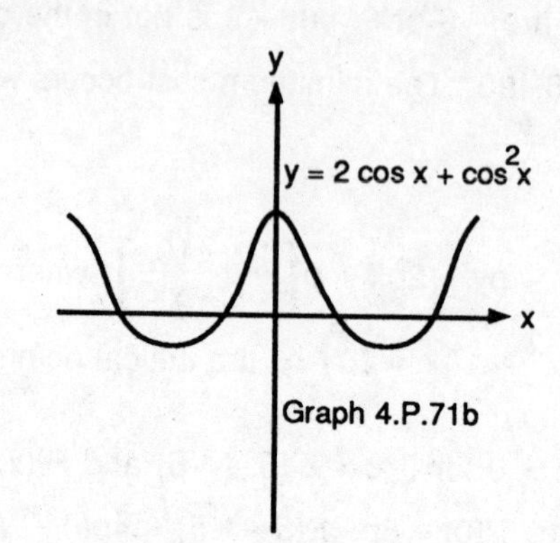


Graph 4.P.71b

# CHAPTER 5

# INTEGRATION

## 5.1 CALCULUS AND AREA

1. $F(x) = \frac{x^3}{3} \Rightarrow F(2) - F(0) = \frac{2^3}{3} - \frac{0^3}{3} = \frac{8}{3}$

3. $F(x) = \frac{2}{3}x^{3/2} \Rightarrow F(4) - F(0) = \frac{2}{3}(4)^{3/2} - \frac{2}{3}(0)^{3/2}$
$= \frac{2}{3}(8) - 0 = \frac{16}{3}$

5. $F(x) = \frac{x^3}{3} \Rightarrow F(1) - F(-1) = \frac{1^3}{3} - \frac{(-1)^3}{3} =$
$\frac{1}{3} + \frac{1}{3} = \frac{2}{3}$

7. $F(x) = \frac{x^4}{4} - x^3 + 4x \Rightarrow F(2) - F(-1) =$
$\left(\frac{2^4}{4} - 2^3 + 4(2)\right) - \left(\frac{(-1)^4}{4} - (-1)^3 + 4(-1)\right)$
$= (4 - 8 + 8) - \left(\frac{1}{4} - (-1) - 4\right) = \frac{27}{4}$

9. $F(x) = \sin x \Rightarrow F(\frac{\pi}{2}) - F(0) = \sin\frac{\pi}{2} - \sin 0 =$
$1 - 0 = 1$

11. $F(x) = -\frac{1}{2}\cos 2x \Rightarrow F(\frac{\pi}{2}) - F(0) = -\frac{1}{2}\cos 2(\frac{\pi}{2})$
$-\left(-\frac{1}{2}\right)\cos 2(0) = -\frac{1}{2}(-1) + \frac{1}{2}(1) = 1$

13. $F(x) = \frac{1}{\pi}\sin \pi x \Rightarrow F(\frac{1}{2}) - F(-\frac{1}{2}) = \frac{1}{\pi}\sin \pi(\frac{1}{2}) -$
$\frac{1}{\pi}\sin \pi(-\frac{1}{2}) = \frac{1}{\pi}(1) - \frac{1}{\pi}(-1) = \frac{2}{\pi}$

15. $F(x) = \tan x \Rightarrow F(\frac{\pi}{3}) - F(-\frac{\pi}{4}) = \tan\frac{\pi}{3} -$
$\tan\left(-\frac{\pi}{4}\right) = \sqrt{3} - (-1) = \sqrt{3} + 1$

17. $F(x) = \frac{x^{n+1}}{n+1}$, $n \neq -1$ fi $F(b) - F(0) =$
$\frac{b^{n+1}}{n+1} - \frac{0^{n+1}}{n+1} = \frac{b^{n+1}}{n+1}$

## SECTION 5.2 FORMULAS FOR FINITE SUMS

1. $\sum_{k=1}^{4} \frac{1}{k} = \frac{1}{1} + \frac{1}{2} + \frac{1}{3} + \frac{1}{4} = \frac{25}{12}$

3. $\sum_{k=1}^{3} (k + 2) = (1 + 2) + (2 + 2) + (3 + 2) = 12$

5. $\sum_{k=0}^{4} \frac{k}{4} = \frac{0}{4} + \frac{1}{4} + \frac{2}{4} + \frac{3}{4} + \frac{4}{4} = \frac{10}{4} = \frac{5}{2}$

7. $\sum_{k=1}^{5} (2k - 2) = (2(1) - 2) + (2(2) - 2) +$
$(2(3) - 2) + (2(4) - 2) + (2(5) - 2) = 20$

9. $\sum_{k=1}^{2} \frac{6k}{k+1} = \frac{6(1)}{1+1} + \frac{6(2)}{2+1} = 7$

11. $\sum_{k=1}^{5} k(k-1)(k-2) = 1(1-1)(2-1) +$

$2(2-1)(2-2) + 3(3-1)(3-2) + 4(4-1)(4-2)$
$+ 5(5-1)(5-2) = 90$

13. $\sum_{k=1}^{4} \cos k\pi = \cos (1)\pi + \cos (2)\pi + \cos (3)\pi$

$+ \cos (4)\pi = 0$

15. $\sum_{k=1}^{4} (-1)^k = (-1)^1 + (-1)^2 + (-1)^3 + (-1)^4 = 0$

17. $\sum_{k=1}^{6} k$

19. $\sum_{k=1}^{4} \frac{1}{2^k}$

21. $\sum_{k=1}^{5} (-1)^{k+1} \frac{1}{k}$

23. $\sum_{k=1}^{5} (-1)^{k+1} \frac{k}{5}$

25. $\sum_{k=1}^{10} k = \frac{10(10+1)}{2} = 55$

27. $\sum_{k=1}^{6} -k^2 = -\sum_{k=1}^{6} k^2 = -\frac{6(6+1)(2(6)+1)}{6} = -91$

29. $\sum_{k=1}^{5} k(k-5) = \sum_{k=1}^{5} k^2 - 5k = \sum_{k=1}^{5} k^2 - 5\left(\sum_{k=1}^{5} k\right) = \frac{5(5+1)(2(5)+1)}{6} - 5\left(\frac{5(5+1)}{2}\right) = -20$

31. $\sum_{k=100}^{100} k^3 = 100^3 = 1\,000\,000$

33. a) $\sum_{k=1}^{6} 2^{k-1} = 2^{1-1} + 2^{2-1} + 2^{3-1} + 2^{4-1} + 2^{5-1} + 2^{6-1} = 1 + 2 + 4 + 8 + 16 + 32$

b) $\sum_{k=0}^{5} 2^k = 2^0 + 2^1 + 2^2 + 2^3 + 2^4 + 2^5 = 1 + 2 + 4 + 8 + 16 + 32$

c) $\sum_{k=-1}^{4} 2^{k+1} = 2^{-1+1} + 2^{0+1} + 2^{1+1} + 2^{2+1} + 2^{3+1} + 2^{4+1} = 1 + 2 + 4 + 8 + 16 + 32$

All of them represent 1 + 2 + 4 + 8 + 16 + 32

35. a) $\sum_{k=1}^{n} 3a_k = 3\left(\sum_{k=1}^{n} a_k\right) = 3(-5) = -15$

b) $\sum_{k=1}^{n} \frac{b_k}{6} = \frac{1}{6}\left(\sum_{k=1}^{n} b_k\right) = \frac{1}{6}(6) = 1$

c) $\sum_{k=1}^{n} (a_k + b_k) = \sum_{k=1}^{n} a_k + \sum_{k=1}^{n} b_k =$
$-5 + 6 = 1$

d) $\sum_{k=1}^{n} (a_k - b_k) = \sum_{k=1}^{n} a_k - \sum_{k=1}^{n} b_k$
$= -5 - 6 = -11$

e) $\sum_{k=1}^{n} (b_k - 2a_k) = \sum_{k=1}^{n} b_k - \sum_{k=1}^{n} 2a_k = \sum_{k=1}^{n} b_k - 2\left(\sum_{k=1}^{n} a_k\right) = -6 - 2(5) = -16$

37. $\sum_{k=1}^{12} k = \frac{12(12+1)}{2} = 6(13) = 78$

39. $\sum_{k=1}^{n} (2k - 1) = \sum_{k=1}^{n} 2k - \sum_{k=1}^{n} 1 = 2\left(\sum_{k=1}^{n} k\right) - \sum_{k=1}^{n} 1 = 2\left(\frac{n(n+1)}{2}\right) - 1(n) = n^2 + n - n = n^2$

41. Answer is in the text

43. Answer is in the text

## SECTION 5.3 DEFINITE INTEGRALS

1. $\int_0^2 x^2\,dx$

3. $\int_{-7}^{5} (x^2 - 3x)\,dx$

5. $\int_2^3 \frac{1}{1-x}\,dx$

7. $\int_0^4 \cos x\,dx$

9. $\int_{-\pi/4}^{0} \sec x\,dx$

11. $\int_{-\pi}^{0} \sin 3x\,dx$

13. $\int_{1/\pi}^{2/\pi} \sin\frac{1}{x}\,dx$

15. $-\int_{\pi/2}^{3\pi/2} \cos x\,dx$

17. $\int_0^5 \sqrt{25 - x^2}\,dx$

19. $y = \sqrt{1 - x^2}$ is the upper half of a circle whose center is (0,0) and radius is 1. ∴ $\int_{-1}^{1} \sqrt{1 - x^2}\,dx = \frac{1}{2}\pi(1)^2 = \frac{1}{2}\pi$

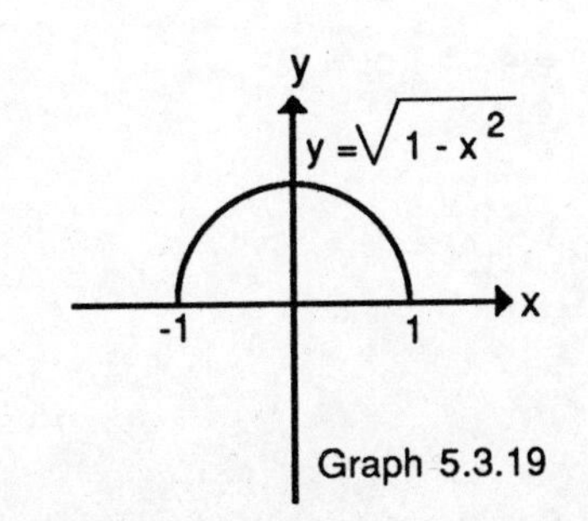

Graph 5.3.19

21.

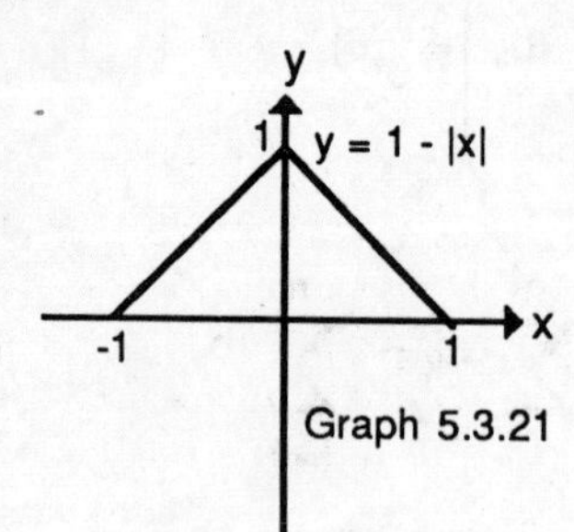

Graph 5.3.21

$y = 1 - |x| \Rightarrow y = \begin{cases} 1 + x, & x < 0 \\ 1 - x, & x \geq 0 \end{cases} \Rightarrow$ the region is a triangle with base = 2 and height = 1. $\therefore \int_{-1}^{1} (1 - |x|)\,dx = \frac{1}{2}(2)(1) = 1$

23. a) $\int_{2}^{2} g(x)\,dx = 0$

b) $\int_{5}^{1} g(x)\,dx = -\int_{1}^{5} g(x)\,dx = -8$

c) $\int_{1}^{2} 3f(x)\,dx = 3\int_{1}^{2} f(x)\,dx = 3(-4)$

$= -12$

d) $\int_{2}^{5} f(x)\,dx = \int_{1}^{5} f(x)\,dx - \int_{1}^{2} f(x)\,dx =$

$6 - (-4) = 10$

e) $\int_{1}^{5} (f(x) - g(x))\,dx =$

$\int_{1}^{5} f(x)\,dx - \int_{1}^{5} g(x)\,dx$

$= 6 - 8 = -2$

f) $\int_{1}^{5} (4f(x) - g(x))\,dx =$

$4\int_{1}^{5} f(x)\,dx - \int_{1}^{5} g(x)\,dx =$

$4(6) - 8 = 16$

25. a) $\int_{1}^{2} f(u)\,du = 5$

b) $\int_{1}^{2} f(z)\,dz = 5$

c) $\int_{2}^{1} f(t)\,dt = -5$

27. Since f is decreasing on [0,1], the maximum of f occurs at 0 and is 1, the minimum of f occurs at 1 and is $\frac{1}{2}$. $\therefore$ the upper bound is $1(1 - 0) = 1$; the lower bound is $0(1 - 0) = 0$.

29. Since f is continuous on [1,2], there exists at least one $c \in (1,2)$ so that $f(c)(2 - 1) = \int_{1}^{2} f(x)\,dx \Rightarrow$

$f(c)(1) = 4 \Rightarrow f(c) = 4$.

31. $\displaystyle\int_0^1 (1 - x)\, dx$

| n | upper endpoint sum | lower endpoint sum |
|---|---|---|
| 4 | 0.625 | 0.375 |
| 10 | 0.55 | 0.45 |
| 20 | 0.525 | 0.475 |
| 50 | 0.51 | 0.49 |

33. $\displaystyle\int_{-\pi}^{\pi} \cos x\, dx$

| n | upper endpoint sum | lower endpoint sum |
|---|---|---|
| 4 | 3.14159551 | −3.14157979 |
| 10 | 1.25664474 | −1.25662538 |
| 20 | 0.628327454 | −0.628307606 |
| 50 | 0.251337007 | −0.251317017 |

35. $\displaystyle\int_{-1}^{1} |x|\, dx$

| n | upper endpoint sum | lower endpoint sum |
|---|---|---|
| 4 | 1.5 | 0.5 |
| 10 | 1.2 | 0.8 |
| 20 | 1.1 | 0.9 |
| 50 | 1.04 | 0.96 |

## SECTION 5.4 THE FUNDAMENTAL THEOREMS OF CALCULUS

1. $\displaystyle\int_1^2 (2x + 5)\, dx = \left[x^2 + 5\right]_1^2 =$

$[4 + 10] - [1 + 5] = 8$

3. $\displaystyle\int_0^3 (4 - x^2)\, dx = \left[4x - \frac{x^3}{3}\right]_0^3 =$

$\left[12 - \frac{27}{3}\right] - [0] = 3$

5. $\displaystyle\int_0^1 (x^2 + \sqrt{x})\, dx = \left[\frac{x^3}{3} + \frac{2}{3}x^{3/2}\right]_0^1 =$

$\left(\frac{1}{3} + \frac{2}{3}\right) - (0) = 1$

7. $\displaystyle\int_1^{32} x^{-6/5}\, dx = \left[-5x^{-1/5}\right]_1^{32} =$

$\left(-\frac{5}{2}\right) - (-5) = \frac{5}{2}$

9. $\displaystyle\int_0^{\pi} \sin x\, dx = [-\cos x]_0^{\pi} =$

$(-\cos \pi) - (-\cos 0) = -(-1) - (-1) = 2$

11. $\displaystyle\int_0^{\pi/3} 2\sec^2 x\, dx = [2\tan x]_0^{\pi/3} =$

$(2\tan(\pi/3)) - (2\tan 0) = 2\sqrt{3} - 0 = 2\sqrt{3}$

13. $\displaystyle\int_{\pi/4}^{3\pi/4} \csc x \cot x\, dx = [-\csc x]_{\pi/4}^{3\pi/4} = (-\csc(3\pi/4)) - (-\csc(\pi/4)) = -\sqrt{2} - (-\sqrt{2}) = 0$

15. $\int_{-1}^{1} (r+1)^2\,dr = \int_{-1}^{1} (r^2+2r+1)\,dr = \left[\frac{r^3}{3}+r^2+r\right]_{-1}^{1} = \left(\frac{1}{3}+1+1\right) - \left(\frac{-1}{3}+1-1\right) = \frac{8}{3}$

17.

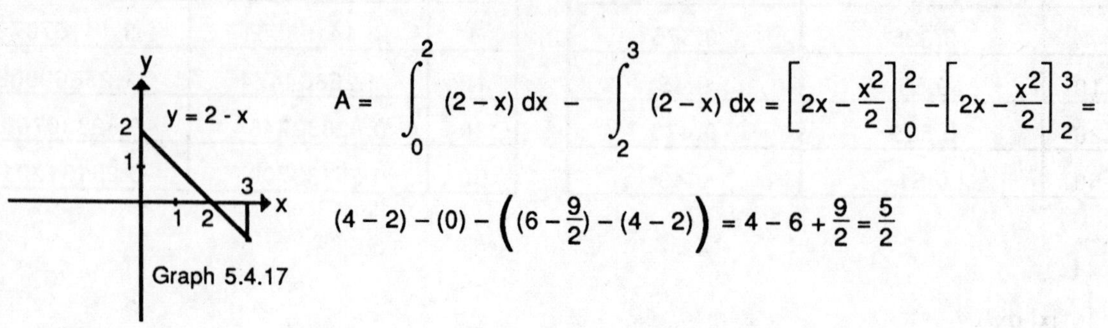

Graph 5.4.17

$A = \int_0^2 (2-x)\,dx - \int_2^3 (2-x)\,dx = \left[2x - \frac{x^2}{2}\right]_0^2 - \left[2x - \frac{x^2}{2}\right]_2^3 =$

$(4-2) - (0) - \left((6-\frac{9}{2}) - (4-2)\right) = 4 - 6 + \frac{9}{2} = \frac{5}{2}$

19.

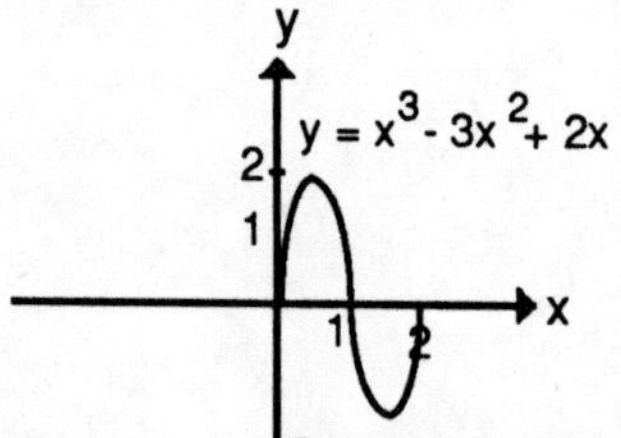

y
$y = x^3 - 3x^2 + 2x$
2
1
1 2
x

Graph 5.4.19

$A = \int_0^1 (x^3 - 3x^2 + 2x)\,dx - \int_1^2 (x^3 - 3x^2 + 2x)\,dx =$

$\left[\frac{x^4}{4} - x^3 + x^2\right]_0^1 - \left[\frac{x^4}{4} - x^3 + x^2\right]_1^2 = \left[\frac{1}{4} - 1 + 1\right] - (0) - \left([4 - 8 + 4] - \left[\frac{1}{4}\right]\right)$

$= \frac{1}{2}$

21. $A = \int_0^1 x^2\,dx + \int_1^2 (2-x)\,dx = \left[\frac{x^3}{3}\right]_0^1 + \left[2x - \frac{x^2}{2}\right]_1^2 = \left(\frac{1}{3}\right) - (0) + (4-2) - \left(2 - \frac{1}{2}\right) = \frac{5}{6}$

23. $A = \int_0^\pi (2 - (1 + \cos x))\,dx = \int_0^\pi (1 - \cos x)\,dx = [x - \sin x]_0^\pi = (\pi - \sin \pi) - (0) = \pi$

25. a) $\frac{d}{dx}\left(\int_0^x \cos t\,dt\right) = \frac{d}{dx}\left([\sin t]_0^x\right) = \frac{d}{dx}(\sin x - \sin 0) = \frac{d}{dx}(\sin x) = \cos x$

b) $\frac{d}{dx}\left(\int_0^x \cos t\,dt\right) = \cos x$

27. $\frac{dy}{dx} = \sqrt{1+x^2}$

29. $\frac{dy}{dx} = \sin\left(\sqrt{x}\right)^2\left(\frac{d}{dx}\left(\sqrt{x}\right)\right) = \frac{1}{2}x^{-1/2}\sin x$

31. d, since $y' = \frac{1}{x}$ and $y(\pi) = \int_\pi^\pi \frac{1}{t}\,dt - 3 = -3$

33. b, since $y' = \sec x$ and

$y(0) = \int_0^0 \sec t\,dt + 4 = 4$

35.

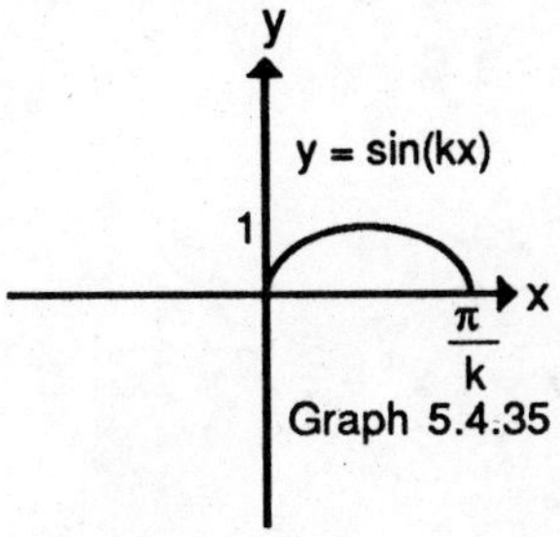

y

y = sin(kx)

1

x

$\frac{\pi}{k}$

Graph 5.4.35

$$A = \int_0^{\pi/k} \sin kx\, dx = \left[-\frac{1}{k}\cos kx\right]_0^{\pi/k} = \left[-\frac{1}{k}\cos\left(k\left(\frac{\pi}{k}\right)\right)\right] - \left[-\frac{1}{k}\cos\left(k(0)\right)\right] = \left(-\frac{1}{k}\cos\pi\right) - \left(-\frac{1}{k}\cos 0\right) = \frac{2}{k}$$

37. a) $\displaystyle\int_1^{100} \frac{1}{2\sqrt{x}}\,dx = \int_1^{100} \frac{1}{2}x^{-1/2}\,dx = \left[x^{1/2}\right]_1^{100} = 100^{1/2} - 1^{1/2} = \$9.00$

b) $\displaystyle\int_{100}^{400} \frac{1}{2\sqrt{x}}\,dx = \left[x^{1/2}\right]_{100}^{400} = 400^{1/2} - 100^{1/2} = \$10.00$

39. $\displaystyle I_{av} = \frac{1}{30-0}\int_0^{30}\left(450 - \frac{x^2}{2}\right)dx = \frac{1}{30}\left[450x - \frac{x^3}{6}\right]_0^{30} = \frac{1}{30}\left[\left(450(30) - \frac{(30)^3}{6}\right) - (0)\right] = 300$

Average holding cost = $I_{av}(\$0.02) = 300(\$0.02) = \$6.00$

41. x = a if f is defined at x = a

43. $\displaystyle\frac{d}{dx}\left(\int_0^x f(t)\,dt\right) = \frac{d}{dx}(x\cos\pi x) = \cos\pi x - \pi x\sin\pi x \Rightarrow f(4) = \cos 4\pi - 4\pi\sin 4\pi = 1$

## SECTION 5.5 INDEFINITE INTEGRALS

1. $\int x^3\,dx = \frac{x^4}{4} + C$

3. $\int (x+1)\,dx = \frac{x^2}{2} + x + C$

5. $\int 3\sqrt{x}\,dx = 2x^{3/2} + C$

7. $\int x^{-1/3}\,dx = \frac{3}{2}x^{2/3} + C$

9. $\int (5x^2 + 2x)\,dx = \frac{5x^3}{3} + x^2 + C$

11. $\int (2x^3 - 5x + 7)\,dx = \frac{1}{2}x^4 - \frac{5}{2}x^2 + 7x + C$

13. $\int 2\cos x\,dx = 2\sin x + C$

15. $\int \sin\frac{x}{3}\,dx = -3\cos\frac{x}{3} + C$

17. $\int 3\csc^2 x\,dx = -3\cot x + C$

19. $\int \frac{\csc x \cot x}{2}\,dx = -\frac{1}{2}\csc x + C$

21. $\int (4\sec x \tan x - 2\sec^2 x)\,dx =$

$4\sec x - 2\tan x + C$

23. $\int (\sin 2x - \csc^2 x)\,dx =$

$-\frac{1}{2}\cos 2x + \cot x + C$

25. $\int 4\sin^2 y\,dy = \int 4\left(\frac{1}{2} - \frac{1}{2}\cos 2y\right)dy =$

$\int (2 - 2\cos 2y)\,dy = 2y - \sin 2y + C$

27. $\int \sin x \cos x\,dx = \int \frac{1}{2}\sin 2x\,dx =$

$-\frac{1}{4}\cos 2x + C$

29. $\int (1 + \tan^2\theta)\,d\theta = \int \sec^2\theta\,d\theta =$

$\tan\theta + C$

31. $\frac{d}{dx}\left(\frac{(7x-2)^4}{28} + C\right) = \frac{4(7x-2)^3(7)}{28} = (7x-2)^3$

33. $\frac{d}{dx}\left(\frac{-1}{x+1} + C\right) = (-1)(-1)(x+1)^{-2} = \frac{1}{(x+1)^2}$

35. a) $\frac{d}{dx}\left(\frac{x^2}{2}\sin x + C\right) = \frac{2x}{2}\sin x + \frac{x^2}{2}\cos x =$

$x\sin x + \frac{x^2}{2}\cos x$ Wrong

b) $\frac{d}{dx}(-x\cos x + C) = -\cos x + x\sin x$ Wrong

c) $\frac{d}{dx}(-x\cos x + \sin x + C) = -\cos x + x\sin x + \cos x = x\sin x$ Right

37. $y = \int 3\sqrt{x}\,dx = \int 3x^{1/2}\,dx = 2x^{3/2} + C \quad \therefore y(9) = 4 \Rightarrow 4 = 2(9)^{3/2} + C \Rightarrow 4 = 54 + C \Rightarrow -50 = C$

$\therefore y = 2x^{3/2} - 50$

39. $y = \int -\pi \sin \pi x \, dx = (\pi)\left(\frac{1}{\pi}\right)\cos \pi x + C = \cos \pi x + C$; $y(0) = 0 \Rightarrow 0 = \cos \pi(0) + C \Rightarrow 0 = 1 + C \Rightarrow$
$C = -1 \ \therefore\ y = \cos(\pi x) - 1$

41. $\frac{dy}{dx} = \int 0 \, dx = C_1$; $\frac{dy}{dx}(0) = 2 \Rightarrow C_1 = 2 \ \therefore\ \frac{dy}{dx} = 2 \Rightarrow y = \int 2 \, dx = 2x + C_2$; $y(0) = 0 \Rightarrow$
$0 = 2(0) + C_2 \Rightarrow C_2 = 0 \ \therefore\ y = 2x$

43. $\frac{dy}{dx} = \int \frac{3}{8} x \, dx = \frac{3}{16}x^2 + C_1$; $\frac{dy}{dx}(4) = 3 \Rightarrow 3 = \frac{3}{16}(4)^2 + C_1 \Rightarrow 3 = 3 + C_1 \Rightarrow C_1 = 0 \ \therefore\ \frac{dy}{dx} = \frac{3}{16}x^2$
$y = \int \frac{3}{16}x^2 \, dx = \frac{3}{16}\left(\frac{x^3}{3}\right) + C_2 = \frac{x^3}{16} + C_2$; $\frac{dy}{dx}(4) = 4 \Rightarrow 4 = \frac{4^3}{16} + C_2 \Rightarrow 4 = 4 + C_2 \Rightarrow C_2 = 0 \ \therefore\ y = \frac{x^3}{16}$

45. Step 1: $\frac{ds}{dt} = \int -k \, dt = -kt + C_1$ ; $\frac{ds}{dt}(0) = 88 \Rightarrow 88 = -k(0) + C_1 \Rightarrow C_1 = 88 \ \therefore\ \frac{ds}{dt} = -kt + 88$
$s = \int (-kt + 88) \, dt = \frac{-kt^2}{2} + 88t + C_2$; $s(0) = 0 \Rightarrow 0 = \frac{-k(0)^2}{2} + 88(0) + C_2 \Rightarrow C_2 = 0 \ \therefore\ s = \frac{-kt^2}{2} + 88t$
Step 2: If $\frac{ds}{dt} = 0$, then $0 = -kt + 88 \Rightarrow t = \frac{88}{k}$
Step 3: $242 = \frac{-k\left(\frac{88}{k}\right)^2}{2} + 88\left(\frac{88}{k}\right) \Rightarrow 242 = \frac{-k(88)^2}{2k^2} + \frac{(88)^2}{k} \Rightarrow 242 = \frac{(88)^2}{k}\left(-\frac{1}{2} + 1\right) \Rightarrow 242 = \frac{(88)^2}{2k}$
$\Rightarrow 484\,k = (88)^2 \Rightarrow k = 16$

47. $\frac{ds}{dt} = \int 5.2 \, dt = 5.2t + C_1$ $\frac{ds}{dt}(0) = 0 \Rightarrow 0 = 5.2(0) + C_1 \Rightarrow C_1 = 0 \ \therefore\ \frac{ds}{dt} = 5.2t$
$s = \int 5.2t \, dt = 2.6t^2 + C_2$ $s(0) = 0 \Rightarrow 0 = 2.6(0)^2 + C_2 \Rightarrow C_2 = 0 \ \therefore\ s = 2.6t^2$
If $s = 4$, $4 = 2.6t^2 \Rightarrow t^2 = \frac{4}{2.6} \Rightarrow t = \pm\sqrt{\frac{4}{2.6}} \Rightarrow t = \sqrt{\frac{4}{2.6}} \approx 1.24$ sec since $t > 0$

## SECTION 5.6 INTEGRATION BY SUBSTITUTION -- RUNNING THE CHAIN RULE BACKWARD

1. $\int \sin 3x\, dx = \int \frac{1}{3}\sin u\, du = -\frac{1}{3}\cos u + C = -\frac{1}{3}\cos 3x + C$

$u = 3x$

$du = 3dx$

$\frac{1}{3}du = dx$

3. $\int \sec 2x \tan 2x\, dx = \int \frac{1}{2}\sec u \tan u\, du = \frac{1}{2}\sec u + C = \frac{1}{2}\sec 2x + C$

$u = 2x$

$du = 2dx$

$\frac{1}{2}du = dx$

5. $\int 28(7x-2)^3\, dx = \int \frac{1}{7}(28)u^3\, du = \int 4u^3\, du = u^4 + C = (7x-2)^4 + C$

$u = 7x - 2 \Rightarrow du = 7dx \Rightarrow \frac{1}{7}du = dx$

7. $\int \frac{9r^2\, dr}{\sqrt{1-r^3}} = \int -3u^{-1/2}\, du = -3(2)u^{1/2} + C = -6\left(1-r^3\right)^{1/2} + C$

$u = 1 - r^3$

$du = -3r^2dr$

$-3du = 9r^2dr$

9. a) $\int \csc^2 2\theta \cot 2\theta\, d\theta = -\int \frac{1}{2}u\, du$

$u = \cot 2\theta \qquad = -\frac{1}{2}\left(\frac{u^2}{2}\right) + C$

$du = -2\csc^2 2\theta\, d\theta \qquad = -\frac{u^2}{4} + C$

$\frac{1}{2}du = -\csc^2 2\theta\, d\theta \qquad = -\frac{\cot^2 2\theta}{4} + C$

b) $\int \csc^2 2\theta \cot 2\theta\, d\theta = \int -\frac{1}{2}u\, du$

$u = \csc 2\theta \qquad = -\frac{1}{2}\left(\frac{u^2}{2}\right) + C$

$du = -2\csc 2\theta \cot 2\theta\, d\theta \qquad = -\frac{u^2}{4} + C$

$-\frac{1}{2}du = \csc 2\theta \cot 2\theta\, d\theta \qquad = -\frac{\csc^2 2\theta}{4} + C$

11. $\int_0^{1/2} \frac{dx}{(2x+1)^3} = \int_1^2 \frac{1}{2}u^{-3}\, du = \left[\frac{1}{2}\left(\frac{u^{-2}}{-2}\right)\right]_1^2 = \left[\frac{-1}{4u^2}\right]_1^2 = \left[\frac{-1}{4(2)^2}\right] - \left[\frac{-1}{4(1)^2}\right] = \frac{-1}{16} + \frac{1}{4} = \frac{3}{16}$

$u = 2x + 1 \Rightarrow du = 2dx \Rightarrow \frac{1}{2}du = dx$ and $x = 0 \Rightarrow u = 1,\ x = \frac{1}{2} \Rightarrow u = 2$

13. $$\int_0^{\pi/6} \frac{\sin 2x}{\cos^2 2x}\,dx = \frac{1}{2}\int_1^{1/2} -u^{-2}\,du = \frac{1}{2}\left[\frac{1}{u}\right]_1^{1/2} = \frac{1}{2}\left[\frac{1}{\frac{1}{2}}\right] - \frac{1}{2}\left[\frac{1}{1}\right] = \frac{1}{2}$$

$u = \cos 2x \Rightarrow du = -2\sin 2x\,dx \Rightarrow -\frac{1}{2}du = \sin 2x\,dx$ and $x = 0 \Rightarrow u = 1$, $x = \frac{\pi}{6} \Rightarrow u = \frac{1}{2}$

15. $$\int_{-1}^{1} x\sqrt{1 - x^2}\,dx = \int_0^0 -\frac{1}{2}u^{1/2}\,du = 0$$

$u = 1 - x^2 \Rightarrow du = -2xdx \Rightarrow -\frac{1}{2}du = xdx$ and $x = -1 \Rightarrow u = 0$, $x = 1 \Rightarrow u = 0$

17. $$\int_{-\pi/2}^{\pi/2} \frac{\cos x}{(2 + \sin x)^2}\,dx = \int_1^3 u^{-2}\,du = \left[-\frac{1}{u}\right]_1^3 = \left[-\frac{1}{3}\right] - \left[-\frac{1}{1}\right] = \frac{2}{3}$$

$u = 2 + \sin x \Rightarrow du = \cos x\,dx$ and $x = -\frac{\pi}{2} \Rightarrow u = 1$, $x = \frac{\pi}{2} \Rightarrow u = 3$

19. $$\int \frac{dx}{(1 - x)^2} = \int -u^{-2}\,du = \frac{1}{u} + C = \frac{1}{1 - x} + C$$

$u = 1 - x$

$du = -dx$

$-du = dx$

21. $$\int \sec^2(x + 2)\,dx = \int \sec^2 u\,du = \tan u + C = \tan(x + 2) + C$$

$u = x = 2$

$du = dx$

23. $$\int 8r(r^2 - 1)^{1/3}\,dr = \int 4u^{1/3}\,du = 4\left(\frac{3}{4}\right)u^{4/3} + C = 3u^{4/3} + C = 3(r^2 - 1)^{4/3} + C$$

$u = r^2 - 1$

$du = 2r\,dr$

$4\,du = 8r\,dr$

25. $$\int \sec\left[\theta + \frac{\pi}{2}\right]\tan\left[\theta + \frac{\pi}{2}\right]d\theta = \int \sec u \tan u\,du = \sec u + C = \sec\left[\theta + \frac{\pi}{2}\right] + C$$

$u = \theta + \frac{\pi}{2}$

$du = d\theta$

27. $\int \frac{6x^3}{\sqrt[4]{1+x^4}}\,dx = \int \frac{3}{2}u^{-1/4}\,du = \frac{3}{2}\left(\frac{4}{3}\right)u^{3/4} + C = \frac{3}{2}\left(\frac{4}{3}\right)u^{3/4} + C = 2u^{3/4} + C = 2(1+x^4)^{3/4} + C$

$u = 1 + x^4$

$du = 4x^3\,dx$

$\frac{3}{2}du = 6x^3\,dx$

29. a) $\int_0^3 \sqrt{y+1}\,dy = \int_1^4 u^{1/2}\,du = \left[\frac{2}{3}u^{3/2}\right]_1^4$

$u = y + 1$ $\quad = \left[\frac{2}{3}(4)^{3/2}\right] - \left[\frac{2}{3}(1)^{3/2}\right]$

$du = dy$ $\quad = \left(\frac{2}{3}(8)\right) - \left(\frac{2}{3}(1)\right) = \frac{14}{3}$

$x = 0 \Rightarrow u = 1,\ x = 3 \Rightarrow u = 4$

b) $\int_{-1}^1 \sqrt{y+1}\,dy = \int_0^1 u^{1/2}\,du = \left[\frac{2}{3}u^{3/2}\right]_0^1$

$= \left[\left(\frac{2}{3}(1)^{3/2}\right) - (0)\right] = \frac{2}{3}$

Use same substitution for u as in part a)

$x = -1 \Rightarrow u = 0,\ x = 0 \Rightarrow u = 1$

31. a) $\int_0^{\pi/4} \tan x \sec^2 x\,dx = \int_0^1 u\,du = \left[\frac{u^2}{2}\right]_0^1$

$u = \tan x$ $\quad = \frac{1^2}{2} - 0 = \frac{1}{2}$

$du = \sec^2 x\,dx$

$x = 0 \Rightarrow u = 0,\ x = \frac{\pi}{4} \Rightarrow u = 1$

b) $\int_{-\pi/4}^0 \tan x \sec^2 x\,dx = \int_{-1}^0 u\,du = \left[\frac{u^2}{2}\right]_{-1}^0$

Use same substitution as in part a) $\quad = (0) - \left(\frac{1}{2}\right)$

$x = -\frac{\pi}{4} \Rightarrow u = -1,\ x = 0 \Rightarrow u = 0$ $\quad = -\frac{1}{2}$

33. a) $\int_0^1 \frac{x^3}{\sqrt{x^4+9}}\,dx = \int_9^{10} \frac{1}{4}u^{-1/2}\,du$

$u = x^4 + 9$ $\quad = \left[\frac{1}{4}(2)u^{1/2}\right]_9^{10}$

$du = 4x^3\,dx$ $\quad = \left(\frac{1}{2}(10)^{1/2}\right) - \left(\frac{1}{2}(9)^{1/2}\right)$

$\frac{1}{4}du = x^3\,dx$ $\quad = \frac{\sqrt{10} - 3}{2}$

$x = 0 \Rightarrow u = 9,\ x = 1 \Rightarrow u = 10$

b) $\int_{-1}^0 \frac{x^3}{\sqrt{x^4+9}}\,dx = \int_{10}^{9} \frac{1}{4}u^{-1/2}\,du =$

Use same substitution as in part a)

$x = -1 \Rightarrow u = 10,\ x = 0 \Rightarrow u = 9$

$= -\int_9^{10} \frac{1}{4}u^{-1/2}\,du = \frac{3 - \sqrt{10}}{2}$

35. a) $\int_0^{\sqrt{7}} x(x^2+1)^{1/3}\,dx = \int_1^8 \frac{1}{2}u^{1/3}\,du$

$u = x^2 + 1 \qquad = \left[\frac{1}{2}\left(\frac{3}{4}\right)u^{4/3}\right]_1^8$

$du = 2x\,dx \qquad = \left(\frac{3}{8}(8)^{4/3}\right) - \left(\frac{3}{8}(1)^{4/3}\right)$

$\frac{1}{2}du = x\,dx \qquad = \frac{45}{8}$

$x = 0 \Rightarrow u = 1,\ x = \sqrt{7} \Rightarrow u = 8$

b) $\int_{-\sqrt{7}}^{0} x(x^2+1)^{1/3}\,dx = \int_8^1 \frac{1}{2}u^{1/3}\,du$

Use same substitution as in part a)

$x = -\sqrt{7} \Rightarrow u = 8,\ x = 0 \Rightarrow u = 1$

$= -\int_1^8 \frac{1}{2}u^{1/3}\,du = -\frac{45}{8}$

37. a) $\int_0^{\pi/6} (1-\cos 3x)\sin 3x\,dx = \int_0^1 \frac{1}{3}u\,du$

$u = 1 - \cos 3x \qquad = \left[\frac{1}{3}\left(\frac{u^2}{2}\right)\right]_0^1$

$du = 3\sin 3x\,dx \qquad = \left[\frac{1}{6}(1)^2\right] - \left[\frac{1}{6}(0)^2\right]$

$\frac{1}{3}du = \sin 3x\,dx \qquad = \frac{1}{6}$

$x = 0 \Rightarrow u = 0 \qquad x = \frac{\pi}{6} \Rightarrow u = 1$

b) $\int_{\pi/6}^{\pi/3} (1-\cos 3x)\sin 3x\,dx = \int_1^2 \frac{1}{3}u\,du$

$= \left[\frac{1}{3}\left(\frac{u^2}{2}\right)\right]_1^2 = \left(\frac{1}{6}(2)^2\right) - \left(\frac{1}{6}(1)^2\right) = \frac{1}{2}$

Use same substitution as in part a)

$x = \frac{\pi}{6} \Rightarrow u = 1,\ x = \frac{\pi}{3} \Rightarrow u = 2$

39. a) $\int_0^{2\pi} \frac{\cos x}{\sqrt{2+\sin x}}\,dx = \int_2^2 u^{-1/2}\,du = 0$

$u = 2 + \sin x$

$du = \cos x\,dx$

$x = 0 \Rightarrow u = 2,\ x = 2\pi \Rightarrow u = 2$

b) $\int_{-\pi}^{\pi} \frac{\cos x}{\sqrt{2+\sin x}}\,dx = \int_2^2 u^{-1/2}\,du = 0$

Use same substitution as in part a)

$x = -\pi \Rightarrow u = 2$

$x = \pi \Rightarrow u = 2$

41. $\int_0^1 \sqrt{t^5+2t}\ (5t^4+2)\,dt = \int_0^3 u^{1/2}\,du = \left[\frac{2}{3}u^{3/2}\right]_0^3 = \left(\frac{2}{3}(3)^{3/2}\right) - \left(\frac{2}{3}(0)^{3/2}\right) = 2\sqrt{3}$

$u = t^5 + 2t \Rightarrow du = (5t^4+2)dt$

$t = 0 \Rightarrow u = 0,\ t = 1 \Rightarrow u = 3$

43. $\int_0^{\pi/2} \cos^3 2x \sin 2x\,dx = \int_1^{-1} -\frac{1}{2}u^3\,du = \left[-\frac{1}{2}\left(\frac{u^4}{4}\right)\right]_1^{-1} = \left(-\frac{(-1)^4}{8}\right) - \left(-\frac{1^4}{8}\right) = 0$

$u = \cos 2x \Rightarrow du = -2\sin 2x\,dx \Rightarrow -\frac{1}{2}du = \sin 2x\,dx$

$x = 0 \Rightarrow u = 1,\ x = \frac{\pi}{2} \Rightarrow u = -1$

45. $$\int_0^{\pi} \frac{8\sin t}{\sqrt{5-4\cos t}}\,dt = \int_1^9 2u^{-1/2}\,du = \left[2(2u^{1/2})\right]_1^9 = \left(4(9)^{1/2}\right) - \left(4(1)^{1/2}\right) = 8$$

$u = 5 - 4\cos t \Rightarrow du = 4\sin t\,dt \Rightarrow 2du = 8\sin t\,dt$

$t = 0 \Rightarrow u = 1,\ t = \pi \Rightarrow u = 9$

47. $$\int_0^1 15x^2\sqrt{5x^3+4}\,dx = \int_4^9 u^{1/2}\,du = \left[\frac{2}{3}u^{3/2}\right]_4^9 = \left(\frac{2}{3}(9)^{3/2}\right) - \left(\frac{2}{3}(4)^{3/2}\right) = \frac{38}{3}$$

$u = 5x^3 + 4 \Rightarrow du = 15x^2dx$

$x = 0 \Rightarrow u = 4,\ x = 1 \Rightarrow u = 9$

49. $$A = -\int_{-2}^0 x\sqrt{4-x^2}\,dx + \int_0^2 x\sqrt{4-x^2}\,dx = -\int_0^4 -\frac{1}{2}u^{1/2}\,du + \int_4^0 -\frac{1}{2}u^{1/2}\,du = 2\int_0^4 \frac{1}{2}u^{1/2}\,du$$

$u = 4 - x^2$

$$= \int_0^4 u^{1/2}\,du = \left[\frac{2}{3}u^{3/2}\right]_0^4 = \left(\frac{2}{3}(4)^{3/2}\right) - \left(\frac{2}{3}(0)^{3/2}\right) = \frac{16}{3}$$

$du = -2x\,dx$ $\quad x = -2 \Rightarrow u = 0$

$-\frac{1}{2}du = dx$ $\quad x = 0 \Rightarrow u = 4$ $\quad x = 2 \Rightarrow u = 0$

51. $$s = \int 24t(3t^2-1)^3\,dt = \int 4u^3\,du = \left[\frac{4u^4}{4}\right] + C = u^4 + C = (3t^2-1)^4 + C$$

$u = 3t^2 - 1 \Rightarrow du = 6t\,dt \Rightarrow 4du = 24t\,dt$

$s = (3t^2-1)^4 + C$ and $s(0) = 0 \Rightarrow 0 = (3(0)^2-1)^4 + C \Rightarrow 0 = 1 + C \Rightarrow C = -1 \quad \therefore s = (3t^2-1)^4 - 1$

53. $$s = \int 6\sin(t+\pi)\,dt = \int 6\sin u\,du = -6\cos u + C = -6\cos(t+\pi) + C$$

$u = t + \pi \Rightarrow du = dt$

$s = -6\cos(t+\pi) + C$ and $s(0) = 0 \Rightarrow 0 = -6\cos(0+\pi) + C \Rightarrow 0 = -6\cos\pi + C \Rightarrow 0 = 6 + C \Rightarrow$

$C = -6 \quad \therefore s = -6\cos(t+\pi) - 6$

55. a) $\displaystyle\int_0^{\pi/4} \frac{18\tan^2 x \sec^2 x}{(2+\tan^3 x)^2}\,dx = \int_0^1 \frac{18u^2}{(2+u^3)^2}\,du = \int_0^1 \frac{6}{(2+v)^2}\,dv = \int_2^3 \frac{6}{w^2}\,dw = \int_2^3 6w^{-2}\,dw$

$= \left[-6w^{-1}\right]_2^3 = \left[-\frac{6}{w}\right]_2^3 = \left(-\frac{6}{3}\right) - \left(-\frac{6}{2}\right) = 1$

$u = \tan x$, $du = \sec^2 x\,dx$, $x = 0 \Rightarrow u = 0$, $x = \frac{\pi}{4} \Rightarrow u = 1$

$v = u^3$, $dv = 3u^2\,du$, $6\,dv = 18u^2\,du$, $u = 0 \Rightarrow v = 0$, $u = 1 \Rightarrow v = 1$

$w = 2 + v$, $dw = dv$, $v = 0 \Rightarrow w = 2$, $v = 1 \Rightarrow w = 3$

b) $\displaystyle\int_0^{\pi/4} \frac{18\tan^2 x \sec^2 x}{(2+\tan^3 x)^2}\,dx = \int_0^1 \frac{6}{(2+u)^2}\,du = \int_2^3 \frac{6}{v^2}\,dv = \int_2^3 6v^{-2}\,dv = \left[-\frac{6}{v}\right]_2^3 = 1$

$u = \tan^3 x$

$du = 3\tan^2 x \sec^2 x\,dx$

$6\,du = 18\tan^2 x \sec^2 x\,dx$

$x = 0 \Rightarrow u = 0,\ x = \frac{\pi}{4} \Rightarrow u = 1$

$v = 2 + u$

$dv = du$

$u = 0 \Rightarrow v = 2,\ u = 1 \Rightarrow v = 3$

c) $\displaystyle\int_0^{\pi/4} \frac{18\tan^2 x \sec^2 x}{(2+\tan^3 x)^2}\,dx = \int_2^3 \frac{6}{u^2}\,du = \int_2^3 6u^{-2}\,du = \left[-\frac{6}{u}\right]_2^3 = 1$

$u = 2 + \tan^3 x$

$du = 3\tan^2 x \sec^2 x\,dx$

$6\,du = 18\tan^2 x \sec^2 x\,dx$

$x = 0 \Rightarrow u = 2$

$x = \frac{\pi}{4} \Rightarrow u = 3$

57. Since $\sin^2 x = 1 - \cos^2 x$, $\sin^2 x + C_1 = 1 - \cos^2 x + C_1 \Rightarrow 1 + C_1 = C_2$, the constant in part b)

Since $\sin^2 x = \frac{1}{2} - \frac{1}{2}\cos 2x$, $\sin^2 x + C_1 = \frac{1}{2} - \frac{\cos 2x}{2} + C_1 \Rightarrow \frac{1}{2} + C_1 = C_3$, the constant in part c)

Since $1 + C_1 = C_2$ and $\frac{1}{2} + C_1 = C_3$, $C_2 - 1 = C_3 - \frac{1}{2} \Rightarrow C_2 = C_3 + \frac{1}{2}$

$\therefore$ the integrals are the same since they differ only by a constant.

## SECTION 5.7 NUMERICAL INTEGRATION

1. a) For $n = 4$, $h = \frac{b-a}{n} = \frac{2-0}{4} = \frac{1}{2} \Rightarrow \frac{h}{2} = \frac{1}{4}$

| $x_i$ | $x_i$ | $f(x_i)$ | m | $mf(x_i)$ |
|---|---|---|---|---|
| $x_0$ | 0 | 0 | 1 | 0 |
| $x_1$ | $\frac{1}{2}$ | $\frac{1}{2}$ | 2 | 1 |
| $x_2$ | 1 | 1 | 2 | 2 |
| $x_3$ | $\frac{3}{2}$ | $\frac{3}{2}$ | 2 | 3 |
| $x_4$ | 2 | 2 | 1 | 2 |

$\sum mf(x_i) = 8 \Rightarrow T = \frac{1}{4}(8) = 2$

b) $h = \frac{1}{2} \Rightarrow \frac{h}{3} = \frac{1}{6}$

| $x_i$ | $x_i$ | $f(x_i)$ | m | $mf(x_i)$ |
|---|---|---|---|---|
| $x_0$ | 0 | 0 | 1 | 0 |
| $x_1$ | $\frac{1}{2}$ | $\frac{1}{2}$ | 4 | 2 |
| $x_2$ | 1 | 1 | 2 | 2 |
| $x_3$ | $\frac{3}{2}$ | $\frac{3}{2}$ | 4 | 6 |
| $x_4$ | 2 | 2 | 1 | 2 |

$\sum mf(x_i) = 12 \Rightarrow S = \frac{1}{6}(12) = 2$

c) $\int_0^2 x\,dx = \left[\frac{x^2}{2}\right]_0^2 = \frac{4}{2} - (0) = 2$

3. a) $n = 4 \Rightarrow h = \frac{b-a}{n} = \frac{2-0}{4} = \frac{1}{2} \Rightarrow \frac{h}{2} = \frac{1}{4}$

| $x_i$ | $x_i$ | $f(x_i)$ | m | $mf(x_i)$ |
|---|---|---|---|---|
| $x_0$ | 0 | 0 | 1 | 0 |
| $x_1$ | $\frac{1}{2}$ | $\frac{1}{8}$ | 2 | $\frac{1}{4}$ |
| $x_2$ | 1 | 1 | 2 | 2 |
| $x_3$ | $\frac{3}{2}$ | $\frac{27}{8}$ | 2 | $\frac{27}{4}$ |
| $x_4$ | 2 | 8 | 1 | 8 |

$\sum mf(x_i) = 17 \Rightarrow T = \frac{1}{4}(17) = \frac{17}{4}$

b) $h = \frac{1}{2} \Rightarrow \frac{h}{3} = \frac{1}{6}$

| $x_i$ | $x_i$ | $f(x_i)$ | m | $mf(x_i)$ |
|---|---|---|---|---|
| $x_0$ | 0 | 0 | 1 | 0 |
| $x_1$ | $\frac{1}{2}$ | $\frac{1}{8}$ | 4 | $\frac{1}{2}$ |
| $x_2$ | 1 | 1 | 2 | 2 |
| $x_3$ | $\frac{3}{2}$ | $\frac{27}{8}$ | 4 | $\frac{27}{2}$ |
| $x_4$ | 2 | 8 | 1 | 8 |

$\sum mf(x_i) = 24 \Rightarrow S = \frac{1}{6}(24) = 4$

c) $\int_0^2 x^3\,dx = \left[\frac{x^4}{4}\right]_0^2 = \frac{16}{4} - (0) = 4$

5. a) $n = 4 \Rightarrow h = \frac{b-a}{n} = \frac{4-0}{4} = 1 \Rightarrow \frac{h}{2} = \frac{1}{2}$

| $x_i$ | $x_i$ | $f(x_i)$ | m | $mf(x_i)$ |
|---|---|---|---|---|
| $x_0$ | 0 | 0 | 1 | 0 |
| $x_1$ | 1 | 1 | 2 | 2 |
| $x_2$ | 2 | $\sqrt{2}$ | 2 | $2\sqrt{2}$ |
| $x_3$ | 3 | $\sqrt{3}$ | 2 | $2\sqrt{3}$ |
| $x_4$ | 4 | 2 | 1 | 2 |

$\sum mf(x_i) = 4 + 2\sqrt{2} + 2\sqrt{3} \Rightarrow$

$T = \frac{1}{2}(4 + 2\sqrt{2} + 2\sqrt{3}) = 2 + \sqrt{2} + \sqrt{3}$

$\approx 5.1463$

b) $h = 1 \Rightarrow \frac{h}{3} = \frac{1}{3}$

| $x_i$ | $x_i$ | $f(x_i)$ | m | $mf(x_i)$ |
|---|---|---|---|---|
| $x_0$ | 0 | 0 | 1 | 0 |
| $x_1$ | 1 | 1 | 4 | 4 |
| $x_2$ | 2 | $\sqrt{2}$ | 2 | $2\sqrt{2}$ |
| $x_3$ | 3 | $\sqrt{3}$ | 4 | $4\sqrt{3}$ |
| $x_4$ | 4 | 2 | 1 | 2 |

$\sum mf(x_i) = 6 + 2\sqrt{2} + 4\sqrt{3} \Rightarrow$

$S = \frac{1}{3}(6 + 2\sqrt{2} + 4\sqrt{3}) = 2 + \frac{2}{3}\sqrt{2} + \frac{4}{3}\sqrt{3}$

$\approx 5.252$

c) $\int_0^4 \sqrt{x}\,dx = \left[\frac{2}{3}x^{3/2}\right]_0^4 = \frac{2}{3}(4)^{3/2} - 0 = \frac{16}{3}$

7. $f(x) = \frac{1}{x} \Rightarrow f'(x) = -\frac{1}{x^2} \Rightarrow f''(x) = \frac{2}{x^3}$ which is decreasing on [1,2]. $\therefore |f''(x)| \le 2$ on [1,2] since $f''(1) = 2$ is the maximum value of $f''$ on the interval. $\therefore M = 2$. $n = 10 \Rightarrow h = \frac{b-a}{n} = \frac{2-1}{10} = \frac{1}{10}$

Then $|E_T| \le \frac{2-1}{12}\left(\frac{1}{10}\right)^2(2) = \frac{1}{600} \approx 0.001667$

9. a) $|E_T| \le 10^{-4} \Rightarrow \frac{b-a}{12}(h^2)M \le 10^{-4}$. $h = \frac{b-a}{n} = \frac{2-0}{n} = \frac{2}{n}$. $f(x) = x \Rightarrow f'(x) = 1 \Rightarrow f''(x) = 0 \Rightarrow M = 0$

For $\frac{2}{12}\left(\frac{2}{n}\right)^2(0) \le 10^{-4}$, let $n = 1$

b) $|E_S| \le 10^{-4} \Rightarrow \frac{b-a}{180}(h)^4M \le 10^{-4}$. $h = \frac{2}{n}$; $f^{(4)}(x) = 0 \Rightarrow M = 0$. For $\frac{2}{180}\left(\frac{2}{n}\right)^4(0) \le 10^{-4}$, let $n = 2$

(Remember n must be even for Simpson's Rule)

11. a) $|E_T| \le 10^{-4} \Rightarrow \frac{b-a}{12}(h^2)M \le 10^{-4}$. $h = \frac{b-a}{n} = \frac{2-0}{n} = \frac{2}{n}$; $f(x) = x^3 \Rightarrow f'(x) = 3x^2 \Rightarrow f''(x) = 6x$ which is increasing on [0,2] $\Rightarrow |6x| \le 12$ on [0,2] $\Rightarrow M = 12$. For $|E_T| \le 10^{-4}$, $\frac{2-0}{12}\left(\frac{2}{n}\right)^2(12) \le 10^{-4} \Rightarrow$

$\frac{8}{n^2} \le 10^{-4} \Rightarrow \frac{n^2}{8} \ge 10000 \Rightarrow n^2 \ge 80000 \Rightarrow n \ge \sqrt{80000} \approx 282.8$ $\therefore$ let $n = 283$

b) $|E_S| \le 10^{-4} \Rightarrow \frac{b-a}{180}(h)^4M \le 10^{-4}$. $h = \frac{2}{n}$; $f'''(x) = 6 \Rightarrow f^{(4)}(x) = 0 \Rightarrow M = 0$. For

$\frac{2}{180}\left(\frac{2}{n}\right)^2(0) \le 10^{-4}$, let $n = 2$ (n must be even)

13. a) $|E_T| \le 10^{-4} \Rightarrow \frac{b-a}{12}(h^2)M \le 10^{-4}$. $h = \frac{b-a}{n} = \frac{4-1}{n} = \frac{3}{n}$. $f(x) = \sqrt{x} = x^{1/2} \Rightarrow f'(x) = \frac{1}{2}x^{-1/2} \Rightarrow f''(x)$ $= -\frac{1}{4}x^{-3/2}$ whose absolute value is decreasing on [1,4] $\Rightarrow \left|f''(x)\right| \le \frac{1}{4}$ on [1,4] $\Rightarrow M = \frac{1}{4}$. For $|E_T| \le 10^{-4}$, $\frac{4-1}{12}\left(\frac{3}{n}\right)^2\left(\frac{1}{4}\right) \le 10^{-4} \Rightarrow \left(\frac{9}{16}\right)\left(\frac{1}{n^2}\right) \le 10^{-4} \Rightarrow n^2 \ge \frac{9}{16}(10^4) \Rightarrow n \ge \frac{3}{4}(10^2) \Rightarrow n \ge 75$

$\therefore$ let n = 75

b) $|E_S| \le 10^{-4} \Rightarrow \frac{b-a}{180}(h)^4M \le 10^{-4}$. $h = \frac{3}{n}$. $f'''(x) = \frac{3}{8}x^{-5/2} \Rightarrow f^{(4)}(x) = -\frac{15}{16}x^{-7/2}$ whose absolute value is decreasing on [1,4] $\Rightarrow \left|f^{(4)}(x)\right| \le \frac{15}{16}$ on [1,4] $\Rightarrow M = \frac{15}{16}$. For $|E_S| \le 10^{-4}$,

$\frac{4-1}{180}\left(\frac{3}{n}\right)^4\left(\frac{15}{16}\right) \le 10^{-4} \Rightarrow \left(\frac{81}{64}\right)\left(\frac{1}{n^4}\right) \le 10^{-4} \Rightarrow n^4 \ge \frac{81}{64}(10^4) \Rightarrow n \ge 10.6$ $\therefore$ let n = 12

(n must be even)

15. Using Simpson's Rule, $h = 200 \Rightarrow \frac{h}{3} = \frac{200}{3}$

| $x_i$ | $x_i$ | $f(x_i)$ | m | $mf(x_i)$ |
|---|---|---|---|---|
| $x_0$ | 0 | 0 | 1 | 0 |
| $x_1$ | 200 | 520 | 4 | 2080 |
| $x_2$ | 400 | 800 | 2 | 1600 |
| $x_3$ | 600 | 1000 | 4 | 4000 |
| $x_4$ | 800 | 1140 | 2 | 2280 |
| $x_5$ | 1000 | 1160 | 4 | 4640 |
| $x_6$ | 1200 | 1110 | 2 | 2220 |
| $x_7$ | 1400 | 860 | 4 | 3440 |
| $x_8$ | 1600 | 0 | 1 | 0 |

$\sum mf(x_i) = 20260 \Rightarrow \text{Area} \approx \frac{200}{3}(20260)$ $= 1\,350\,666.667$ ft$^2$. Since the average depth = 20 ft, Volume ≈ 20(Area) ≈ 27 013 333.33 ft$^3$

Number of fish = Volume/1000 = 27013 (to the nearest fish)

Maximum to be caught = 75% of 27013 = 20260

Number of licenses = $\frac{20260}{20} = 1013$

17. Using Simpson's Rule, $h = \frac{b-a}{n} = \frac{24-0}{6} = \frac{24}{6} = 4$

| $x_i$ | $x_i$ | $y_i$ | m | $my_i$ |
|---|---|---|---|---|
| $x_0$ | 0 | 0 | 1 | 0 |
| $x_1$ | 4 | 18.75 | 4 | 75 |
| $x_2$ | 8 | 24 | 2 | 48 |
| $x_3$ | 12 | 26 | 4 | 104 |
| $x_4$ | 16 | 24 | 2 | 48 |
| $x_5$ | 20 | 18.75 | 4 | 75 |
| $x_6$ | 24 | 0 | 1 | 0 |

$\sum my_i = 350 \Rightarrow$

$S = \frac{4}{3}(350) = \frac{1400}{3} \approx 466.7$ in$^2$

19. Using odd numbered hours,

| Hour | Weekday KW | m | m(Weekday KW) | Hour | Weekday KW | m | m(Weekday KW) |
|---|---|---|---|---|---|---|---|
| 1 | 1.88 | 1 | 1.88 | 15 | 1.72 | 2 | 3.44 |
| 3 | 2.02 | 2 | 4.04 | 17 | 1.97 | 2 | 3.94 |
| 5 | 2.25 | 2 | 4.50 | 19 | 2.68 | 2 | 5.36 |
| 7 | 3.60 | 2 | 7.20 | 21 | 2.65 | 2 | 5.30 |
| 9 | 3.05 | 2 | 6.10 | 23 | 2.21 | 2 | 4.42 |
| 11 | 2.38 | 2 | 4.76 | 1 | 1.88 | 1 | 1.88 |
| 13 | 2.02 | 2 | 4.04 | | | | |

$n = 12 \Rightarrow h = \frac{b-a}{n} = \frac{24}{12} = 2 \Rightarrow \frac{h}{2} = 1$

$\sum$ m(Weekday KW) = 56.86 $\Rightarrow$ T = 1(56.86) = 56.86 KW/Weekday

21. a) $|E_S| \le \frac{b-a}{180}(h^4)M$. $n = 4 \Rightarrow h = \frac{\frac{\pi}{2} - 0}{4} = \frac{\pi}{8}$. $|f^{(4)}| \le 1 \Rightarrow M = 1 \therefore |E_S| \le \frac{\frac{\pi}{2} - 0}{180}\left(\frac{\pi}{8}\right)^4(1) \approx 0.0002075$

b) $h = \frac{\pi}{8} \Rightarrow \frac{h}{3} = \frac{\pi}{24}$

| $x_i$ | $x_i$ | $f(x_i)$ | m | $mf(x_i)$ |
|---|---|---|---|---|
| $x_0$ | 0 | 1 | 1 | 1 |
| $x_1$ | $\frac{\pi}{8}$ | 0.974495358 | 4 | 3.897981432 |
| $x_2$ | $\frac{\pi}{4}$ | 0.900316316 | 2 | 1.800632632 |
| $x_3$ | $\frac{3\pi}{8}$ | 0.784213303 | 4 | 3.136853212 |
| $x_4$ | $\frac{\pi}{2}$ | 0.636619772 | 1 | 0.636619772 |

$\sum mf(x_i) = 10.472087048 \Rightarrow S = \frac{\pi}{24}(10.472087048) = 1.370792988$

c) 0.0151372%

Exercises 23 and 25 were done using the Calculus Toolkit Integral Evaluator with n = 50

23. S = 1.08942941

25. 0.828116331

## SECTION 5.8 A BRIEF INTRODUCTION TO LOGARITHMS AND EXPONENTIALS

1. $\ln 1.5 = \ln \frac{3}{2} = \ln 3 - \ln 2$

3. $\ln \frac{4}{9} = \ln 2^2 - \ln 3^2 = 2\ln 2 - 2\ln 3$

5. $\ln \frac{1}{2} = \ln 2^{-1} = -\ln 2$

7. $\ln 4.5 = \ln \frac{9}{2} = \ln 3^2 - \ln 2 = 2\ln 3 - \ln 2$

9. $\ln 3\sqrt{2} = \ln 3\left(2^{1/2}\right) = \ln 3 + \frac{1}{2}\ln 2$

11. $e^{\ln 7} = 7$

13. $\ln e^2 = 2$

15. $e^{2+\ln 3} = e^2 e^{\ln 3} = 3e^2$

17. $\frac{dy}{dx} = \frac{1}{x}$

19. $\frac{dy}{dx} = \frac{2}{x}$

21. $\frac{dy}{dx} = -\frac{1}{x}$

23. $\frac{dy}{dx} = \frac{1}{x+2}$

25. $\frac{dy}{dx} = \frac{\sin x}{2 - \cos x}$

27. $\frac{dy}{dx} = \frac{1}{\ln x}\left(\frac{1}{x}\right) = \frac{1}{x \ln x}$

29. $\frac{dy}{dx} = 2e^x$

31. $\frac{dy}{dx} = -e^{-x}$

33. $\frac{dy}{dx} = \frac{2}{3}e^{2x/3}$

35. $\frac{dy}{dx} = e^x + xe^x - e^x = xe^x$

37. $\frac{dy}{dx} = e^{\sqrt{x}}\left(\frac{1}{2}x^{-1/2}\right) = \frac{e^{\sqrt{x}}}{2\sqrt{x}}$

39. $$\int_0^3 \frac{1}{x+1}\,dx = \int_1^4 \frac{1}{u}\,du = \Big[\ln u\Big]_1^4 = \ln 4 - \ln 1 = \ln 4$$

$u = x + 1$

$du = dx$

$x = 0 \Rightarrow u = 1,\ x = 3 \Rightarrow u = 4$

41. $$\int_{\ln 3}^{\ln 5} e^{2x}\,dx = \int_{2\ln 3}^{2\ln 5} \frac{1}{2}e^u\,du = \left[\frac{1}{2}e^u\right]_{2\ln 3}^{2\ln 5} = \left(\frac{1}{2}e^{2\ln 5}\right) - \left(\frac{1}{2}e^{2\ln 3}\right) = \frac{1}{2}(25) - \frac{1}{2}(9) = 8$$

$u = 2x \Rightarrow du = 2\,dx \Rightarrow \frac{1}{2}du = dx \quad x = \ln 3 \Rightarrow u = 2\ln 3,\ x = \ln 5 \Rightarrow u = 2\ln 5$

43. $$\int_0^1 (1 + e^x)e^x\,dx = \int_2^{1+e} u\,du = \left[\frac{u^2}{2}\right]_2^{1+e} = \left(\frac{(1+e)^2}{2}\right) - \left(\frac{2^2}{2}\right) = \frac{1 + 2e + e^2}{2} - \frac{4}{2} = \frac{e^2 + 2e - 3}{2}$$

$u = 1 + e^x$

$du = e^x\,dx$

$x = 0 \Rightarrow u = 2$

$x = 1 \Rightarrow u = 1 + e$

45. $\displaystyle\int_1^4 \frac{e^{\sqrt{x}}}{2\sqrt{x}}\,dx = \int_1^2 e^u\,du = \left[e^u\right]_1^2 = e^2 - e^1 = e^2 - e$

$u = \sqrt{x}$

$du = \dfrac{1}{2\sqrt{x}}\,dx$

$x = 1 \Rightarrow u = 1,\ x = 4 \Rightarrow u = 2$

47. $\displaystyle\int_1^{e^2} \frac{1}{x}\,dx = \left[\ln x\right]_1^{e^2} = \ln e^2 - \ln 1 = 2$

49. $\displaystyle\int_{\ln 2}^{\ln 3} e^x\,dx = \left[e^x\right]_{\ln 2}^{\ln 3} = e^{\ln 3} - e^{\ln 2} = 3 - 2 = 1$

51. $\displaystyle\int_2^4 \frac{dx}{x+2} = \int_4^6 \frac{1}{u}\,du = \left[\ln u\right]_4^6 = \ln 6 - \ln 4 = \ln\frac{6}{4} = \ln\frac{3}{2}$

$u = x + 2$

$du = dx$

$x = 2 \Rightarrow u = 4,\ x = 4 \Rightarrow u = 6$

53. $\displaystyle\int_{-1}^{1} 2x\,e^{-x^2}\,dx = \int_{-1}^{-1} -e^u\,du = 0$

$u = -x^2$

$du = -2x\,dx$

$-du = 2x\,dx$

$x = -1 \Rightarrow u = -1,\ x = 1 \Rightarrow u = -1$

55. $e^{2k} = 4 \Rightarrow \ln e^{2k} = \ln 4 \Rightarrow 2k = \ln 4$

$\Rightarrow k = \dfrac{\ln 4}{2}$

57. $100\,e^{10k} = 200 \Rightarrow e^{10k} = 2 \Rightarrow \ln e^{10k} = \ln 2$

$\Rightarrow 10k = \ln 2 \Rightarrow k = \dfrac{\ln 2}{10}$

59. $e^t = 1 \Rightarrow \ln e^t = \ln 1 \Rightarrow t = 0$

61. $e^{-0.3t} = 27 \Rightarrow \ln e^{-0.3t} = \ln 27 \Rightarrow -0.3t$

$= \ln 27 \Rightarrow t = \dfrac{\ln 27}{-0.3} = \dfrac{3\ln 3}{-0.3} = -10\ln 3$

63. $4e^{-0.1t} = 20 \Rightarrow e^{-0.1t} = 5 \Rightarrow \ln e^{-0.1t} = \ln 5$

$\Rightarrow -0.1t = \ln 5 \Rightarrow t = \dfrac{\ln 5}{-0.1} = -10\ln 5$

65. $\ln y = 2t + 4 \Rightarrow e^{\ln y} = e^{2t+4} \Rightarrow y = e^{2t+4}$

67. $\ln(y - 40) = 5t \Rightarrow e^{\ln(y-40)} = e^{5t} \Rightarrow$

$y - 40 = e^{5t} \Rightarrow y = e^{5t} + 40$

69. $y = y_0 e^{kt},\ y_0 = 1 \Rightarrow y = e^{kt}$. $\therefore$ when $y = 2$ at $t = 0.5$, $2 = e^{0.5k} \Rightarrow \ln 2 = \ln e^{0.5k} \Rightarrow \ln 2 = 0.5k \Rightarrow k = \dfrac{\ln 2}{0.5}$

or $k = 2\ln 2 = \ln 4 \Rightarrow y = e^{t\ln 4} \Rightarrow y(24) = e^{24\ln 4} = 4^{24} \approx 2.8147497 \times 10^{14}$

71. a) $10000e^{k(1)} = 7500 \Rightarrow e^k = 0.75 \Rightarrow \ln e^k = \ln 0.75 \Rightarrow k = \ln 0.75 \;\therefore\; y = 10000e^{t(\ln 0.75)}$

Then $1000 = 10000e^{t\ln 0.75} \Rightarrow 0.1 = e^{t\ln 0.75} \Rightarrow \ln 0.1 = \ln e^{t\ln 0.75} \Rightarrow \ln 0.1 = t(\ln 0.75) \Rightarrow$

$t = \dfrac{\ln 0.1}{\ln 0.75} \approx 8.00$ years (to the nearest hundredth of a year)

b) $1 = 10000e^{t\ln 0.75} \Rightarrow 0.0001 = e^{t\ln 0.75} \Rightarrow \ln 0.0001 = \ln e^{t\ln 0.75} \Rightarrow \ln 0.0001 = t\ln 0.75 \Rightarrow$

$t = \dfrac{\ln 0.0001}{\ln 0.75} \approx 32.02$ years (to the nearest hundredth of a year)

73. $A(100) = 90000 \Rightarrow 90000 = 1000e^{r(100)} \Rightarrow 90 = e^{100r} \Rightarrow \ln 90 = \ln e^{100r} \Rightarrow \ln 90 = 100r \Rightarrow$

$r = \dfrac{\ln 90}{100} \approx .0450$ or 4.50%

75. $P_0 = 246\,605\,103$ and $k = .609\% \Rightarrow P(t) = 246\,605\,103e^{0.00609t} \Rightarrow P(10) = 246\,605\,103e^{0.00609(10)}$

$\Rightarrow P(10) = 262\,090\,086$

77. $T \approx \dfrac{70}{2.8} = 25$ years or the year 2013

79. For 1990: $p = 116.5e^{0.044(2)} = 127.2 \;\therefore$ Purchasing Power $= \dfrac{100}{127.2} = 0.786$

For 1992: $p = 116.5e^{0.044(4)} = 138.9 \;\therefore$ Purchasing Power $= \dfrac{100}{138.9} = 0.720$

For 1994: $p = 116.5e^{0.044(6)} = 151.7 \;\therefore$ Purchasing Power $= \dfrac{100}{151.7} = 0.659$

81. $p_0e^{8t} = 2p_0 \Rightarrow e^{8t} = 2 \Rightarrow \ln e^{8t} = \ln 2 \Rightarrow 8t = \ln 2 \Rightarrow t = \dfrac{\ln 2}{8} \approx 0.0866$ years or about 32 days.

83. For $P = \frac{1}{5}P_0$, $\frac{1}{5}P_0 = P_0e^{-0.1t} \Rightarrow \frac{1}{5} = e^{-0.1t} \Rightarrow \ln\frac{1}{5} = \ln e^{-0.1t} \Rightarrow \ln\frac{1}{5} = -0.1t \Rightarrow t = \dfrac{\ln\frac{1}{5}}{-0.1} \approx 16.09$ years

85. $p = 1.50p_0 \Rightarrow 1.5p_0 = p_0e^{kt} \Rightarrow 1.5 = e^{kt} \Rightarrow \ln 1.5 = \ln e^{kt} \Rightarrow \ln 1.5 = kt \Rightarrow t = \dfrac{\ln 1.5}{k}$

a) Food: $t = \dfrac{\ln 1.5}{0.03} \approx 13.52$ years

Rent: $t = \dfrac{\ln 1.5}{0.05} \approx 8.11$ years

Medicine: $t = \dfrac{\ln 1.5}{0.064} \approx 6.34$ years

b) Food: $t = \dfrac{\ln 2}{0.03} \approx 23.3$ years

Rent: $t = \dfrac{\ln 2}{0.05} \approx 14.0$ years

Medicine: $t = \dfrac{\ln 2}{0.064} \approx 10.9$ years

87. a) $p(t) = 100\,e^{\int_0^{\tau}(1+1.3\tau)\,d\tau}$

b) $p(1) \Rightarrow t = 1. \;\therefore\; \displaystyle\int_0^1 (1 + 1.3\tau)\,d\tau = \left[\tau + \frac{1.3\tau^2}{2}\right]_0^1 = \left(1 + \frac{1.3(1)^2}{2}\right) - (0) = 1 + .65 = 1.65$

$\therefore p(1) = 100\,e^{\int_0^9(1+1.3\tau)d\tau} = 100\,e^{1.65} = 520.7$

$p(2) \Rightarrow t = 1. \;\therefore\; \displaystyle\int_0^2 (1 + 1.3\tau)\,d\tau = \left[\tau + \frac{1.3\tau^2}{2}\right]_0^2 = \left(2 + \frac{1.3(2)^2}{2}\right) - (0) = 2 + 2.6 = 4.6$

$\therefore p(2) = 100\,e^{\int_0^2(1+1.3\tau)\,d\tau} = 100\,e^{4.6} = 9948.4$. The percentage increases are 420.7% and 9848.4%.

## PRACTICE EXERCISES

1. a) $\sum_{k=1}^{100} k = \frac{100(100+1)}{2} = 5050$  b) $\sum_{k=1}^{100} k^2 = \frac{100(100+1)(2(100)+1)}{6} = 338350$

   c) $\sum_{k=1}^{100} k^3 = \left[\frac{100(100+1)}{2}\right]^2 = 25\,502\,500$

3. a) $\sum_{k=1}^{6}\left(k^2 - \frac{1}{6}\right) = \sum_{k=1}^{6} k^2 - \frac{1}{6}\sum_{k=1}^{6} 1 = \frac{6(6+1)(2(6)+1)}{6} - \frac{1}{6}(6(1)) = 90$

   b) $\sum_{k=1}^{6} k(k+1) = \sum_{k=1}^{6}\left(k^2 + k\right) = \sum_{k=1}^{6} k^2 + \sum_{k=1}^{6} k = \frac{6(6+1)(2(6)+1)}{6} + \frac{6(6+1)}{2} = 112$

5. a) $\sum_{k=1}^{3} 2^{k-1} = 2^0 + 2^1 + 2^2 = 1 + 2 + 4 = 7$  b) $\sum_{k=0}^{4} (-1)^k \cos k\pi = \cos 0 - \cos \pi + \cos 2\pi - \cos 3\pi + \cos 4\pi = 5$

   c) $\sum_{k=-1}^{2} k(k+1) = (-1)(-1+1) + (0)(0+1) + (1)(1+1) + (2)(2+1) = 8$

   d) $\sum_{k=1}^{4} \frac{(-1)^{k+1}}{k(k+1)} = \frac{1}{1(1+1)} - \frac{1}{2(2+1)} + \frac{1}{3(3+1)} - \frac{1}{4(4+1)} = \frac{1}{2} - \frac{1}{6} + \frac{1}{12} - \frac{1}{20} = \frac{11}{30}$

7. $\sum_{k=1}^{6} k^2 = \frac{6(6+1)(2(6)+1)}{6} = 91$  9. $\int_1^2 \frac{1}{x}\,dx$

11. a) True, $\int_5^2 f(x)\,dx = -\int_2^5 f(x)\,dx$

    b) True, $\int_{-2}^{5} (f(x) + g(x))\,dx = \int_{-2}^{5} f(x)\,dx + \int_{-2}^{5} g(x)\,dx = \int_{-2}^{2} f(x)\,dx + \int_{2}^{5} f(x)\,dx + \int_{-2}^{5} g(x)\,dx$

    c) False, $\int_{-2}^{5} f(x)\,dx > \int_{-2}^{5} g(x)\,dx \Rightarrow f(x) \not\leq g(x)$

13. $A = \int_{\pi/4}^{3\pi/4} (2 - \csc^2 x)dx = \Big[2x + \cot x\Big]_{\pi/4}^{3\pi/4} = \left(2\left(\frac{3\pi}{4}\right) + \cot\left(\frac{3\pi}{4}\right)\right) - \left(2\left(\frac{\pi}{4}\right) + \cot\left(\frac{\pi}{4}\right)\right) =$
$\left(\frac{3\pi}{2} - 1\right) - \left(\frac{\pi}{2} + 1\right) = \pi - 2$

15. $A = \int_0^1 (e - e^x)dx = \Big[ex - e^x\Big]_0^1 = (e(1) - e^1) - (e(0) - e^0) = 1$

17. $A = \int_0^4 (4 - x)dx - \int_4^6 (4 - x)dx = \left[4x - \frac{x^2}{2}\right]_0^4 - \left[4x - \frac{x^2}{2}\right]_4^6 = \left(4(4) - \frac{4^2}{2}\right) - (0) =$
$\left[\left(4(6) - \frac{6^2}{2}\right) - \left(4(4) - \frac{4^2}{2}\right)\right] = (16 - 8) - (24 - 18 - 16 + 8) = 10$

19. $\int_{-1}^1 (3x^2 - 4x + 7)dx = \Big[x^3 - 2x^2 + 7x\Big]_{-1}^1 = (1^3 - 2(1)^2 + 7(1)) - ((-1)^3 - 2(-1)^2 + 7(-1) = 16$

21. $\int_1^2 \frac{4}{x^2}dx = \int_1^2 4x^{-2}\,dx = \Big[-4x^{-1}\Big]_1^2 = \left[\frac{-4}{x}\right]_1^2 = \left(\frac{-4}{2}\right) - \left(\frac{-4}{1}\right) = 2$

23. $\int_1^4 \frac{dt}{t\sqrt{t}} = \int_1^4 \frac{dt}{t^{3/2}} = \int_1^4 t^{-3/2}\,dt = \Big[-2t^{-1/2}\Big]_1^4 = \left[\frac{-2}{\sqrt{t}}\right]_1^4 = \frac{-2}{\sqrt{4}} - \frac{-2}{\sqrt{1}} = 1$

25. $\int_0^1 \frac{36\,dx}{(2x + 1)^3} = \int_1^3 18u^{-3}\,du = \left[\frac{18u^{-2}}{-2}\right]_1^3 = \left[\frac{-9}{u^2}\right]_1^3 = \left(\frac{-9}{3^2}\right) - \left(\frac{-9}{1^2}\right) = 8$

$u = 2x + 1 \Rightarrow du = 2dx \Rightarrow 18du = 36dx \quad x = 0 \Rightarrow u = 1,\ x = 1 \Rightarrow u = 3$

27. $\int_0^\pi \sin 5\theta\,d\theta = \left[-\frac{1}{5}\cos 5\theta\right]_0^\pi = -\frac{1}{5}\cos 5(\pi) - \left(-\frac{1}{5}\right)\cos 5(0) = \left(-\frac{1}{5}(-1)\right) - \left(-\frac{1}{5}(1)\right) = \frac{2}{5}$

29. $\int_0^{\pi/3} \sec^2\theta\,d\theta = \Big[\tan\theta\Big]_0^{\pi/3} = \tan\frac{\pi}{3} - \tan 0 = \sqrt{3}$

31. $\int_{\pi}^{3\pi} \cot^2 \frac{x}{6} dx = \int_{\pi/6}^{\pi/2} 6\cot^2 u\, du = \int_{\pi/6}^{\pi/2} (\csc^2 u - 1)du = \left[6(-\cot u - u)\right]_{\pi/6}^{\pi/2} = 6\left(-\cot \frac{\pi}{2} - \frac{\pi}{2}\right) -$

$6\left(-\cot \frac{\pi}{6} - \frac{\pi}{6}\right) = 6\sqrt{3} - 2\pi.$

$u = \frac{x}{6} \Rightarrow du = \frac{1}{6} dx \Rightarrow 6du = dx \quad x = \pi \Rightarrow u = \frac{\pi}{6},\ x = 3\pi \Rightarrow u = \frac{\pi}{2}$

33. $\int_{-\pi/3}^{0} \sec x \tan x\, dx = \left[\sec x\right]_{-\pi/3}^{0} = \sec 0 - \sec \frac{\pi}{3} = -1$

35. $\int_{0}^{\pi/2} 5(\sin x)^{3/2} \cos x\, dx = \int_{0}^{1} 5u^{3/2} du = \left[5\left(\frac{2}{5}\right)u^{5/2}\right]_{0}^{1} = \left[2u^{5/2}\right]_{0}^{1} = 2(1)^{5/2} - 2(0)^{5/2} = 2$

$u = \sin x \Rightarrow du = \cos x\, dx \quad x = 0 \Rightarrow u = 0,\ x = \frac{\pi}{2} \Rightarrow u = 1$

37. $\int_{4}^{8} \frac{1}{t} dt = \left[\ln t\right]_{4}^{8} = \ln 8 - \ln 4 = \ln \frac{8}{4} = \ln 2$

39. $\int_{0}^{2} \frac{x\, dx}{x^2 + 5} = \int_{5}^{9} \frac{1}{2}\left(\frac{1}{u}\right) du \left[\frac{1}{2} \ln u\right]_{5}^{9} = \frac{1}{2} \ln 9 - \frac{1}{2} \ln 5 = \frac{1}{2} \ln \frac{9}{5}$

$u = x^2 + 5 \Rightarrow du = 2x\, dx \Rightarrow \frac{1}{2} du = x\, dx$

$x = 0 \Rightarrow u = 5,\ x = 2 \Rightarrow u = 9$

41. $\int_{\ln 3}^{\ln 4} e^x dx = \left[e^x\right]_{\ln 3}^{\ln 4} = e^{\ln 4} - e^{\ln 3} = 4 - 3 = 1$

43. $\int_{0}^{\pi/4} e^{\tan x} \sec^2 x\, dx = \int_{0}^{1} e^u du = \left[e^u\right]_{0}^{1} = e^1 - e^0 = e - 1$

$u = \tan x \Rightarrow du = \sec^2 x\, dx$

$x = 0 \Rightarrow u = 0,\ x = \frac{\pi}{4} \Rightarrow u = 1$

45. $\int_2^3 \left(t-\frac{2}{t}\right)\left(t+\frac{2}{t}\right)dt = \int_2^3 \left(t^2-\frac{4}{t^2}\right)dt = \int_2^3 (t^2-4t^{-2})dt = \left[\frac{t^3}{3}+4t^{-1}\right]_2^3 =$

$\left[\frac{t^3}{3}+\frac{4}{t}\right]_2^3 = \left(\frac{3^3}{3}+\frac{4}{3}\right)-\left(\frac{2^3}{3}+\frac{4}{2}\right) = \left(9+\frac{4}{3}\right)-\left(\frac{8}{3}+2\right) = \frac{17}{3}$

47. $\int_{-4}^0 |x|\,dx = \int_{-4}^0 -x\,dx = \left[-\frac{x^2}{2}\right]_{-4}^0 = -\frac{0^2}{2}-\left(-\frac{(-4)^2}{2}\right) = 8$

49. $\int_{-\pi/2}^{\pi/2} 15\sin^4 3x\cos 3x\,dx = \int_1^{-1} 5u^4\,du = [u^5]_1^{-1} = (-1)^5-(1)^5 = -2$

$u = \sin 3x \Rightarrow du = 3\cos 3x\,dx \Rightarrow 5du = 15\cos 3x\,dx$

$x = -\frac{\pi}{2} \Rightarrow u = 1,\ x = \frac{\pi}{2} \Rightarrow u = -1$

51. $\int_0^{\pi/2} \frac{3\sin x\cos x}{\sqrt{1+3\sin^2 x}}dx = \int_1^4 \frac{1}{2}\left(\frac{1}{\sqrt{u}}\right)du = \int_1^4 \frac{1}{2}u^{-1/2}\,du = [u^{1/2}]_1^4 = 4^{1/2}-1^{1/2} = 1$

$u = 1+3\sin^2 x \Rightarrow du = 6\sin x\cos x\,dx \Rightarrow$

$\frac{1}{2}du = 3\sin x\cos x\,dx$

$x = 0 \Rightarrow u = 1,\ x = \frac{\pi}{2} \Rightarrow u = 4$

53. $\frac{dy}{dx} = \frac{1}{\sqrt{x}}\left(\frac{1}{2}\right)x^{-1/2} = \frac{1}{2x}$

55. $\frac{dy}{dx} = \frac{1}{3x^2+6}(6x) = \frac{6x}{3x^2+6} = \frac{2x}{x^2+2}$

57. $y = \frac{1}{e^x} = e^{-x} \Rightarrow \frac{dy}{dx} = -e^{-x}$

59. $y = e^{1+\ln x} = e(e^{\ln x}) = ex \Rightarrow \frac{dy}{dx} = e$

61. $\frac{d}{dx}\left(\frac{1}{2}(\ln 5x)^2+C\right) = \frac{1}{2}(2)(\ln 5x)\left(\frac{1}{5x}\right)(5) = \frac{\ln 5x}{x}$

63. $\frac{d}{dx}\left(\frac{e^x}{2}(\sin x-\cos x)+C\right) = \frac{e^x}{2}(\sin x-\cos x)+\frac{e^x}{2}(\cos x+\sin x) = \frac{e^x}{2}(2\sin x) = e^x\sin x$

65. a) $\ln e^{2x} = 2x$

b) $\ln 2e = \ln 2+\ln e = (\ln 2)+1$

c) $\ln\frac{1}{e} = \ln 1-\ln e = 0-1 = -1$

67. $\ln(y^2+y)-\ln y = x,\ y>0 \Rightarrow \ln\left(\frac{y^2+y}{y}\right) = x \Rightarrow \ln(y+1) = x \Rightarrow e^{\ln(y+1)} = e^x \Rightarrow y+1 = e^x \Rightarrow y = e^x-1$

69. $e^{2y} = 4x^2,\ x > 0 \Rightarrow \ln e^{2y} = \ln 4x^2 \Rightarrow 2y = \ln 4 + \ln x^2 \Rightarrow 2y = 2\ln 2 + 2\ln x \Rightarrow y = \ln 2 + \ln x$

71. Calculator Problem

73. a) Yes b) No c) Yes

75. $A = \int_0^x f(x)\,dx = \sin x$ on $[0,1] \Rightarrow f(x) = \cos x$

77. $\int_0^1 \sqrt{1 + x^4}\,dx = F(1) - F(0)$

79. $\frac{d^2s}{dt^2} = \pi^2 \cos \pi t \Rightarrow v(t) = \int \pi^2 \cos \pi t\,dt = \pi \sin \pi t + C_1$. $v(0) = 8 \Rightarrow \pi \sin \pi(0) + C_1 = 8 \Rightarrow C_1 = 8$

$\therefore v(t) = \pi \sin \pi t + 8$. Then $s(t) = \int (\pi \sin \pi t + 8)dt = -\cos \pi t + 8t + C_2$. $s(0) = 0 \Rightarrow$

$-\cos \pi(0) + 8(0) + C_2 = 0 \Rightarrow C_2 = 1$. $\therefore s(t) = -\cos \pi t + 8t + 1 \Rightarrow s(1) = -\cos \pi(1) + 8(1) + 1 = 10$

81. Step 1: $v(t) = \int \frac{d^2s}{dt^2}\,dt = \int -k\,dt = -kt + C$. $v(0) = 44 \Rightarrow 44 = -k(0) + C \Rightarrow C = 44$. $\therefore v(t) = -kt + 44$

$s(t) = \int \frac{ds}{dt}\,dt = \int (-kt + 44)dt = \frac{-kt^2}{2} + 44t + C$. $s(0) = 0 \Rightarrow 0 = \frac{-k(0)^2}{2} + 44(0) + C \Rightarrow C = 0$

$\therefore s(t) = -\frac{kt^2}{2} + 44t$

Step 2: $-kt^* + 44 = 0 \Rightarrow t^* = \frac{44}{k}$

Step 3: $\frac{-k\left(\frac{44}{k}\right)^2}{2} + 44\left(\frac{44}{k}\right) = 45 \Rightarrow -k\left(\frac{44^2}{k^2}\right) + 2\left(\frac{44^2}{k}\right) = 90 \Rightarrow -\left(\frac{44^2}{k}\right) + 2\left(\frac{44^2}{k}\right) = 90 \Rightarrow \frac{44^2}{k} = 90 \Rightarrow$

$90k = 44^2 \Rightarrow k = \frac{44^2}{90} \approx 21.5$ ft/sec$^2$

83. a) Average Value of $V^2 = \frac{1}{b - a} \int_a^b V_{max}^2 \sin^2 120\pi t\,dt$. Let $a = x$, $b = x + 1$. Then Average Value of

$$V^2 = \frac{1}{(x + 1) - x} \int_x^{x+1} V_{max}^2 \sin^2 120\pi t\,dt = \int_x^{x+1} V_{max}^2\left(\frac{1}{2} - \frac{1}{2}\cos 240\pi t\right)dt = \frac{1}{2} V_{max}^2 \int_x^{x+1} (1 - \cos 240\pi t)dt$$

$$= \frac{1}{2} V_{max}^2\left[t - \frac{1}{240\pi} \sin 240\pi t\right]_x^{x+1} = \frac{1}{2} V_{max}^2\left[\left((x + 1) - \frac{1}{240\pi} \sin 240\pi(x + 1)\right) - \left(x - \frac{1}{240\pi} \sin 240\pi x\right)\right] =$$

$$\frac{1}{2} V_{max}^2[(x + 1 - 0) - (x - 0)] = \frac{1}{2} V_{max}^2 \quad \therefore V_{rms} = \sqrt{(V^2)_{av}} = \sqrt{\frac{V_{max}^2}{2}} = \frac{|V_{max}|}{\sqrt{2}} = \frac{V_{max}}{\sqrt{2}}$$

b) $V_{max} = (240)\sqrt{2} \approx 339$ volts

85. $f'_{av} = \frac{1}{b - a} \int_a^b f'(x)\,dx = \frac{1}{b - a}\left[f(x)\right]_a^b = \frac{1}{b - a}(f(b) - f(a)) = \frac{f(b) - f(a)}{b - a}$

87. $|E_s| \le \frac{3-1}{180}(h)^4 M$ where $h = \frac{3-1}{n} = \frac{2}{n}$. $f(x) = \frac{1}{x} = x^{-1} \Rightarrow f'(x) = -x^{-2} \Rightarrow f''(x) = 2x^{-3} \Rightarrow f'''(x) = -6x^{-4}$

$\Rightarrow f^{(4)}(x) = 24x^{-5}$ which is decreasing on [1,3] $\Rightarrow$ maximum of $f^{(4)}(x)$ on [1,3] is $f^{(4)}(1) = 24 \Rightarrow M = 24$

For $|E_s| \le 0.0001$, $\frac{3-1}{180}\left(\frac{2}{n}\right)^4(24) \le 0.0001 \Rightarrow \frac{768}{180}\left(\frac{1}{n^4}\right) \le 0.0001 \Rightarrow \frac{1}{n^4} \le (0.0001)\frac{180}{768} \Rightarrow$

$n^4 \ge 10000\left(\frac{768}{180}\right) \Rightarrow n \ge 14.37 \Rightarrow n \ge 16$ (n must be even) $\Rightarrow h \le \frac{2}{16} \Rightarrow h \le \frac{1}{8}$

89. $h = \frac{b-a}{n} = \frac{\pi - 0}{6} = \frac{\pi}{6} \Rightarrow \frac{h}{2} = \frac{\pi}{12}$

| $x_i$ | $x_i$ | $f(x_i)$ | m | $mf(x_i)$ |
|---|---|---|---|---|
| $x_0$ | 0 | 0 | 1 | 0 |
| $x_1$ | $\pi/6$ | 1/2 | 2 | 1 |
| $x_2$ | $\pi/3$ | 3/2 | 2 | 3 |
| $x_3$ | $\pi/2$ | 2 | 2 | 4 |
| $x_4$ | $2\pi/3$ | 3/2 | 2 | 3 |
| $x_5$ | $5\pi/6$ | 1/2 | 2 | 1 |
| $x_6$ | $\pi$ | 0 | 1 | 0 |

$\sum_{i=0}^{6} mf(x_i) = 12$

$T = \frac{\pi}{12}(12) = \pi$

| $x_i$ | $x_i$ | $f(x_i)$ | m | $mf(x_i)$ |
|---|---|---|---|---|
| $x_0$ | 0 | 0 | 1 | 0 |
| $x_1$ | $\pi/6$ | 1/2 | 4 | 2 |
| $x_2$ | $\pi/3$ | 3/2 | 2 | 3 |
| $x_3$ | $\pi/2$ | 2 | 4 | 8 |
| $x_4$ | $2\pi/3$ | 3/2 | 2 | 3 |
| $x_5$ | $5\pi/6$ | 1/2 | 4 | 2 |
| $x_6$ | $\pi$ | 0 | 1 | 0 |

$\sum_{i=0}^{6} mf(x_i) = 18$, $\frac{h}{3} = \frac{\pi}{18}$

$\therefore S = \frac{\pi}{18}(18) = \pi$

91. Cost is given at $2.10/ft$^2$

| $x_i$ | $x_i$ | $f(x_i)$ | m | $mf(x_i)$ |
|---|---|---|---|---|
| $x_0$ | 0 | 0 | 1 | 0 |
| $x_1$ | 15 | 36 | 2 | 72 |
| $x_2$ | 30 | 54 | 2 | 108 |
| $x_3$ | 45 | 51 | 2 | 102 |
| $x_4$ | 60 | 49.5 | 2 | 99 |
| $x_5$ | 75 | 54 | 2 | 108 |
| $x_6$ | 90 | 64.4 | 2 | 128.8 |
| $x_7$ | 105 | 67.5 | 2 | 135 |
| $x_8$ | 120 | 42 | 1 | 42 |

$\sum_{i=0}^{8} mf(x_i) = 794.8$ $h = 15 \Rightarrow \frac{h}{2} = \frac{15}{2}$

$\therefore T = \frac{15}{2}(794.8) = 5961 \text{ ft}^2$

Total cost = \$2.10(Area) = \$2.10(5961) = \$12,518.10 (approximately)

The job cannot be done for \$11,000.

93. $L(x) = L_0e^{-kx} \Rightarrow \frac{1}{2}L_0 = L_0e^{-k(18)} \Rightarrow \frac{1}{2} = e^{-18k} \Rightarrow \ln\frac{1}{2} = \ln e^{-18k} \Rightarrow \ln\frac{1}{2} = -18k \Rightarrow k = \frac{\ln\frac{1}{2}}{-18} \approx 0.0385$

$\therefore L(x) = L_0e^{-0.0385x} \Rightarrow \frac{1}{10}L_0 = L_0e^{-0.0385x} \Rightarrow \frac{1}{10} = e^{-0.0385x} \Rightarrow \ln\frac{1}{10} = \ln e^{-0.0385x} \Rightarrow$

$\ln\frac{1}{10} = -0.0385x \Rightarrow x = \frac{\ln\frac{1}{10}}{-0.0385} \approx 59.8$ ft

95. a) CPI = 175.8 in 1980 $\Rightarrow p(t) = 175.8e^{kt}$. CPI = 211.9 in 1986 $\Rightarrow 211.9 = 175.8\, e^{k(6)} \Rightarrow$

$\frac{211.9}{175.8} = e^{6k} \Rightarrow \ln\left(\frac{211.9}{175.8}\right) = \ln e^{6k} \Rightarrow \ln\left(\frac{211.9}{175.8}\right) = 6k \Rightarrow k = \frac{\ln\left(\frac{211.9}{175.8}\right)}{6} \approx 0.0311$ or 3.11%

b) $p(t) = 175.8e^{0.0311t} \Rightarrow p(12) = 175.8e^{0.0311(12)} \approx 255.3$

97. $A_0e^{-kt} = \frac{1}{2}A_0 \Rightarrow e^{-kt} = \frac{1}{2} \Rightarrow \ln e^{-kt} = \ln\frac{1}{2} \Rightarrow -kt = \ln\frac{1}{2} \Rightarrow -kt = \ln 1 - \ln 2 \Rightarrow -kt = -\ln 2 \Rightarrow kt = \ln 2 \Rightarrow$

$t = \frac{\ln 2}{k}$

# CHAPTER 6

# APPLICATIONS OF DEFINITE INTEGRALS

## 6.1 AREAS BETWEEN CURVES

1. $A = \int_1^e \frac{1}{x} - \frac{-1}{x}\,dx = 2\int_1^e \frac{1}{x}\,dx = 2\,[\ln(x)\,]_1^e = 2$

3. $A = \int_1^{\ln 3} e^{2x} - e^x\,dx = \frac{e^{2x}}{2} - e^x \Big|_0^{\ln 3} = 2$

5. $A = \int_0^\pi 1 - \cos^2 x\,dx = \int_0^\pi \sin^2 x\,dx = \frac{1}{2}\int_0^\pi 1 - \cos 2x\,dx = \frac{1}{2}\left[x - \frac{\sin 2x}{2}\right]_0^\pi = \frac{\pi}{2}$

7. $A = 2\int_0^2 2 - (x^2 - 2)\,dx = 2\int_0^2 4 - x^2\,dx = 2\left[4x - \frac{x^3}{3}\right]_0^2 = \frac{32}{3}$

9. $A = 2\int_0^2 4 - y^2\,dy = \frac{32}{3}$, see question 7

11. $A = \int_0^1 x - x^2\,dx = \left[\frac{x^2}{2} - \frac{x^3}{3}\right]_0^1 = \frac{1}{6}$

13. $A = \int_{-1}^2 (y + 2) - y^2\,dy = \left[\frac{y^2}{2} + 2y - \frac{y^3}{3}\right]_{-1}^2 = \frac{9}{2}$

15. $A = \int_0^3 x - (x^2 - 2x)\,dx = \left[\frac{3x^2}{2} - \frac{x^3}{3}\right]_0^3 = \frac{9}{2}$

17. $A = 2\int_0^1 3 - 2y^2 - y^2\,dy = 6\left[y - \frac{y^3}{3}\right]_0^1 = 4$

19. $A = \int_1^2 2 - (x - 2)^2 - x\,dx = \left[-2x + \frac{3x^2}{2} - \frac{x^3}{3}\right]_1^2 = \frac{1}{6}$

21. $A = 2\int_0^1 (1 - x^2) - \cos(\pi x/2)\,dx = 2\left[x - \frac{x^3}{3} - \frac{2}{\pi}\sin(\pi x/2)\right]_0^1 = \frac{4}{3} - \frac{4}{\pi}$

23. $A = \int_0^1 x^{1/3} - x^{1/2}\,dx = \left[\frac{3x^{4/3}}{4} - \frac{2x^{3/2}}{3}\right]_0^1 = \frac{1}{12}$

25. $A = \int_0^1 12(y^2 - y^3)\,dy = 12\left[\frac{y^3}{3} - \frac{y^4}{4}\right]_0^1 = 1$

27. $A = \int_0^{\pi/4} \cos x - \sin x\,dx = [\sin x + \cos x]_0^{\pi/4} = \sqrt{2} - 1$

29. a) $\int_0^c \sqrt{y}\,dy = \frac{1}{2}\int_0^4 \sqrt{y}\,dy \Rightarrow \left[\frac{2y^{3/2}}{3}\right]_0^c = \frac{1}{2}\left[\frac{2y^{3/2}}{3}\right]_0^4 \Rightarrow \frac{2}{3}c^{3/2} = \frac{1}{2}\left[\frac{16}{3}\right] \Rightarrow c^{3/2} = 4 \Rightarrow c = 4^{2/3}$

b) $\int_0^{\sqrt{c}} c - x^2\,dx = \frac{1}{2}\int_0^2 4 - x^2\,dx \Rightarrow \left[cx - \frac{x^3}{3}\right]_0^{\sqrt{c}} = \frac{1}{2}\left[4x - \frac{x^3}{3}\right]_0^2 \Rightarrow \frac{2c^{3/2}}{3} = \frac{8}{3} \Rightarrow c = 4^{2/3}$

31. $A = \int_a^b 2f(x)\,dx - \int_a^b f(x)\,dx = 2\int_a^b f(x)\,dx - \int_a^b f(x)\,dx = \int_a^b f(x)\,dx = 4$

## 6 2 VOLUMES OF SOLIDS OF REVOLUTION--DISKS AND WASHERS

1. $V = \pi\int_0^2 (2-x)^2\,dx = \pi\left[4x - 2x^2 + \frac{x^3}{3}\right]_0^2 = \frac{8\pi}{3}$

3. $V = 2\pi\int_0^3 (\sqrt{9-x^2})^2\,dx = 2\pi\left[9x - \frac{x^3}{3}\right]_0^3 = 36\pi$

5. $V = \pi\int_0^2 (x^3)^2\,dx = \left[\frac{\pi x^7}{7}\right]_0^2 = \frac{128\pi}{7}$

7. $V = \pi\int_0^{\pi/2} (\sqrt{\cos x})^2\,dx = [\pi \sin x]_0^{\pi/2} = \pi$

9. $V = \pi\int_0^2 (2y)^2\,dy = \left[\frac{4\pi y^3}{3}\right]_0^2 = \frac{32\pi}{3}$

11. $V = 2\pi\int_0^1 (\sqrt{5}y^2)^2\,dy = \left[\frac{10\pi y^5}{5}\right]_0^1 = 2\pi$

13. $V = \pi\int_0^2 (y^{3/2})^2\,dy = \left[\frac{\pi y^4}{4}\right]_0^2 = 4\pi$

15. $V = \pi\int_0^3 \left(\frac{2}{\sqrt{y+1}}\right)^2 dy = 4\pi\int_0^3 \frac{1}{y+1}\,dy = 4\pi[\ln(y+1)]_0^3 = 4\pi\ln 4$

17. $V = \pi\int_0^1 1^2 - x^2\,dx = \pi\left[x - \frac{x^3}{3}\right]_0^1 = \frac{2\pi}{3}$

19. $V = \pi\int_0^2 4^2 - (x^2)^2\,dx = \pi\left[16x - \frac{x^5}{5}\right]_0^2 = \frac{128\pi}{5}$

21. $V = \pi\int_{-1}^2 (x+3)^2 - (x^2+1)^2\,dx = \pi\left[-\frac{x^5}{5} - \frac{x^3}{3} + 3x^2 + 8x\right]_{-1}^2 = \frac{117\pi}{5}$

23. $V = 2\pi\int_0^{\pi/4} (\sqrt{2})^2 - \sec^2 x\,dx = 2\pi[2x - \tan x]_0^{\pi/4} = \pi(\pi - 2)$

25. $V = \pi\int_0^1 (y+1)^2 - 1^2\,dy = \pi\left[\frac{y^3}{3} + y^2\right]_0^1 = \frac{4\pi}{3}$

27. $V = \pi\int_0^4 2^2 - (\sqrt{y})^2\,dy = \pi\left[4y - \frac{y^2}{2}\right]_0^4 = 8\pi$

29. $V = 2\pi\int_0^5 (\sqrt{25-y^2})^2\,dy = 2\pi\left[25y - \frac{y^3}{3}\right]_0^5 = \frac{500\pi}{3}$

31. $V = 2\pi\int_0^{\pi/2} 1^2 - \cos x\, dx = 2\pi[x - \sin x]_0^{\pi/2} = \pi^2 - 2\pi$

33. $V = \pi\int_0^1 1^2 - (\sqrt{1 - y^2})^2\, dy = \left[\frac{\pi y^3}{3}\right]_0^1 = \frac{\pi}{3}$

35. a) $V = \pi\int_0^4 2^2 - (\sqrt{x})^2\, dx = \pi\left[4x - \frac{x^2}{2}\right]_0^4 = 8\pi$

b) $V = \pi\int_0^2 (y^2)^2\, dy = \left[\frac{\pi y^5}{5}\right]_0^2 = \frac{32\pi}{5}$

c) $V = \pi\int_0^4 (2 - \sqrt{x})^2\, dx = \pi\left[4x - \frac{8}{3}x^{3/2} + \frac{x^2}{2}\right]_0^4 = \frac{8\pi}{3}$

d) $V = \pi\int_0^2 4^2 - (4 - y^2)^2\, dy = \pi\left[\frac{8y^3}{3} - \frac{y^5}{5}\right]_0^2 = \frac{224\pi}{15}$

37. a) $V = 2\pi\int_0^1 (1 - x^2)^2\, dx = 2\pi\left[x - \frac{2}{3}x^3 + \frac{1}{5}x^5\right]_0^1 = \frac{16\pi}{15}$

b) $V = 2\pi\int_0^1 (2 - x^2)^2 - 1^2\, dx = 2\pi\left[3x - \frac{4x^3}{3} + \frac{x^5}{5}\right]_0^1 = \frac{56\pi}{15}$

c) $V = 2\pi\int_0^1 2^2 - (1 + x^2)^2\, dx = 2\pi\left[3x - \frac{2x^3}{3} - \frac{x^5}{5}\right]_0^1 = \frac{64\pi}{15}$

39. $V = \pi\int_0^\pi (c - \sin x)^2\, dx = \pi\int_0^\pi c^2 - 2c\sin x + \sin^2 x\, dx = \pi\int_0^\pi c^2 - 2c\sin x + \frac{1 - \cos 2x}{2}\, dx =$

$\pi\left[c^2 x + 2c\cos x + \frac{x}{2} - \frac{\sin 2x}{4}\right]_0^\pi = \pi\left(c^2\pi - 4c + \frac{\pi}{2}\right)$. If $V(c) = \pi\left(c^2\pi - 4c + \frac{\pi}{2}\right)$, then $V'(c) =$

$\pi(2c\pi - 4) \Rightarrow$ that $c = \frac{2}{\pi}$ is the critical point. $V''(c) = 2\pi^2 > 0 \Rightarrow$ that $c = \frac{2}{\pi}$ minimizes the volume.

41. $V = \pi\int_{-16}^{-7} (\sqrt{256 - y^2})^2\, dy = \pi\left[256y - \frac{y^3}{3}\right]_{-16}^{-7} = 1053\pi\ \text{cm}^3$

## 6.3 CYLINDRICAL SHELLS--AN ALTERNATIVE TO WASHERS

1. $V = 2\pi\int_0^2 \left(x - \left(-\frac{x}{2}\right)\right)(x)\, dx = \left[\pi x^3\right]_0^2 = 8\pi$

3. $V = 2\pi\int_0^1 (x^2 + 1)x\, dx = 2\pi\left[\frac{x^4}{4} + \frac{x^2}{2}\right]_0^1 = \frac{3\pi}{2}$

5. $V = 2\pi\int_{1/2}^2 \left(\frac{1}{x}\right)(x)\, dx = [2\pi x]_{1/2}^2 = 3\pi$

7. $V = 2\pi\int_0^1 (2y)(y)\, dy = 4\pi\left[\frac{y^3}{3}\right]_0^1 = \frac{4\pi}{3}$

9. $V = 2\pi\int_0^2 \left((y+2) - y^2\right)y\,dy = 2\pi\left[\frac{y^3}{3} + y^2 - \frac{y^4}{4}\right]_0^2 = \frac{16\pi}{3}$

11. $V = 2\pi\int_0^2 (2y - y^2)y\,dy = 2\pi\left[\frac{2y^3}{3} - \frac{y^4}{4}\right]_0^2 = \frac{8\pi}{3}$

13. $V = 2\pi\int_0^{\sqrt{3}} x\sqrt{x^2+1}\,dx = \pi\int_0^{\sqrt{3}} (x^2+1)^{1/2}(2x)\,dx = \frac{2\pi}{3}\left[(x^2+1)^{3/2}\right]_0^{\sqrt{3}} = \frac{14\pi}{3}$

15. $V = 2\pi\int_0^1 12(y^2 - y^3)y\,dy = 24\pi\left[\frac{y^4}{4} - \frac{y^5}{5}\right]_0^1 = \frac{6\pi}{5}$

17. a) $V = 2\pi\int_1^2 y(y-1)\,dy = 2\pi\left[\frac{y^3}{3} - \frac{y^2}{2}\right]_1^2 = \frac{5\pi}{3}$

b) $V = 2\pi\int_1^2 x(2-x)\,dx = 2\pi\left[x^2 - \frac{x^3}{3}\right]_1^2 = \frac{4\pi}{3}$

19. a) $V = 2\pi\int_0^1 y\left(1 - (y - y^3)\right)dy = 2\pi\left[\frac{y^2}{2} - \frac{y^3}{3} + \frac{y^5}{5}\right]_0^1 = \frac{11\pi}{15}$

b) $V = \pi - \pi\int_0^1 (y - y^3)^2\,dy = \pi - \pi\left[\frac{y^3}{3} - \frac{2y^5}{5} + \frac{y^7}{7}\right]_0^1 = \pi - \frac{8\pi}{105} = \frac{97\pi}{105}$

c) $V = \pi\int_0^1 \left[1 - (y - y^3)\right]^2 dy = \pi\left[y - y^2 + \frac{y^4}{2} + \frac{y^3}{3} - \frac{2y^5}{5} + \frac{y^7}{7}\right]_0^1 = \frac{121\pi}{210}$

d) $V = 2\pi\int_0^1 (1-y)(1 - y + y^3)\,dy = 2\pi\left[y - y^2 + \frac{y^3}{3} + \frac{y^4}{4} - \frac{y^5}{5}\right]_0^1 = \frac{23\pi}{30}$

21. a) $V = \pi\int_0^2 (4x)^2 - (x^3)^2\,dx = \pi\left[\frac{16x^3}{3} - \frac{x^7}{7}\right]_0^2 = \frac{512\pi}{21}$

b) $V = \pi\int_0^2 (8 - x^3)^2 - (8 - 4x)^2\,dx = \pi\int_0^2 64x - 16x^2 - 16x^3 + x^6\,dx =$

$\pi\left[32x^2 - \frac{16x^3}{3} - 4x^4 + \frac{x^7}{7}\right]_0^2 = \frac{832\pi}{21}$

23. a) $V = 2\pi\int_0^1 \left((2x - x^2) - x\right)(x)\,dx = 2\pi\left[\frac{x^3}{3} - \frac{x^4}{4}\right]_0^1 = \frac{\pi}{6}$

b) $V = 2\pi\int_0^1 (1-x)\left((2x - x^2) - x\right)dx = 2\pi\left[\frac{x^2}{2} - \frac{2x^3}{3} + \frac{x^4}{4}\right]_0^1 = \frac{\pi}{6}$

## 6.4 LENGTHS OF CURVES IN THE PLANE

1. $L = \int_0^3 \sqrt{1 + \left(\frac{dy}{dx}\right)^2}\, dx = \int_0^3 \sqrt{1 + x^2(x^2 + 2)}\, dx = \int_0^3 \sqrt{(x^2 + 1)^2}\, dx = \frac{1}{3}\left[x^3 + x\right]_0^3 = 12.$

3. $L = \int_0^3 \sqrt{1 + \left(\frac{dx}{dy}\right)^2}\, dy = \int_0^3 \sqrt{1 + y}\, dy = \left[\frac{2(1 + y)^{3/2}}{3}\right]_0^3 = \frac{14}{3}.$

5. $y = \frac{x^3}{3} + \frac{1}{4x},\ 1 \le x \le 3 \Rightarrow \frac{dy}{dx} = y^2 - \frac{1}{4x^2} \Rightarrow 1 + \left(\frac{dy}{dx}\right)^2 = x^4 + \frac{1}{2} + \frac{1}{16x^4} = \left(x^2 + \frac{1}{4x^2}\right)^2.$

   $L = \int_1^3 \sqrt{1 + \left(\frac{dy}{dx}\right)^2}\, dx = \int_1^3 \left(x^2 + \frac{1}{4x^2}\right) dx = \left[\frac{x^3}{3} - \frac{1}{4x}\right]_1^3 = \frac{53}{6}.$

7. $x = \frac{y^4}{4} + \frac{1}{8y^2} \Rightarrow \frac{dx}{dy} = y^3 - \frac{1}{4y^3} \Rightarrow 1 + \left(\frac{dx}{dy}\right)^2 = 1 + y^6 - \frac{1}{2} + \frac{1}{16y^6} = y^6 + \frac{1}{2} + \frac{1}{16y^6} = \left(y^3 + \frac{1}{4y^3}\right)^2.$

   $L = \int_1^2 \sqrt{1 + \left(\frac{dx}{dy}\right)^2}\, dy = \int_1^2 \left|y^3 + \frac{1}{4y^3}\right| dy = \int_1^2 y^3 + \frac{1}{4y^3}\, dy = \left[\frac{y^4}{4} - \frac{1}{8y^2}\right]_1^2 = \frac{123}{32}.$

9. $y = \frac{e^x + e^{-x}}{2} \Rightarrow \frac{dy}{dx} = \frac{e^x - e^{-x}}{2} \Rightarrow 1 + \left(\frac{dy}{dx}\right)^2 = \left(\frac{e^x + e^{-x}}{2}\right)^2.\ L = \int_{-\ln 2}^{\ln 2} \sqrt{1 + \left(\frac{dy}{dx}\right)^2}\, dx =$

   $\frac{1}{2}\int_{\ln(1/2)}^{\ln 2} e^x + e^{-x}\, dx = \frac{1}{2}\left[e^x - e^{-x}\right]_{\ln(1/2)}^{\ln 2} = \frac{3}{2}.$

11. $x^{2/3} + y^{2/3} = 1 \Rightarrow \frac{dy}{dx} = -\frac{y^{1/3}}{x^{1/3}} = -\frac{(1 - x^{2/3})^{1/2}}{x^{1/3}} \Rightarrow 1 + \left(\frac{dy}{dx}\right)^2 = \frac{1}{x^{2/3}}.\ L = 4\int_0^1 \sqrt{1 + \left(\frac{dy}{dx}\right)^2}\, dx =$

   $4\int_0^1 x^{-1/3}\, dx = \left[6x^{2/3}\right]_0^1 = 6.$

13. $\frac{dy}{dx} = \sqrt{\frac{1}{4x}} = \frac{1}{2}x^{-1/2}$ at $(0,0) \Rightarrow y = x^{1/2} + C$ at $(0,0) \Rightarrow C = 0$ and $y = x^{1/2}$.

15. $\frac{dy}{dx} = \frac{1}{x} = x^{-1}$ at $(1,0) \Rightarrow y = \ln(x) + C$ at $(1,0) \Rightarrow C = 0 \Rightarrow y = \ln x$.

17. $y = \ln(1 - x^2) \Rightarrow \frac{dy}{dx} = \frac{-2x}{1 - x^2} \Rightarrow 1 + \left(\frac{dy}{dx}\right)^2 = \left(\frac{1 + x^2}{1 - x^2}\right)^2.\ L = \int_0^{1/2} \sqrt{1 + \left(\frac{dy}{dx}\right)^2}\, dx = \int_0^{1/2} \frac{1 + x^2}{1 - x^2}\, dx.$

19. Use the Calculus Tool Kit. $L = \int_0^{20} \sqrt{1 + \left(\frac{3\pi}{20}\cos\left(\frac{3\pi x}{20}\right)\right)^2}\, dx$ where $\frac{3\pi}{20} \approx 0.47124$.

   $L \approx S10 = 21.067928 \approx 21.07.$

## 6.5 AREAS OF SURFACES OF REVOLUTION

1. $S = 2\pi\int_0^4 \frac{x}{2}\sqrt{1+\left(\frac{1}{2}\right)^2}\,dx = \frac{\pi\sqrt{5}}{2}\int_0^4 x\,dx = \frac{\pi\sqrt{5}}{2}\left[\frac{x^2}{2}\right]_0^4 = 4\pi\sqrt{5}.$

3. $S = 2\pi\int_1^3 \frac{1}{2}(x+1)\sqrt{1+\left(\frac{1}{2}\right)^2}\,dx = \frac{\pi\sqrt{5}}{2}\int_1^3 (x+1)\,dx = \frac{\pi\sqrt{5}}{2}\left[\frac{x^2}{2}+x\right]_1^3 = 3\pi\sqrt{5}.$

5. $S = 2\pi\int_0^2 \frac{x^3}{9}\sqrt{1+\left(\frac{x^2}{3}\right)^2}\,dx = \frac{\pi}{54}\int_0^2 (9+x^4)^{1/2}(4x^3)\,dx = \frac{\pi}{81}\left[(9+x^4)^{3/2}\right]_0^2 = \frac{98\pi}{81}.$

7. $S = 2\pi\int_0^2 \sqrt{2x-x^2}\sqrt{1+\left(\frac{1-x}{\sqrt{2x-x^2}}\right)^2}\,dx = 2\pi\int_0^2 \sqrt{2x-x^2+1-2x+x^2}\,dx =$

$2\pi\int_0^2 dx = [2\pi x]_0^2 = 4\pi.$

9. $S = 2\pi\int_0^1 \frac{y^3}{3}\sqrt{1+(y^2)^2}\,dy = \frac{\pi}{9}\left[(1+y^4)^{3/2}\right]_0^1 = \frac{\pi(\sqrt{8}-1)}{9}.$

11. $S = 2\pi\int_{1/2}^1 \sqrt{2y-1}\sqrt{1+\frac{1}{2y-1}}\,dy = 2\pi\sqrt{2}\int_{1/2}^1 y^{1/2}\,dy = \frac{2^{5/2}\pi}{3}\left[y^{3/2}\right]_{1/2}^1 = \frac{2\pi(2\sqrt{2}-1)}{3}.$

13. $S = 2\pi\int_1^2 y\,ds = 2\pi\int_1^2 y\sqrt{dx^2+dy^2} = 2\pi\int_1^2 y\sqrt{(y^3-y^{-3}/4)^2+1}\,dy = 2\pi\int_1^2 y^4+\frac{1}{4}y^{-2}\,dy =$

$2\pi\left[\frac{y^5}{5}-\frac{y^{-1}}{4}\right]_1^2 = \frac{253\pi}{20}.$

15. $S = 4\pi\int_0^1 \sqrt{1-x^2}\sqrt{1+x^2(1-x^2)^{-1}}\,dx = 4\pi\int_0^1 dx = 4\pi.$

17. $\frac{S}{2} = 2\pi\int_0^1 y\sqrt{dx^2+dy^2} \Rightarrow S = 4\pi\int_0^1 y\sqrt{\frac{x^{2/3}}{y^{2/3}}+1}\,dy = 4\pi\int_0^1 y^{2/3}\,dy = \frac{12\pi}{5}\left[y^{5/3}\right]_0^1 = \frac{12\pi}{5}.$

19. $x^2+y^2=r^2 \Rightarrow x\,dx+y\,dy=0 \Rightarrow dy = -\frac{x}{y}dx \Rightarrow ds = \sqrt{dx^2+dy^2} = \sqrt{dx^2+\left(-\frac{x}{y}dx\right)^2} =$

$\sqrt{\frac{x^2+y^2}{y^2}}\,dx = \frac{r}{y}\,dx$. The crust of the indicated slice is $2\pi\int_a^{a+h} y\,ds = 2\pi\int_a^{a+h} y\frac{r}{y}\,dx =$

$2r\pi\int_a^{a+h} dx = 2\pi rh$ a constant. $\therefore$ equal widths imply equal amounts of crust.

## 6.6 WORK

1. $W = \int_0^{20} 40 - 2x\,dx = \left[40x - x^2\right]_0^{20} = 400$ ft · lb.

3. $W = \int_0^{50} (0.74)(50 - x)\,dx = (0.74)\left[50x - \frac{x^2}{2}\right]_0^{50} = 925$ N · m.

5. $W = \int_0^{180} 4(180 - x)\,dx = 4\left[180x - \frac{x^2}{2}\right]_0^{180} = 64800$ ft · lb.

7. $W = \int_0^{0.4} 15x\,dx = \left[\frac{15x^2}{2}\right]_0^{0.4} = 1.2$ N · m.

9. a) $W = \int_0^{1/2} 10000x\,dx = \left[\frac{10000x^2}{2}\right]_0^{1/2} = 1250$ in · lb $= 104\frac{1}{6}$ ft · lb.

   b) $W = \int_{1/2}^{1} 10000x\,dx = \left[\frac{1000x^2}{2}\right]_{1/2}^{1} = 3750$ in · lb $= 312\frac{1}{2}$ ft · lb.

11. a) $W = 62.5\int_0^{20} (20 - y)(10)(12)\,dy = 120(62.5)\int_0^{20} 20 - y\,dy =$

    $7500\left[20y - \frac{y^2}{2}\right]_0^{20} = 1500000$ ft · lb.

    b) $\frac{1500000}{250}$ sec = 6000 sec = 100 min = 1 hr and 40 min.

    c) $W = 7500\int_{10}^{20} 20 - y\,dy = 7500\left[20y - \frac{y^2}{2}\right]_{10}^{20} = 375000$ ft · lb, the work required to drop the

    water level 10 ft. The time needed to lower the water 10 feet is $\dfrac{37500 \text{ ft} \cdot \text{lb}}{\frac{60 \text{ sec}}{1 \text{ min}}\,\frac{250 \text{ ft} \cdot \text{lb}}{1 \text{ sec}}} = 25$ min.

13. $W = 62.5\int_0^{10} 25\pi y\,dy = 1562.5\pi\left[\frac{y^2}{2}\right]_0^{10} \approx 245436.93$ ft · lb.

15. $W = \omega\int_0^{30} (30 - y)\pi 10^2\,dy = 100\pi\omega\left[30y - \frac{y^2}{2}\right]_0^{30} = 45000\pi\omega \approx 7238229.48$ ft · lb.

17. a) $W = \frac{64.5\pi}{4}\int_0^{8} (10 - y)y^2\,dy = \frac{64.5\pi}{4}\left[\frac{10y^3}{3} - \frac{y^4}{4}\right]_0^{8} \approx 34582.65$ ft · lb.

    b) $W = \frac{57\pi}{4}\int_0^{8} (13 - y)y^2\,dy = \frac{57\pi}{4}\left[\frac{13y^3}{3} - \frac{y^4}{4}\right]_0^{8} \approx 53482.5$ ft · lb.

19. $W = \omega\int_0^{16} \pi(\sqrt{y})^2(16 - y)\,dy = \omega\pi\int_0^{16} 16y - y^2\,dy = \omega\pi\left[8y^2 - \frac{y^3}{3}\right]_0^{16} = \frac{2048\omega\pi}{3}$ N · m $\approx$

    21446605.85 N · m.

21. $W = \omega\int_0^{10}\left(\sqrt{100-y^2}\right)^2\pi(12-y)\,dy = \omega\pi\int_0^{10} 1200 - 100y - 12y^2 + y^3\,dy =$

$\omega\pi\left[1200y - 50y^2 - 4y^3 + \frac{y^4}{4}\right]_0^{10} = 5500\omega\pi \approx 967610.54$ ft · lb.

The cost is (967610.54)(0.005) ≈ \$4838.05.

23. Let a = 6370000 and b = 35780000. $\int_a^b \frac{mMG}{r^2}\,dr = mMG\int_a^b r^{-2}\,dr = -mMG\left[\frac{1}{r}\right]_a^b = mMG\left(\frac{b-a}{ab}\right) =$

$(1000)(5.975)(10^{24})(6.6720)(10^{-11})\left(\frac{29410000}{(35780000)(6370000)}\right) = 0.000005144 \times 10^{16} =$

$5.144 \times 10^{10}$ N · m.

## 6.7 FLUID PRESSURES AND FLUID FORCES

1. $F = \omega\int_0^3 (7-y)(2y)\,dy = 2\omega\left[\frac{7y^2}{2} - \frac{y^3}{3}\right]_0^3 = 45\omega = 2812.5$ lb.

3. $F = 2\omega\int_0^3\left(\frac{2y}{3}\right)(3-y)\,dy = \frac{4\omega}{3}\left[\frac{3y^2}{2} - \frac{y^3}{3}\right]_0^3 = 6\omega = 375$ lb. No, the length is of no concern.

5. $F = 2\omega\int_{-5}^{-1}\left(\frac{y+5}{2}\right)(-y)\,dy = -\omega\left[\frac{y^3}{3} + \frac{5y^2}{2}\right]_{-5}^{-1} = \frac{56\omega}{3} \approx 1166.67$ lb.

7. $F = 2\omega\int_{-1}^0 -y\sqrt{1-y^2}\,dy = \frac{2\omega}{3}\left[(1-y^2)^{3/2}\right]_{-1}^{0} = \frac{2\omega}{3} \approx 41.67$ lb.

9. $F = \omega\int_0^{33}(33.5-y)63\,dy = 63\omega\left[33.5y - \frac{y^2}{2}\right]_0^{33} = 35343\omega = \frac{2261952}{1728}$ lb = 1309 lb.

11. $F = 2\omega\int_{-3}^0 -y\sqrt{9-y^2}\,dy = \omega\int_{-3}^0 (9-y^2)^{(1/2)}(-2y)\,dy = \frac{2\omega}{3}\left[(9-y^2)^{3/2}\right]_{-3}^{0} = 18\omega = 1161$ lb.

13. Let h represent the depth of the water. $6667 \text{ lb} = 62.5\int_0^h (h-y)\left(\frac{2y}{5}\right)(2)\,dy = 50\int_0^h hy - y^2\,dy =$

$50\left[\frac{hy^2}{2} - \frac{y^3}{3}\right]_0^h = \frac{50h^3}{6} \Rightarrow h^3 = \frac{6(6667)}{50} \Rightarrow h \approx 9.28$ ft. The area of the end is $\left(\frac{1}{2}\right)\left(\frac{4h}{5}\right)(h) = \frac{2h^2}{5}$.

The volume of the water allowed in the tank is $\left(\frac{2h^2}{5}\right)(30) = 12h^2 \approx 1034$ ft$^3$.

15. Consider a rectangular plate of length l and width w. The length is parallel with the surface of the water. The force on one side of the plate is $\omega\int_{-w}^0 (-y)(l)\,dy = -\omega l\left[\frac{y^2}{2}\right]_{-w}^{0} = \frac{\omega l w^2}{2}$. The average force on one side of the plate is $\frac{\omega}{w}\int_{-w}^0 (-y)\,dy = \frac{\omega}{w}\left[-\frac{y^2}{2}\right]_{-w}^{0} = \frac{\omega w}{2}$. Notice that the force $\frac{\omega l w^2}{2} =$

$\left(\frac{\omega w}{2}\right)(lw)$ = (the average pressure up and down)(the area of the plate).

## 6.8 CENTERS OF MASS

1. $\overline{x} = \frac{100x + (5)(80)}{180} = 0 \Rightarrow x = -4$. The child is 4 ft from the fulcrum.

3. $M_o = \int_0^2 4x\,dx = \left[2x^2\right]_0^2 = 8$, $M = \int_0^2 4\,dx = [4x]_0^2 = 8 \Rightarrow \overline{x} = \frac{8}{8} = 1$.

5. $M_o = \int_0^4 x\left(1 + \frac{x}{4}\right)^2 dx = \int_0^4 x + \frac{x^2}{2} + \frac{x^3}{16}\,dx = \left[\frac{x^2}{2} + \frac{x^3}{6} + \frac{x^4}{64}\right]_0^4 = \frac{68}{3}$, $M = \int_0^4 \left(1 + \frac{x}{4}\right)^2 dx =$

$\int_0^4 1 + \frac{x}{2} + \frac{x^2}{16}\,dx = \left[x + \frac{x^2}{4} + \frac{x^3}{48}\right]_0^4 = \frac{28}{3} \Rightarrow \overline{x} = \frac{17}{7}$.

7. $M_x = \delta\int_0^2 y\left(\frac{2-y}{2} - \frac{y-2}{2}\right) dy = \delta\int_0^2 2y - y^2\,dy = \delta\left[y^2 - \frac{y^3}{3}\right]_0^2 = \frac{4\delta}{3}$, $M = 2\delta\int_{-1}^0 2x + 2\,dx =$

$2\delta\left[x^2 + 2x\right]_{-1}^0 = 2\delta \Rightarrow \overline{y} = \frac{2}{3}$, and by symmetry $\overline{x} = 0$.

9. $M_x = \delta\int_0^1 y(y - y^3)\,dy = \delta\left[\frac{y^3}{3} - \frac{y^5}{5}\right]_0^1 = \frac{2\delta}{15}$, $M_y = \delta\int_0^1 \frac{x}{2}(y - y^3)\,dy = \delta\int_0^1 \frac{(y - y^3)^2}{2}\,dy =$

$\frac{\delta}{2}\left[\frac{y^3}{3} - \frac{2y^5}{5} + \frac{y^7}{7}\right]_0^1 = \frac{4\delta}{105}$, $M = \delta\int_0^1 y - y^3\,dy = \delta\left[\frac{y^2}{2} - \frac{y^4}{4}\right]_0^1 = \frac{\delta}{4} \Rightarrow \overline{x} = \frac{16}{105}$ and $\overline{y} = \frac{8}{15}$.

11. $M_x = \delta\int_0^2 y(y - y^2 + y)\,dy = \delta\int_0^2 2y^2 - y^3\,dy = \delta\left[\frac{2y^3}{3} - \frac{y^4}{4}\right]_0^2 = \frac{4\delta}{3}$, $M_y = \frac{\delta}{2}\int_0^2 y^2 - (y^2 - y)^2\,dy =$

$\frac{\delta}{2}\int_0^2 2y^3 - y^4\,dy = \frac{\delta}{2}\left[\frac{y^4}{2} - \frac{y^5}{5}\right]_0^2 = \frac{4\delta}{5}$, $M = \delta\int_0^2 y - (y^2 - y)\,dy = \delta\left[y^2 - \frac{y^3}{3}\right]_0^2 = \frac{4\delta}{3} \Rightarrow$
$\overline{x} = \frac{3}{5}$ and $\overline{y} = 1$.

13. $M_x = \delta\int_{-\pi/2}^{\pi/2} \frac{\cos^2 x}{2}\,dx = \frac{\delta}{4}\int_{-\pi/2}^{\pi/2} 1 + \cos x\,dx = \frac{\delta}{4}\left[x + \frac{\sin 2x}{2}\right]_{-\pi/2}^{\pi/2} = \frac{\pi\delta}{4}$, $M = \delta\int_{-\pi/2}^{\pi/2} \cos x\,dx =$

$\delta[\sin x]_{-\pi/2}^{\pi/2} = 2\delta \Rightarrow \overline{y} = \frac{\pi}{8}$ and by symmetry $\overline{x} = 0$.

15. $M_x = \delta\int_0^2 \frac{(2x - x^2)^2 - (2x^2 - 4x)^2}{2}\,dx = \frac{\delta}{2}\int_0^2 -3x^4 + 12x^3 - 12x^2\,dx =$

$-\frac{3\delta}{2}\left[\frac{x^5}{5} - x^4 + \frac{4x^3}{3}\right]_0^2 = -\frac{8\delta}{5},\ M_y = \delta\int_0^2 x\left[(2x - x^2) - (2x^2 - 4x)\right]dx = -\delta\int_0^2 3x^3 - 6x^2\,dx =$

$-\delta\left[\frac{3x^4}{4} - 2x^3\right]_0^2 = 4\delta,\ M = \delta\int_0^2 (2x - x^2) - (2x^2 - 4x)\,dx = \delta\left[3x^2 - x^3\right]_0^2 = 4\delta \Rightarrow$

$\overline{x} = 1$ and $\overline{y} = -\frac{2}{5}$.

17. $M_y = \delta\int_0^3 x(3 - \sqrt{9 - x^2})\,dx = \delta\int_0^3 3x\,dx + \frac{\delta}{2}\int_0^3 \sqrt{9 - x^2}(-2x)\,dx =$

$\delta\left[\frac{3x^2}{2}\right]_0^3 + \frac{\delta}{3}\left[(9 - x^2)^{3/2}\right]_0^3 = \frac{9\delta}{2},\ M = \delta\int_0^3 3 - \sqrt{9 - x^2}\,dx = \delta\int_0^3 3\,dx - \delta\int_0^3 \sqrt{9 - x^2}\,dx$ (interpret

integral as area of a quarter circle) $= \delta[3x]_0^3 - \delta\left(\frac{9\pi}{4}\right) = \frac{\delta(36 - 9\pi)}{4} \Rightarrow$

$\overline{x} = \frac{2}{4 - \pi}$ and by symmetry $\overline{y} = \overline{x}$.

19. $\overline{y} = \frac{1}{3}(3) = 1$ and by symmetry $\overline{x} = 0$.

21. By symmetry $\overline{x} = \overline{y}$. The centroid is located at the intersection of medians: $y = x$ and $y = -\frac{x}{2} + \frac{a}{2}$.

$\therefore\ \overline{x} = \overline{y} = \frac{a}{3}$.

23. $$\overline{x} = \frac{M_y}{M} = \frac{\int_1^2 x\left(\frac{2}{x^2}\right)x^3\,dx}{\int_1^2 x^3\left(\frac{2}{x^2}\right)dx} = \frac{\left[\frac{2x^3}{3}\right]_1^2}{\left[x^2\right]_1^2} = \frac{14}{9},\ \overline{y} = \frac{M_x}{M} = \frac{\int_1^2 \frac{x^3}{2}\left(\frac{2}{x^2}\right)^2 dx}{\int_1^2 x^3\left(\frac{2}{x^2}\right)dx} = \frac{\left[2\ln x\right]_1^2}{\left[x^2\right]_1^2} = \frac{\ln 4}{3}.$$

25. The centroid of the square is located at (2,2). $V = (2\pi)(\overline{y})(A) = (2\pi)(2)(8) = 32\pi$, $S = (2\pi)(\overline{y})(L) = (2\pi)(20)(4\sqrt{8}) = 32\sqrt{2}\pi$.

27. The centroid is located at (2,0). $V = (2\pi)(\overline{y})(A) = (2\pi)(2)(\pi) = 4\pi^2$.

29. $S = 2\pi\overline{y}L \Rightarrow 4\pi a^2 = (2\pi\overline{y})(\pi a) \Rightarrow \overline{y} = \frac{2a}{\pi}$. By symmetry $\overline{x} = 0$.

31. $V = 2\pi\overline{y}A \Rightarrow \frac{4}{3}\pi a^3 = (2\pi\overline{y})\left(\frac{\pi a^2}{2}\right) \Rightarrow \overline{y} = \frac{4a}{3\pi}$. By symmetry $\overline{x} = 0$.

33. $V = 2\pi \bar{y} A$ = (area of the region)(distance from the centroid,$(0,4a/3\pi)$, to the line $y = x - a$)$(2\pi)$. We must find the distance from $(0,4a/3\pi)$ to $y = x - a$. The line containing the centroid and perpendicular to $y = x - a$ has a slope of $-1$ and contains the point $(0,4a/3\pi)$. This line is $y = -x + \frac{4a}{3\pi}$. The intersection of $y = -x - a$ and $y = -x + \frac{4a}{3\pi}$ is $\left(\frac{4a + 3a\pi}{6\pi}, \frac{4a - 3a\pi}{6\pi}\right)$. The distance from the centroid to the line $y = x - a$ is $\sqrt{\left(\frac{4a + 3a\pi}{6\pi}\right)^2 + \left(\frac{4a}{3\pi} - \frac{4a}{6\pi} + \frac{3a\pi}{6\pi}\right)^2} = \frac{\sqrt{2}(4a + 3a\pi)}{6\pi}$.

$\therefore V = \left(\frac{\pi a^2}{2}\right)\left(\frac{\sqrt{2}(4a + 3a\pi)}{6\pi}\right)(2\pi) = \frac{\sqrt{2}\pi a^3(4 + 3\pi)}{6}$.

## 6.9 THE BASIC IDEA. OTHER MODELING APPLICATIONS

1. Area of the cross section is $2x$. $V = \int_0^4 2x\,dx = \left[x^2\right]_0^4 = 16$.

3. Area of the cross section is $4(1 - x^2)$. $V = 2\int_0^1 4(1 - x^2)\,dx = 8\left[x - \frac{x^3}{3}\right]_0^1 = \frac{16}{3}$.

5. Area of the cross section is $\frac{\pi}{x}$. $V = \int_1^2 \frac{\pi}{x}\,dx = \pi[\ln x]_1^2 = \pi\ln 2$.

7. Area of the cross section is $\frac{1}{2}(2\sqrt{1 - y^2})(2\sqrt{1 - y^2}) = 2(1 - y^2)$. $V = 2\int_0^1 2(1 - y^2)\,dy =$ $4\left[y - \frac{y^3}{3}\right]_0^1 = \frac{8}{3}$.

9. $\int_0^h s^2\,dx = s^2h$. The volume does not depend on the number of revolutions.

11. a)

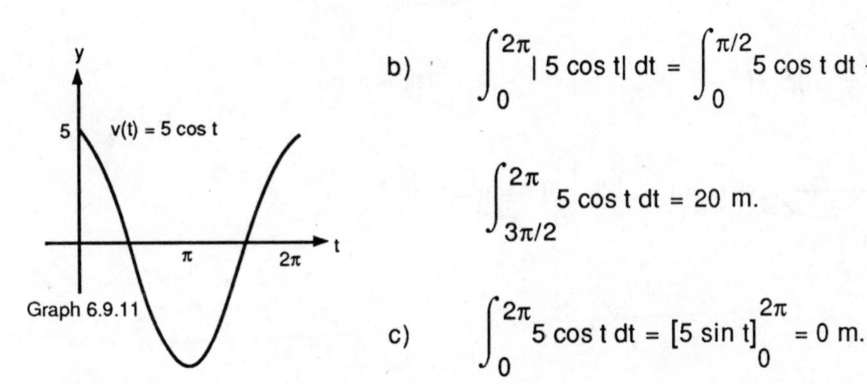

Graph 6.9.11

b) $\int_0^{2\pi} |5\cos t|\,dt = \int_0^{\pi/2} 5\cos t\,dt - \int_{\pi/2}^{3\pi/2} 5\cos t\,dt + \int_{3\pi/2}^{2\pi} 5\cos t\,dt = 20$ m.

c) $\int_0^{2\pi} 5\cos t\,dt = [5\sin t]_0^{2\pi} = 0$ m.

13. a)

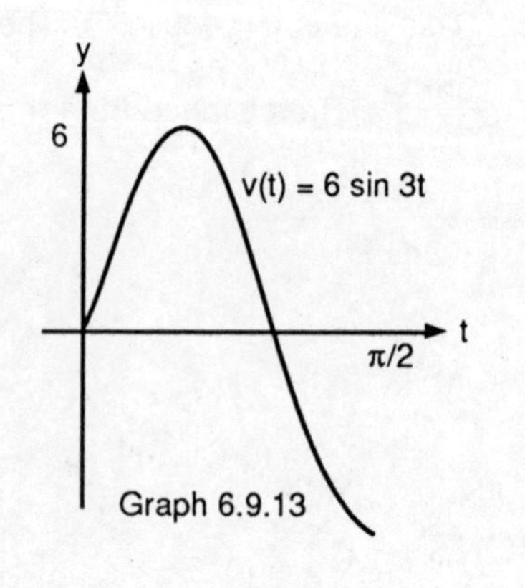

Graph 6.9.13

b) $\int_0^{\pi/2} |6 \sin 3t|\, dt = \int_0^{\pi/2} 6 \sin 3t\, dt - \int_{\pi/3}^{\pi/2} 6 \sin 3t\, dt = 6$ m.

c) $\int_0^{\pi/2} 6 \sin 3t\, dt = 2$ m.

15. a)

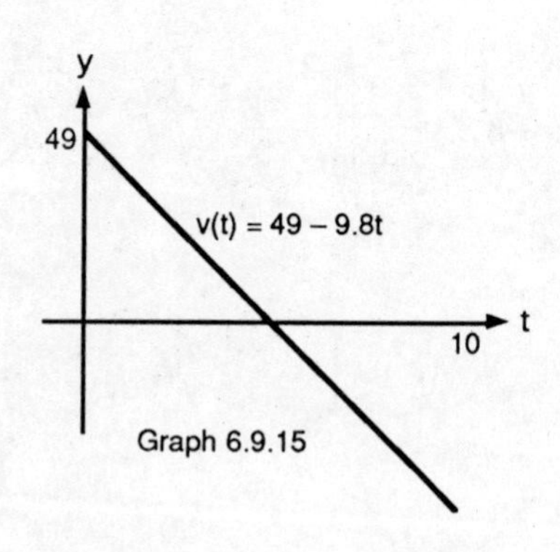

Graph 6.9.15

b) $\int_0^{10} |49 - 9.8t|\, dt = \int_0^{5} 49 - 9.8t\, dt -$

$\int_5^{10} 49 - 9.8t\, dt = 245$ m.

c) $\int_0^{10} 49 - 9.8t\, dt = 0$ m.

17. a)

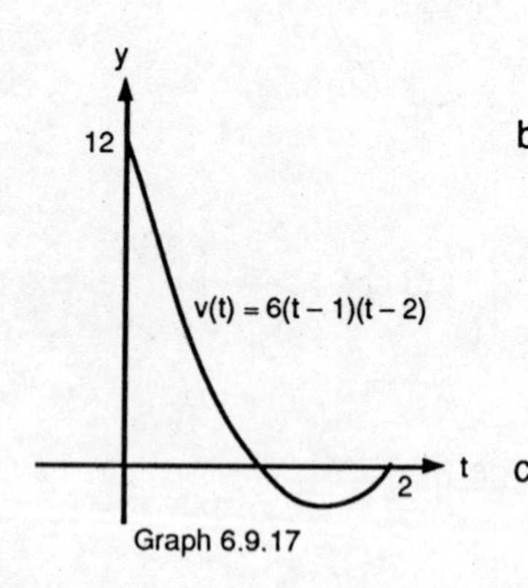

Graph 6.9.17

b) $6\int_0^{2} |t^2 - 3t + 2|\, dt = 6\int_0^{1} t^2 - 3t + 2\, dt -$

$6\int_1^{2} t^2 - 3t + 2\, dt = 6$ m.

c) $6\int_0^{2} t^2 - 3t + 2\, dt = 4$ m.

19. a) distance = 2, shift = 2 b) distance = 4, shift = 0

c) distance = 4, shift = 4 d) distance = 2, shift = 2

21. Out of 450 squares, about 136 are not shrimp-colored. Consequently, around 70% of the granular material are shrimp-colored.

23. $S = \int_0^{\sqrt{3}} 2\pi \frac{x}{\sqrt{3}}\, dx = \frac{2\pi}{\sqrt{3}}\left[\frac{x^2}{2}\right]_0^{\sqrt{3}} = \sqrt{3}\pi.$

## PRACTICE EXERCISES

1. $A = \int_1^2 x - \frac{1}{x^2}\, dx = \left[\frac{x^2}{2} + \frac{1}{x}\right]_1^2 = 1.$

2. $A = \int_1^2 x - x^{-1/2}\, dx = \left[\frac{x^2}{2} - 2x^{1/2}\right]_1^2 = \frac{7 - 4\sqrt{2}}{2}.$

3. $A = \int_{-2}^1 (3 - x^2) - (x + 1)\, dx = \left[-\frac{x^3}{3} - \frac{x^2}{2} + 2x\right]_{-2}^1 = \frac{9}{2}.$

5. $A = \int_0^3 2y^2\, dy = \left[\frac{2y^3}{3}\right]_0^3 = 18.$

7. $A = \int_{-1}^2 \left(\frac{y+2}{4}\right) - \left(\frac{y^2}{4}\right) dx = \left[\frac{y^2}{8} + \frac{y}{2} - \frac{y^3}{12}\right]_{-1}^2 = \frac{9}{8}.$

9. $A = \int_0^{\pi/4} (-\sin x + x)\, dx = \left[\cos x + \frac{x^2}{2}\right]_0^{\pi/4} = \frac{-32 + 16\sqrt{2} + \pi^2}{32} \approx 0.497.$

11. $A = \int_0^{\pi} (2\sin x - \sin 2x)\, dx = \left[-2\cos x + \frac{\cos 2x}{2}\right]_0^{\pi} = 4.$

13. $A = \int_1^2 \sqrt{y} - (2 - y)\, dy = \left[\frac{2y^{3/2}}{3} - 2y + \frac{y^2}{2}\right]_1^2 = \frac{8\sqrt{2} - 7}{2}.$

15. $A = \int_{-\ln 2}^{\ln 2} 2e^{-x}\, dx = -2\left[e^{-x}\right]_{-\ln 2}^{\ln 2} = 3.$

17. a) $V = 2\pi \int_0^1 (3x^4)^2\, dx = 18\pi\left[\frac{x^9}{9}\right]_0^1 = 2\pi.$

b) $V = 2\pi \int_0^1 x(3x^4)\, dx = 6\pi\left[\frac{x^6}{6}\right]_0^1 = \pi.$

19. a) $V = \pi \int_1^5 (\sqrt{x-1})^2\, dx = \pi\left[\frac{x^2}{2} - x\right]_1^5 = 8\pi.$

b) $V = 2\left(50\pi - \pi \int_0^2 (y^2+1)^2\, dy\right) = 100\pi - 2\pi \int_0^2 y^4 + 2y^2 + 1\, dy =$

$100\pi - 2\pi\left[\frac{y^5}{5} + \frac{2y^3}{3} + y\right]_0^2 = \frac{1088\pi}{15} \approx 227.87.$

c) $V = 2\pi \int_0^2 \left(5 - (y^2+1)\right)^2 dy = 2\pi\left[16y - \frac{8y^3}{3} + \frac{y^5}{5}\right]_0^2 = \frac{512\pi}{15} \approx 107.233.$

21. $V = \pi \int_0^{\pi/3} \tan^2 x\, dx = \pi \int_0^{\pi/3} \sec^2 x - 1\, dx = \pi[\tan x - x]_0^{\pi/3} = \frac{\pi(3\sqrt{3} - \pi)}{3}.$

23. $V = \pi \int_1^{16} \left(\frac{1}{\sqrt{x}}\right)^2 dx = \pi[\ln x]_1^{16} = \pi \ln 16.$

25. The volume of an end is $V = \pi \int_1^2 \left(\sqrt{4 - x^2}\right)^2 dx = \pi\left[4x - \frac{x^3}{3}\right]_1^2 = \frac{5\pi}{3}$. The volume of the cylinder is $2(3\pi)$. The total volume is $2\left(\frac{5\pi}{3}\right) + 6\pi = \frac{10\pi}{3} + 6\pi = \frac{28\pi}{3} \approx 29.32.$

27. $y = x^{1/2} - \frac{x^{3/2}}{3} \Rightarrow y' = \frac{1}{2}x^{-1/2} - \frac{1}{2}x^{1/2} \Rightarrow 1 + (y')^2 = \left(\frac{x^{-1/2} + x^{1/2}}{2}\right)^2$. $L = \int_0^3 \sqrt{1 + (y')^2}\, dx =$

$\frac{1}{2}\int_0^3 x^{-1/2} + x^{1/2}\, dx = \frac{1}{2}\left[2x^{1/2} + \frac{2}{3}x^{3/2}\right]_0^3 = 2\sqrt{3}.$

29. $y = \frac{x^2}{8} - \ln x \Rightarrow y' = \frac{x}{4} - \frac{1}{x} \Rightarrow 1 + (y')^2 = \frac{1}{16}\left(x + \frac{4}{x}\right)^2$. $L = \int_4^8 \sqrt{1 + (y')^2}\, dy =$

$\frac{1}{4}\int_4^8 x + \frac{4}{x}\, dy = \frac{1}{4}\left[\frac{x^2}{2} + 4\ln x\right]_4^8 = 6 + \ln 2.$

31. $y = \sqrt{2x+1} \Rightarrow 1 + (y')^2 = \frac{2x+2}{2x+1}$. $S = \int_0^{12} 2\pi\sqrt{2x+1}\,\frac{\sqrt{2x+2}}{\sqrt{2x+1}}\, dx = \pi \int_0^{12} (2x+2)^{1/2}(2)\, dx =$

$\frac{(2\pi)}{3}\left[(2x+2)^{3/2}\right]_0^{12} = \frac{2^{5/2}\pi(13^{3/2} - 1)}{3} \approx 271.739$

33. $y = \frac{x^{3/2}}{3} - x^{1/2} \Rightarrow 1 + (y')^2 = \frac{1}{4}\left(x^{1/2} + x^{-1/2}\right)^2$. $S = -2\pi \int_0^3 \left(\frac{x^{3/2}}{3} - x^{1/2}\right)\sqrt{\left(\frac{x^{1/2} + x^{-1/2}}{2}\right)^2}\, dx =$

$-\pi\left[\frac{x^3}{9} + \frac{x^2}{6} - \frac{x^2}{2} - x\right]_0^3 = 3\pi.$

35. $3y = (x^2+2)^{1/2} \Rightarrow dy = x\sqrt{x^2+2}\, dx$. $S = 2\pi \int_0^1 x\sqrt{dx^2 + \left(x\sqrt{x^2+2}\right)^2 dx^2} =$

$2\pi \int_0^1 x^3 + x\, dx = 2\pi\left[\frac{x^4}{4} + \frac{x^2}{2}\right]_0^1 = \frac{3\pi}{2}.$

37. Weight of the equipment is (10 kg)(9.8 m/sec$^2$) = 98 newtons. The work needed to pull up the equipment is (98 N)(40 m) = 3920 N · m. The work needed to pull up the rope is W = $\int_0^{40} (0.8)(40 - x)\,dx = 0.8\left[40x - \frac{x^2}{2}\right]_0^{40} = 640$ N · m. The total amount of work is

3920 N · m + 640 N · m = 4560 N · m.

39. The work it takes to stretch the spring 1 ft is W = $\int_0^1 20x\,dx = \left[10x^2\right]_0^1 = 10$ ft · lb. The work needed to stretch the spring an additional foot is W = $\int_1^2 20x\,dx = \left[10x^2\right]_1^2 = 30$ ft · lb.

41. $W = \omega\int_0^8 (14 - y)(\pi)\left(\frac{5y}{4}\right)^2 dx = \frac{25\omega\pi}{16}\left[\frac{14}{3}x^3 - \frac{1}{4}x^4\right]_0^8 = \frac{6400\omega\pi}{3} \approx 418879.02$ ft · lb.

43. $F = 2\omega\int_0^2 (2 - y)(2y)\,dy = 4\omega\left[y^2 - \frac{y^3}{3}\right]_0^2 = \frac{16\omega}{3} \approx 333.33$ lb.

45. $F = 2\omega\int_0^4 (9 - y)\left(\frac{\sqrt{y}}{2}\right)dy = \omega\left[6y^{3/2} - \frac{2}{5}y^{5/2}\right]_0^4 = \frac{176\omega}{5} = 2200$ lb.

47. $M_x = \int_{-1}^1 \left(\frac{3 - x^2 + 2x^2}{2}\right)(3 - x^2 - 2x^2)\,dx = \frac{3}{2}\left[3x - \frac{2x^3}{3} - \frac{x^5}{5}\right]_{-1}^1 = \frac{32}{5}$, $M = \int_{-1}^1 3 - 3x^2\,dx =$ $\left[3x - x^3\right]_{-1}^1 = 4$. $\overline{y} = \frac{M_x}{M} = \frac{8}{5}$, and by symmetry $\overline{x} = 0$.

49. $\overline{x} = \dfrac{\omega\int_0^4 x\left(4 - \frac{x^2}{4}\right)dx}{\omega\int_0^4\left(4 - \frac{x^2}{4}\right)dx} = \dfrac{\left[2x^2 - \frac{x^4}{16}\right]_0^4}{\left[4x - \frac{x^3}{12}\right]_0^4} = \dfrac{16}{16\left(\frac{2}{3}\right)} = \frac{3}{2}$. $\overline{y} = \dfrac{\omega\int_0^4 16 - \frac{x^4}{16}\,dx}{2\omega\int_0^4\left(4 - \frac{x^2}{4}\right)dx} = \dfrac{\left[16x - \frac{x^5}{80}\right]_0^4}{16\left(\frac{2}{3}\right)} = \frac{12}{5}$.

51. $\overline{y} = \dfrac{\omega\int_0^2 1^2 - 0^2\,dx + \omega\int_2^3 1^2 - (x - 2)^2\,dx}{2\omega\left[2 + \frac{1}{2}(1)(1)\right]} = \dfrac{2\omega + \omega\left[-3x - \frac{x^3}{3} + 2x^2\right]_2^3}{5\omega} = \frac{8}{15}$, and by symmetry $\overline{x} = 0$.

53. $\overline{x} = \dfrac{\omega\int_1^{16} x\left(\frac{1}{\sqrt{x}}\right)dx}{\omega\int_1^{16}\frac{1}{\sqrt{x}}\,dx} = \dfrac{\left[\frac{2}{3}x^{3/2}\right]_1^{16}}{\left[2x^{1/2}\right]_1^{16}} = \dfrac{\frac{126}{3}}{6} = 7$, $\overline{y} = \dfrac{\int_1^{16}\frac{1}{x}\,dx}{2\omega\int_1^{16}\frac{1}{\sqrt{x}}\,dx} = \dfrac{\left[\ln x\right]_1^{16}}{2 \cdot 6} = \frac{4\ln 2}{12} = \frac{\ln 2}{3}$.

55. $V = \frac{\pi}{4}\int_0^1 x - 2x^{5/2} + x^4\,dx = \frac{\pi}{4}\left[\frac{x^2}{2} - \frac{4}{7}x^{7/2} + \frac{x^5}{5}\right]_0^1 = \frac{9\pi}{280} \approx 0.10097$.

57. $V = \int_{\pi/4}^{5\pi/4} \pi(\sin x - \cos x)^2\,dx = \pi\int_{\pi/4}^{5\pi/4} 1 - 2\sin x\cos x\,dx = \pi\Big[x - \sin^2 x\Big]_{\pi/4}^{5\pi/4} = \pi^2.$

$dx = 2(u - \sqrt{6})du.$

59. $V = 2\int_0^3 \left(\frac{1}{2}(9 - y)^2\right) dy = \left[9y - \frac{y^3}{3}\right]_0^3 = 18.$

61. a)

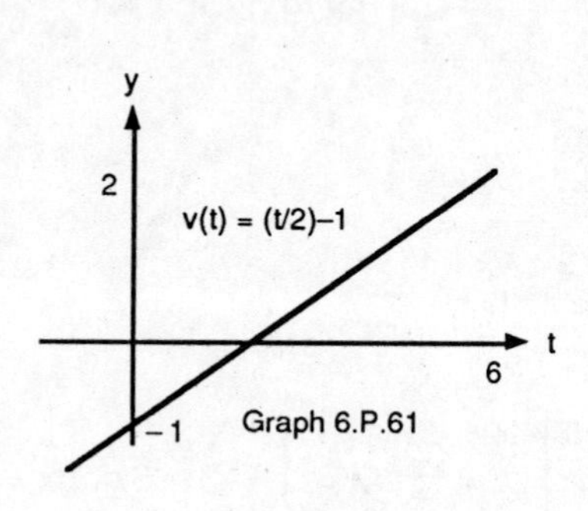

Graph 6.P.61

b) $\int_0^6 \left|\frac{t}{2} - 1\right| dt = \int_0^2 1 - \frac{t}{2}\,dt + \int_2^6 \frac{t}{2} - 1\,dt =$

$\left[t - \frac{t^2}{4}\right]_0^2 + \left[\frac{t^2}{4} - t\right]_2^6 = 5$ ft.

c) $\int_0^6 \frac{t}{2} - 1\,dt = \left[\frac{t^2}{4} - t\right]_0^6 = 3$ ft.

63. a)

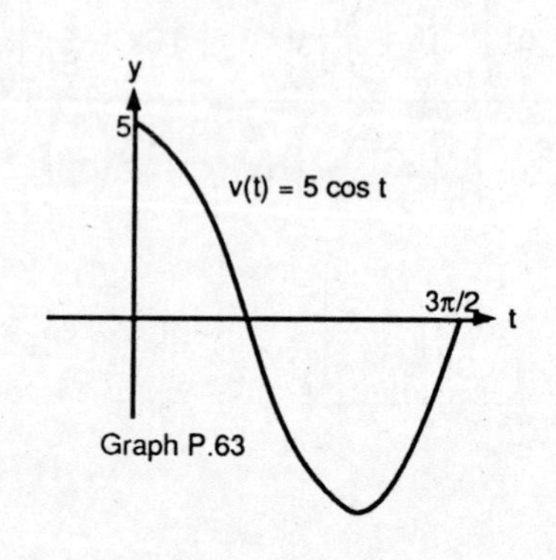

Graph P.63

b) $\int_0^{3\pi/2} |5\cos t|\,dt = \int_0^{\pi/2} 5\cos t\,dt - \int_{\pi/2}^{5\pi/2} 5\cos t\,dt =$

$[5\sin t]_0^{\pi/2} - [5\sin t]_2^{3\pi/2} = 15$ ft.

c) $\int_0^{3\pi/2} 5\cos t\,dt = [5\sin t]_0^{3\pi/2} = -5$ ft.

# CHAPTER 7

# CALCULUS OF TRANSCENDENTAL FUNCTIONS

## 7.1 INVERSE FUNCTIONS AND THEIR DERIVATIVES

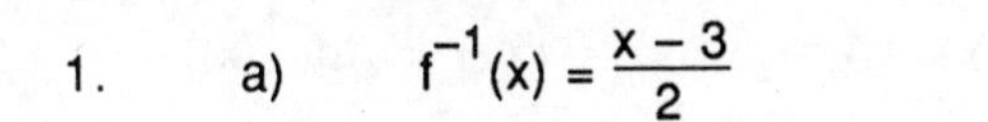

1. a) $f^{-1}(x) = \dfrac{x-3}{2}$

b)

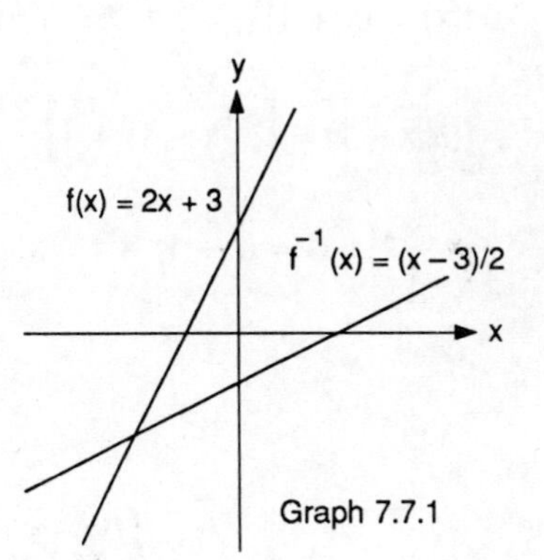

Graph 7.7.1

c) $\left.\dfrac{df}{dx}\right|_{x=-1} = 2\Big|_{x=-1} = 2$

$\left.\dfrac{df^{-1}}{dx}\right|_{x=1} = \dfrac{1}{2}\Big|_{x=1} = \dfrac{1}{2}$

3. a) $f^{-1}(x) = 5x - 35$

b)

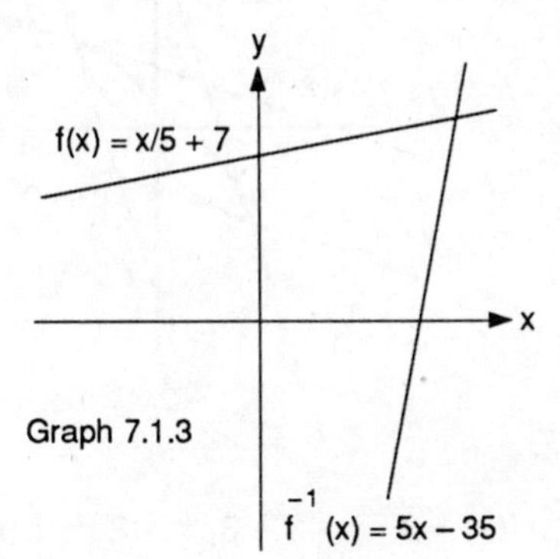

Graph 7.1.3

c) $\left.\dfrac{df}{dx}\right|_{x=-1} = \dfrac{1}{5}\Big|_{x=-1} = \dfrac{1}{5}$

$\left.\dfrac{df^{-1}}{dx}\right|_{x=34/5} = 5\Big|_{x=34/5} = 5$

5. $f(x) = x^2 + 1,\ x \geq 0 \Rightarrow y = x^2 + 1.$ $x = y^2 + 1 \Rightarrow y = \pm\sqrt{x-1}$, and (2,5) is on the graph of $y = f(x) \Rightarrow$ (5,2) is on the graph of $y = f^{-1}(x) \Rightarrow f^{-1}(x) = \sqrt{x-1}$.

7. $f(x) = x^3 - 1 \Rightarrow y = x^3 - 1.$ $x = y^3 - 1 \Rightarrow y = \sqrt[3]{x+1}$.

9. $f(x) = x^5 \Rightarrow y = x^5.$ $x = y^5 \Rightarrow y = \sqrt[5]{x} \Rightarrow f^{-1}(x) = \sqrt[5]{x}.$ $f(f^{-1}(x)) = f(\sqrt[5]{x}) = \left(\sqrt[5]{x}\right)^5 = x$ and $f^{-1}(f(x)) = f^{-1}(x^5) = \sqrt[5]{x^5}) = x$.

11. $f(x) = x^3 + 1 \Rightarrow y = x^3 + 1.$ $x = y^3 + 1 \Rightarrow y = \sqrt[3]{x-1} \Rightarrow f^{-1}(x) = \sqrt[3]{x-1}.$ $f(f^{-1}(x)) = f(\sqrt[3]{x-1}) = \left(\sqrt[3]{x-1}\right)^3 + 1 = x - 1 + 1 = x$ and $f^{-1}(f(x)) = f^{-1}(x^3 + 1) = \sqrt[3]{(x^3+1) - 1} = x$.

13. $f(x) = \frac{1}{x^2}, x > 0 \Rightarrow y = \frac{1}{x^2}.\ x = \frac{1}{y^2} \Rightarrow y = \frac{1}{\sqrt{x}} \Rightarrow f^{-1}(x) = \frac{1}{\sqrt{x}}, x > 0.\ f(f^{-1}(x)) = f\left(\frac{1}{\sqrt{x}}\right) = \frac{1}{\left(\frac{1}{\sqrt{x}}\right)^2} = \frac{1}{\frac{1}{x}} = x$

and $f^{-1}(f(x)) = f^{-1}\left(\frac{1}{x^2}\right) = \frac{1}{\sqrt{\frac{1}{x^2}}} = \frac{1}{\frac{1}{\sqrt{x^2}}} = \sqrt{x^2} = x, x > 0.$

15. $f(x) = (x + 1)^2, x \geq 1 \Rightarrow y = (x + 1)^2.\ x = (y + 1)^2 \Rightarrow y = \sqrt{x} - 1 \Rightarrow f^{-1}(x) = \sqrt{x} - 1.\ f(f^{-1}(x)) =$

$f(\sqrt{x} - 1) = \left[(\sqrt{x} - 1) + 1\right]^2 = \left[\sqrt{x}\right]^2 = x$ and $f^{-1}(f(x)) = f^{-1}\left((x + 1)^2\right) = \sqrt{(x + 1)^2} - 1 =$

$|x + 1| - 1 = (x + 1) - 1 = x,\ x \geq 1.$

17. a)

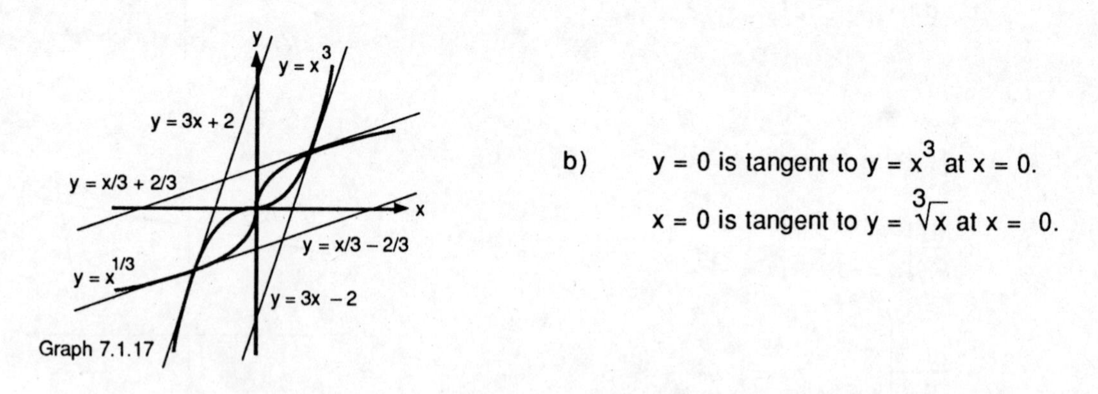


Graph 7.1.17

b) $y = 0$ is tangent to $y = x^3$ at $x = 0$.

$x = 0$ is tangent to $y = \sqrt[3]{x}$ at $x = 0$.

19. $f(x) = x^2 - 4x - 3, x > 2 \Rightarrow f'(x) = 2x - 4.\ \left.\frac{df^{-1}}{dx}\right|_{x=-3} = \left.\frac{1}{2x - 4}\right|_{x=4} = \frac{1}{4}.$

## 7.2 Ln x, $e^x$, AND LOGARITHMIC DIFFERENTIATION

1. $y = \sqrt{x(x+1)} = (x(x+1))^{1/2} \Rightarrow \ln y = \frac{1}{2}\ln(x(x + 1)) \Rightarrow 2\ln y = \ln(x) + \ln(x + 1) \Rightarrow \frac{2y'}{y} = \frac{1}{x} + \frac{1}{x + 1} \Rightarrow$

$y' = \left(\frac{1}{2}\right)\sqrt{x(x + 1)}\left(\frac{1}{x} + \frac{1}{x + 1}\right)$

3. $y = \sqrt{x + 3}\sin x = (x + 3)^{1/2}\sin x \Rightarrow \ln y = \left(\frac{1}{2}\right)\ln(x+3) + \ln(\sin x) \Rightarrow$

$y' = \sqrt{x + 3}\,(\sin x)\left(\frac{1}{2(x + 3)} + \cot x\right)$

5. $y = x(x + 1)(x + 2) \Rightarrow \ln y = \ln(x) + \ln(x + 1) + \ln(x + 2) \Rightarrow \frac{y'}{y} = \frac{1}{x} + \frac{1}{x + 1} + \frac{1}{x + 2} \Rightarrow$

$y' = x(x + 1)(x + 2)\left[\frac{1}{x} + \frac{1}{x + 1} + \frac{1}{x + 2}\right]$

7. $y = \dfrac{x+5}{x \cos x} \Rightarrow \ln y = \ln(x+5) - \ln(x) - \ln(\cos x) \Rightarrow \dfrac{y'}{y} = \dfrac{1}{x+5} - \dfrac{1}{x} + \dfrac{\sin x}{\cos x} \Rightarrow$

$y' = \dfrac{x+5}{x \cos x}\left[\dfrac{1}{x+5} - \dfrac{1}{x} + \tan x\right]$

9. $y = \dfrac{x\sqrt{x^2+1}}{(x+1)^{2/3}} \Rightarrow \ln y = \ln(x) + \dfrac{1}{2}\ln(x^2+1) - \dfrac{2}{3}\ln(x+1) \Rightarrow \dfrac{y'}{y} = \dfrac{1}{x} + \dfrac{x}{x^2+1} - \dfrac{2}{3(x+1)} \Rightarrow$

$y' = \dfrac{x\sqrt{x^2+1}}{(x+1)^{2/3}}\left[\dfrac{1}{x} + \dfrac{x}{x^2+1} - \dfrac{2}{3(x+1)}\right]$

11. $y = \sqrt[3]{\dfrac{x(x-2)}{x^2+1}} \Rightarrow \ln y = \dfrac{1}{3}\left[\ln(x) + \ln(x-2) - \ln(x^2+1)\right] \Rightarrow y' = \dfrac{1}{3}\sqrt[3]{\dfrac{x(x-2)}{x^2+1}}\left[\dfrac{1}{x} + \dfrac{1}{x-2} - \dfrac{2x}{x^2+1}\right]$

13. $\displaystyle\int_{-3}^{-2} \frac{1}{x}\,dx = [\ln|x|]_{-3}^{-2} = \ln(2) - \ln(3) = \ln\left(\frac{2}{3}\right)$

15. $\displaystyle\int_{-1}^{0} \frac{3}{3x-2}\,dx = [\ln|3x-2|]_{-1}^{0} = \ln\left(\frac{2}{5}\right)$

17. $\displaystyle\int_{3}^{4} \frac{1}{x-5}\,dx = [\ln|x-5|]_{3}^{4} = \ln(1) - \ln(2) = -\ln 2$

19. $\displaystyle\int_{0}^{3} \frac{2x}{x^2-25}\,dx = \left[\ln\left|x^2-25\right|\right]_{0}^{3} = \ln(16) - \ln(25) = \ln\left(\frac{16}{25}\right)$

21. $\displaystyle\int_{0}^{\pi} \frac{\sin x}{2-\cos x}\,dx = \left[\ln|2-\cos x|\right]_{0}^{\pi} = \ln(3) - \ln(1) = \ln 3$, or

$\displaystyle\int_{0}^{\pi} \frac{\sin x}{2-\cos x}\,dx = \int_{1}^{3} \frac{1}{u}\,du = [\ln u]_{1}^{3} = \ln 3$, where $u = 2 - \cos x$

23. $\displaystyle\int_{1}^{2} \frac{2\ln x}{x}\,dx = \int_{0}^{\ln 2} 2u\,du = \left[u^2\right]_{0}^{\ln 2} = (\ln 2)^2$, where $u = \ln x$

25. $\displaystyle\int_{-1}^{1} \frac{e^x}{1+e^x}\,dx = \int_{1+e^{-1}}^{1+e} \frac{1}{u}\,du = [\ln|u|]_{1+e^{-1}}^{1+e} = \ln|1+e| - \ln\left|1+e^{-1}\right| =$

$\ln|1+e| - \ln|1+e| + \ln e = 1$, where $u = 1 + e^x$

27. $\displaystyle\int_{0}^{\pi/2} \tan\left(\frac{x}{2}\right)dx = -2\int_{0}^{\pi/2} \frac{\left(-\frac{1}{2}\right)\sin\left(\frac{x}{2}\right)}{\cos\left(\frac{x}{2}\right)}\,dx = -2\left[\ln\left|\cos\left(\frac{x}{2}\right)\right|\right]_{0}^{\pi/2} = \ln 2$

29. $\displaystyle\int_{\pi/2}^{\pi} 2\cot\left(\frac{x}{3}\right)dx = 6\int_{\pi/2}^{\pi} \frac{\left(\frac{1}{3}\right)\cos\left(\frac{x}{3}\right)}{\sin\left(\frac{x}{3}\right)}\,dx = 6\left[\ln\left|\sin\left(\frac{x}{3}\right)\right|\right]_{\pi/2}^{\pi} = \ln 27$

31. $\displaystyle\lim_{x\to\infty} \ln\left(\frac{1}{x}\right) = -\infty$

33. $\displaystyle\lim_{x\to\infty} e^{-x} = \lim_{x\to\infty} \frac{1}{e^x} = 0$

35. $\displaystyle\lim_{x\to-\infty} \ln\left(2+e^x\right) = \ln\left[\lim_{x\to-\infty} \left(2+e^x\right)\right] = \ln 2$

37. $A = \int_0^{\pi/3} \tan x \, dx = -\int_0^{\pi/3} \frac{-\sin x}{\cos x} dx = [\ln|\sec x|]_0^{\pi/3} = \ln 2$

39. $f(x) = e^x \Rightarrow f'(x) = e^x \Rightarrow f'(0) = 1$, at (0,1) the linearization is $y - 1 = f'(0)(x - 0) \Rightarrow L(x) = x + 1$

41. $f(x) = \ln(x+1) \Rightarrow f'(x) = \frac{1}{x+1} \Rightarrow f'(0) = 1$, at (0,0) the linearization is $y - 0 = f'(0)(x - 0) \Rightarrow L(x) = x$

43. $f(x) = x^2 \ln\left(\frac{1}{x}\right), x > 0 \Rightarrow f'(x) = 2x\ln\left(\frac{1}{x}\right) + x^2(x)\left(-\frac{1}{x^2}\right) = x\left[2\ln\left(\frac{1}{x}\right) - 1\right]$. The critical points are 0 and $e^{-1/2}$, but 0 is not in the domain. Since $f'(x) > 0$ for $0 < x < e^{-1/2}$ and $f'(x) < 0$ when $e^{-1/2} < x$ there is a maximum at $x = e^{-1/2}$ and it is $\frac{1}{2e}$.

45. $\frac{dy}{dx} = (\cos x)\, e^{(\sin x)}$ at $(0,0) \Rightarrow y = e^{(\sin x)} + C$ at $(0,0) \Rightarrow C = -1 \Rightarrow y = e^{(\sin x)} - 1$

47. $\frac{d^2s}{dt^2} = 4(4-t)^{-2}$ and $v(0) = 2 \Rightarrow \frac{ds}{dt} = 4(4-t)^{-1} + C$, at $(0,2) \Rightarrow C = 1 \Rightarrow v(t) = 4(4-t)^{-1} + 1$.

The total distance $= \int_1^2 \left|4(4-t)^{-1} + 1\right| dt = \int_1^2 \frac{4}{4-t} + 1 \, dt = [-4\ln|4-t| + t]_1^2 = \ln\left(\frac{81}{16}\right) + 1$.

49. $f(x) = \ln(x) - 1 \Rightarrow f'(x) = \frac{1}{x} \Rightarrow x_{n+1} = x_n - \frac{\ln(x_n) - 1}{\frac{1}{x_n}} \Rightarrow x_{n+1} = x_n[2 - \ln(x_n)]$. When $x_1 = 2 \Rightarrow$ $x_2 = 2.61370564$, $x_3 = 2.71624393$ and $x_5 = 2.71828183$.

51. $e^x e^{-x} = e^{(x-x)} = e^0 = 1 \Rightarrow e^{-x} = \frac{1}{e^x}$ for all x

## 7.3 THE OTHER EXPONENTIAL AND LOGARITHMIC FUNCTIONS

1. $y = x^{\pi} \Rightarrow \frac{dy}{dx} = \pi x^{(\pi - 1)}$

3. $y = x^{-\sqrt{2}} \Rightarrow \frac{dy}{dx} = -\sqrt{2}\, x^{-\sqrt{2} - 1}$

5. $y = 2^x \Rightarrow \ln y = (\ln 2)x \Rightarrow \frac{y'}{y} = \ln 2 \Rightarrow \frac{dy}{dx} = 2^x \ln 2$

7. $y = 2^{-x} \Rightarrow \ln y = (-\ln 2)x \Rightarrow \frac{dy}{dx} = -2^{-x}(\ln 2)$

9. $y = 2^{\sec x} \Rightarrow \ln y = (\ln 2)\sec x \Rightarrow \frac{y'}{y} = (\ln 2)(\sec x)(\tan x) \Rightarrow \frac{dy}{dx} = 2^{\sec x}(\ln 2)(\sec x)(\tan x)$

11. $y = x^{\ln x}, x > 0 \Rightarrow \ln y = (\ln x)^2 \Rightarrow \frac{y'}{y} = 2(\ln x)\left(\frac{1}{x}\right) \Rightarrow \frac{dy}{dx} = \left(x^{\ln x}\right)\left(\frac{\ln x^2}{x}\right)$

13. $y = (\sin x)^{(\tan x)} \Rightarrow \ln y = (\tan x)(\ln(\sin x)) \Rightarrow \frac{y'}{y} = \left(\sec^2 x\right)(\ln \sin x) + (\tan x)\left(\frac{1}{\sin x}(\cos x)\right) \Rightarrow$ $\frac{dy}{dx} = (\sin x)^{(\tan x)}\left((\sec^2 x)(\ln \sin x) + 1\right)$

15. $\int_0^1 3x^{\sqrt{3}} dx = \left[\frac{3\, x^{\sqrt{x} + 1}}{\sqrt{3} + 1}\right]_0^1 = \frac{3}{\sqrt{3} + 1}$

17. $\displaystyle\int_0^3 (\sqrt{2}+1)\,x^{\sqrt{2}}\,dx = \Big[x^{\sqrt{2}+1}\Big]_0^3 = 3^{\sqrt{2}+1}$

19. $\displaystyle\int_0^1 5^x\,dx = \int_0^1 e^{(\ln 5)x}\,dx = \left(\frac{1}{\ln 5}\right)\int_0^1 e^{(\ln 5)x}(\ln 5)\,dx = \left(\frac{1}{\ln 5}\right)\Big[e^{(\ln 5)x}\Big]_0^1 = \frac{4}{\ln 5}$

21. $\displaystyle\int_0^1 \frac{1}{2^x}\,dx = -\int_0^{-1} 2^u\,du = \int_{-1}^0 2^u\,du = \frac{1}{\ln 4}$, where $u = -x$. See question 20 for evaluation of the last integral.

23. $\displaystyle\int_{-1}^0 4^{-x}\ln 2\,dx = \int_{-1}^0 2^{-2x}\ln 2\,dx = -\frac{1}{2}\int_{-1}^0 e^{(-2(\ln 2))x}(-2\ln 2)\,dx = -\frac{1}{2}\Big[e^{\ln 2^{-2x}}\Big]_{-1}^0 = \frac{3}{2}$

25. $\displaystyle\int_1^{\sqrt{2}} x\,2^{x^2}\,dx = \frac{1}{2\ln 2}\int_1^{\sqrt{2}} e^{(\ln 2)x^2}(2\ln 2)x\,dx = \frac{1}{\ln 4}\Big[2^{x^2}\Big]_1^{\sqrt{2}} = \frac{1}{\ln 2}$

27. $\log_4 16 = \dfrac{\ln 16}{\ln 4} = \dfrac{2\ln 4}{\ln 4} = 2$

29. $\log_5 0.04 = \log_5 \dfrac{1}{25} = -5$

31. $\log_2 4 = \dfrac{2\ln 2}{\ln 2} = 2$

33. $\dfrac{\log_2 x}{\log_3 x} = \dfrac{\ln x}{\ln 2} \div \dfrac{\ln x}{\ln 3} = \dfrac{\ln 3}{\ln 2}$

35. $\dfrac{\log_9 x}{\log_3 x} = \dfrac{\ln x}{2\ln 3} \div \dfrac{\ln x}{\ln 3} = \dfrac{1}{2}$

37. $3^{\log_3 7} + 2^{\log_2 5} = 5^{\log_5 x} \Rightarrow 7 + 5 = x \Rightarrow x = 12$

39. $y = \log_4 x = \dfrac{\ln x}{\ln 4} \Rightarrow \dfrac{dy}{dx} = \dfrac{1}{(\ln 4)x}$

41. $y = \log_{10} e^x = \dfrac{x}{\ln 10} \Rightarrow \dfrac{dy}{dx} = \dfrac{1}{\ln 10}$

43. $y = (\ln 2)(\log_2 2) = \dfrac{(\ln 2)(\ln x)}{\ln 2} \Rightarrow \dfrac{dy}{dx} = \dfrac{1}{x}$

45. $y = \log_{10}\sqrt{x+1} = \dfrac{\ln(x+1)}{2\ln 10} \Rightarrow \dfrac{dy}{dx} = \dfrac{1}{(\ln 100)(x+1)}$

47. $y = \dfrac{1}{\log_2 x} = (\ln 2)(\ln x)^{-1} \Rightarrow \dfrac{dy}{dx} = -(\ln 2)(\ln x)^{-2}\left(\dfrac{1}{x}\right) = \dfrac{-\ln 2}{x(\ln x)^2}$

49. $y = \log_5 (x+1)^2 = \dfrac{2\ln(x+1)}{\ln 5} \Rightarrow \dfrac{dy}{dx} = \dfrac{2}{(\ln 5)(x+1)}$

51. $\displaystyle\int_1^{10} \frac{\log_{10} x}{x}\,dx = \frac{1}{\ln 10}\int_1^{10} (\ln x)\left(\frac{1}{x}\right)dx = \frac{1}{\ln 10}\left[\frac{(\ln x)^2}{2}\right]_1^{10} = \frac{\ln 10}{2}$

53. $\displaystyle\int_1^4 \frac{(\ln 2)(\log_2 x)}{x}\,dx = \int_1^4 \ln x\left(\frac{1}{x}\right)dx = \left[\frac{(\ln x)^2}{2}\right]_1^4 = 2(\ln 2)^2$

55. $\displaystyle\int_0^2 \frac{\log_2 (x+2)}{x+2}\,dx = \frac{1}{\ln 2}\int_0^2 \ln(x+2)\left(\frac{1}{x+2}\right)dx = \frac{1}{\ln 2}\left[\frac{\ln(x+2)^2}{2}\right]_0^2 = \frac{3\ln 2}{2}$

57. $\displaystyle\int_0^9 \frac{2\log_{10}(x+1)}{x+1}\,dx = \frac{2}{\ln 10}\int_0^9 \ln(x+10)\left(\frac{1}{x+1}\right)dx = \frac{2}{\ln 10}\left[\frac{(\ln x+1)^2}{2}\right]_0^9 = \ln 10$

59. a) $\displaystyle\lim_{x\to\infty} \log_2 x = \lim_{x\to\infty} \frac{\ln x}{\ln 2} = \infty$ b) $\displaystyle\lim_{x\to\infty} \log_2 x^{-1} = \lim_{x\to\infty} -\frac{\ln x}{\ln 2} = -\infty$

61. a) $\displaystyle\lim_{x\to\infty} 3^x = \left(\lim_{x\to\infty} e^x\right)^{\ln 3} = \infty$ b) $\displaystyle\lim_{x\to\infty} 3^{-x} = \left(\lim_{x\to\infty} e^{-x}\right)^{\ln 3} = 0$

63. a) $y = 2^{\ln x} \Rightarrow \ln y = (\ln 2)(\ln x) \Rightarrow \frac{y'}{y} = (\ln 2)\left(\frac{1}{x}\right) \Rightarrow \frac{dy}{dx} = 2^{\ln x}\left(\frac{\ln 2}{x}\right)$

b) $y = \ln\left(2^x\right) = (\ln 2)x \Rightarrow \frac{dy}{dx} = \ln 2$ c) $y = \ln\left(x^2\right) = 2\ln x \Rightarrow \frac{dy}{dx} = \frac{2}{x}$

d) $y = (\ln x)^2 \Rightarrow \frac{dy}{dx} = 2(\ln x)\left(\frac{1}{x}\right) = \frac{2\ln x}{x}$

65. Let $\left[H_3O^+\right] = x$. Solve $7.37 = -\log_{10} x$ and $7.44 = -\log_{10} x$. The solutions of these equations are $10^{-7.37}$ and $10^{-7.44}$. Consequently, the bounds for $\left[H_3O^+\right]$ are $\left[10^{-7.44}, 10^{-7.37}\right]$.

67. Let O = original sound level = $10\log_{10}(I \times 10^{12})$ db. Solving $O + 10 = 10\log_{10}(kI \times 10^{12})$ for $k \Rightarrow$ $10\log_{10}(I \times 10^{12}) + 10 = 10\log_{10}(kI \times 10^{12}) \Rightarrow 1 = \log_{10} k \Rightarrow k = 10$.

69. $\log_b x = \frac{\ln x}{\ln b} = \left(\frac{\ln x}{\ln b}\right)\left(\frac{\ln a}{\ln a}\right) = \left(\frac{\ln a}{\ln b}\right)\left(\frac{\ln x}{\ln a}\right) = \frac{\ln a}{\ln b}\log_a x$.

## 7.4 GROWTH AND DECAY

1. a) $k = \frac{\ln .99}{1000} \approx -0.00001$

b) $0.9 = e^{kt} \Rightarrow kt = \ln(0.9) \Rightarrow t = \frac{\ln(0.9)}{k} \approx 10536$ yr

c) $y = y_o e^{(20000)k} \approx y_o(0.82) \Rightarrow 82\%$

3. $\frac{dy}{dt} = -0.6y \Rightarrow y = y_o e^{-0.6t}$. If $y_o = 100$, $t = 1$ and $y = y_o e^{-0.6t}$, then $y = 100e^{-0.6} \approx 54.88$ grams.

5. $y = y_o e^{-0.18t}$ represents the decay equation; solving $(0.9)y_o = y_o e^{-0.18t} \Rightarrow t = \frac{\ln(0.9)}{-0.18} \approx 0.585$ days

7. a) $\frac{v_o m}{k} = \frac{(22)(5)}{1/5} = (22)(25) = 550$ ft

b) $1 = 22e^{-t/25} \Rightarrow \ln 1 = \ln(22) - \frac{t}{25} \Rightarrow t = 25\ln 22 \approx 77.28$ sec

9. $T - T_s = (T_o - T_s)e^{-kt}$, $T_o = 90°C$, $T_s = 20°C \Rightarrow 60 - 20 = 70e^{-10k} \Rightarrow k = \frac{\ln(7/4)}{10} \approx 0.05596$

a) $35 - 20 = 70e^{-0.05596t} \Rightarrow t \approx 27.5$ min the total time; it will take $27.5 - 10 = 17.5$ min to reach to 35°C

b) $T - T_s = (T_o - T_s)e^{-kt}$, $T_o = 90°C$, $T_s = -15°C \Rightarrow 35 + 15 = 105e^{-0.05596t} \Rightarrow t \approx 13.26$ min

11. $T - T_s = (T_o - T_s)e^{-kt}$, $39 - T_s = (46 - T_s)e^{-10k}$ and $33 - T_s = (46 - T_s)e^{-20k} \Rightarrow \frac{39 - T_s}{46 - T_s} = e^{-10k}$ and $\frac{33 - T_s}{46 - T_s} = e^{-20k} = \left(e^{-10k}\right)^2 \Rightarrow \frac{33 - T_s}{46 - T_s} = \left(\frac{39 - T_s}{46 - T_s}\right)^2 \Rightarrow (33 - T_s)(46 - T_s) = (39 - T_s)^2 \Rightarrow$ $T_s = -3°C$

13. $V(t) = V_o e^{-t/40} \Rightarrow 0.9V_o = V_o e^{-t/40} \Rightarrow t = -40\ln(0.9) \approx 4.2$ sec

15. $\frac{c_o}{2} = c_o e^{-k(5700)} \Rightarrow k = \frac{\ln 2}{5700} \approx 0.00012 \Rightarrow c(t) = c_o e^{-0.00012t} \Rightarrow (0.445)c_o = c_o e^{-0.00012t} \Rightarrow$ $t \approx 6658.3$ yr

17. from exercise 15 we have $c(t) = c_o e^{-0.00012t} \Rightarrow (0.995)c_o = c_o e^{-0.00012t} \Rightarrow t \approx 41.22$ yr

## 7.5 INDETERMINATE FORMS AND L'HOPITAL'S RULE

1. $\lim_{x \to 2} \frac{x-2}{x^2-4} = \lim_{x \to 2} \frac{1}{2x} = \frac{1}{4}$

3. $\lim_{x \to 1} \frac{x^3-1}{4x^3-x-3} = \lim_{x \to 1} \frac{3x^2}{12x^2-1} = \frac{3}{11}$

5. $\lim_{t \to 0} \frac{\sin t^2}{t} = \lim_{t \to 0} \frac{(\cos t^2)(2t)}{1} = 0$

7. $\lim_{x \to 0} \frac{3^{\sin x}-1}{x} = \lim_{x \to 0} \frac{3^{\sin x}(\ln 3)\cos x}{1} = \ln 3$

9. $\lim_{x \to \infty} \frac{5x^2-3x}{7x^2+1} = \lim_{x \to \infty} \frac{10x-3}{14x} = \lim_{x \to \infty} \frac{10}{14} = \frac{5}{7}$

11. $\lim_{x \to \pi/2} \frac{1-\sin x}{1+\cos 2x} = \lim_{x \to \pi/2} \frac{-\cos x}{-2\ \text{sun}\ 2x} = \lim_{x \to \pi/2} \frac{\sin x}{-4\cos 2x} = \frac{1}{4}$

13. $\lim_{x \to 0} \frac{2\sqrt{x}}{\sqrt{x+7/2}} = 0$

15. $\lim_{t \to 0} \frac{10(\sin t - t)}{t^3} = \lim_{t \to 0} \frac{10(\cos t - 1)}{3t^2} = \lim_{t \to 0} \frac{10(-\sin t)}{6t} = \lim_{t \to 0} -\frac{10\cos t}{6} = -\frac{5}{3}$

17. $\lim_{x \to 0} \left(\frac{1}{\sin x} - \frac{1}{x}\right) = \lim_{x \to 0} \frac{x-\sin x}{x\sin x} = \lim_{x \to 0} \frac{1-\cos x}{\sin x + x\cos x} = \lim_{x \to 0} \frac{\sin x}{\cos x + \cos x - x\sin x} = 0$

19. $\lim_{x \to 0^+} x^{1/\ln x} = e$, let $y = x^{1/\ln x} \Rightarrow \ln y = \frac{\ln x}{\ln x} = 1 \Rightarrow \lim_{x \to 0^+} \ln y = 1 \Rightarrow \lim_{x \to 0^+} e^{1/\ln} = \lim_{x \to o^+} e^y = e^1 = e$

21. $\lim_{x \to 1^+} x^{1/(x-1)} = e$, let $y = x^{1/(x-1)} \Rightarrow \ln y = \frac{\ln x}{x-1} \Rightarrow \lim_{x \to 1^+} \frac{\ln x}{x-1} = \lim_{x \to 1^+} \frac{1/x}{1} = 1 \Rightarrow$

$\lim_{x \to 1^+} x^{1/(x-1)} = e^1 = e$

23. b, part a is neither in the $\frac{0}{0}$ nor $\frac{\infty}{\infty}$ form and one may not use L'Hopital's rule

25. $A(t) = \int_0^t e^{-x}\,dx = 1 - \frac{1}{e^t}$, $V(t) = -\frac{\pi}{2}\int_0^t e^{-2x}(-2)\,dx = \frac{\pi}{2}\left[1 - \frac{2}{e^{2t}}\right]$

a) $\lim_{t \to \infty} A(t) = \lim_{t \to \infty} 1 - \frac{1}{e^t} = 1$

b) $\lim_{t \to \infty} \frac{V(t)}{A(t)} = \lim_{t \to \infty} \left(\frac{\left(\frac{\pi}{2}\right)\left(1 - \frac{1}{e^{2t}}\right)}{1 - \frac{1}{e^t}}\right) = \frac{\pi}{2}$

c) $\lim_{t \to 0^+} \frac{V(t)}{A(t)} = \lim_{t \to 0^+} \left(\frac{\left(\frac{\pi}{2}\right)\left(1 - \frac{1}{e^{2t}}\right)}{1 - \frac{1}{e^t}}\right) = \left(\frac{\pi}{2}\right) \lim_{t \to 0^+} \frac{2e^{-2t}}{e^{-t}} = \left(\frac{\pi}{2}\right) \lim_{t \to 0^+} \frac{2e^t}{e^{2t}} = \pi$

27.

| $n$ | $10$ | $10^3$ | $10^5$ | $10^7$ | $10^9$ |
|---|---|---|---|---|---|
| $\left(1+\frac{1}{n}\right)^n$ | 2.59374246 | 2.716923932 | 2.718268237 | 2.718281693 | 2.718281827 |

29. a) $\lim_{k\to\infty}\left(1+\frac{r}{k}\right)^{kt} = e^{rt}$, let $y = \left(1+\frac{r}{k}\right)^{kt} \Rightarrow \lim_{k\to\infty} \ln y = \lim_{k\to\infty} \frac{\ln\left(1+\frac{r}{k}\right)}{\frac{1}{kt}} =$

$\lim_{k\to\infty} \frac{rt}{1 + r/k} = rt. \therefore \lim_{k\to\infty} A_o\left(1+\frac{r}{k}\right)^{kt} = A_o e^{rt}$ b) 106.184

## 7.6 THE RATES AT WHICH FUNCTIONS GROW

1. a) Yes, $\lim_{x\to\infty} \frac{x+3}{e^x} = \lim_{x\to\infty} \frac{1}{e^x} = 0$

b) Yes, $\lim_{x\to\infty} \frac{x^3 - 3x + 1}{e^x} = \lim_{x\to\infty} \frac{3x^2 - 3}{e^x} = \lim_{x\to\infty} \frac{6x}{e^x} = \lim_{x\to\infty} \frac{6}{e^x} = 0$

c) Yes, $\lim_{x\to\infty} \frac{\sqrt{x}}{e^x} = \lim_{x\to\infty} \frac{(1/2)x^{(-1/2)}}{e^x} = 0$ d) No, $\lim_{x\to\infty} \frac{4^x}{e^x} = \lim_{x\to\infty} \left(\frac{4}{e}\right)^x = \infty$

e) Yes, $\lim_{x\to\infty} \frac{\left(\frac{5}{2}\right)^x}{e^x} = \lim_{x\to\infty} \left(\frac{5}{2e}\right)^x = 0$ f) Yes, $\lim_{x\to\infty} \frac{\ln x}{e^x} = \lim_{x\to\infty} \frac{1}{xe^x} = 0$

g) Yes, $\lim_{x\to\infty} \frac{\log_{10} x}{e^x} = \left(\frac{1}{\ln 10}\right) \lim_{x\to\infty} \frac{\ln x}{e^x} = \left(\frac{1}{\ln 10}\right)(0) = 0$

h) Yes, $\lim_{x\to\infty} \frac{e^{-x}}{e^x} = \lim_{x\to\infty} \frac{1}{e^{2x}} = 0$ i) No, $\lim_{x\to\infty} \frac{e^{x+1}}{e^x} = \lim_{x\to\infty} e = e \neq 0$

j) No, $\lim_{x\to\infty} \frac{(1/2)e^x}{e^x} = \lim_{x\to\infty} \frac{1}{2} = \frac{1}{2} \neq 0$

3. a) Yes, $\lim_{x\to\infty} \frac{\log_3 x}{\ln x} = \lim_{x\to\infty} \frac{\ln x}{(\ln 3)(\ln x)} = \frac{1}{\ln 3}$ b) Yes, $\lim_{x\to\infty} \frac{\log_2 x^2}{\ln x} = \lim_{x\to\infty} \frac{2 \ln x}{(\ln 2)(\ln x)} = \frac{2}{\ln 2}$

c) Yes, $\lim_{x\to\infty} \frac{\log_{10}\sqrt{x}}{\ln x} = \lim_{x\to\infty} \frac{\ln x}{2(\ln 10)(\ln x)} = \frac{1}{\ln 100}$

d) No, $\lim_{x\to\infty} \frac{1/x}{\ln x} = \lim_{x\to\infty} -\frac{1}{x} = 0$ e) No, $\lim_{x\to\infty} \frac{1/\sqrt{x}}{\ln x} = \lim_{x\to\infty} -\frac{1}{2\sqrt{x}} = 0$

f) No, $\lim_{x \to \infty} \frac{e^{-1}}{\ln x} = \lim_{x \to \infty} \frac{1}{e^x \ln x} = 0$ g) No, $\lim_{x \to \infty} \frac{x}{\ln x} = \lim_{x \to \infty} x = \infty$

h) Yes, $\lim_{x \to \infty} \frac{5 \ln x}{\ln x} = 5$ i) No, $\lim_{x \to \infty} \frac{2}{\ln x} = 0$

j) No, $0 = \lim_{x \to \infty} \frac{-1}{\ln x} \le \lim_{x \to \infty} \frac{\sin x}{\ln x} < \lim_{x \to \infty} \frac{1}{\ln x} = 0 \Rightarrow \lim_{x \to \infty} \frac{\sin x}{\ln x} = 0$

5. $\lim_{x \to \infty} \frac{\sqrt{10x+1}}{\sqrt{x}} = \sqrt{\lim_{x \to \infty} \frac{10x+1}{x}} = \sqrt{10}$ and $\lim_{x \to \infty} \frac{\sqrt{x+1}}{\sqrt{x}} = \sqrt{\lim_{x \to \infty} \frac{x+1}{x}} = \sqrt{1} = 1$. Since the growth rate is transitive, we may conclude that $\sqrt{10x+1}$ and $\sqrt{x+1}$ have the same growth rate.

7. a) False b) False c) True d) True e) True f) True g) False h) True

9. $\lim_{x \to \infty} \frac{x^n}{e^x} = \lim_{x \to \infty} \frac{n x^{n-1}}{e^x} = \cdots = \lim_{x \to \infty} \frac{n!}{e^x} = 0 \Rightarrow x^n = o(e^x)$ for any non-negative integer n

11. a) $\lim_{x \to \infty} \frac{x^{1/n}}{\ln x} = \lim_{x \to \infty} \frac{x^{(1-n)/n}}{n\,(1/x)} = \left(\frac{1}{n}\right) \lim_{x \to \infty} x^{1/n} = \infty \Rightarrow \ln x = o(x^{1/n})$ for any positive integer n

b) $\ln\left(e^{17000000}\right) = 17000000 < \left(e^{17 \times 10^6}\right)^{1/10^6} = e^{17} \approx 24154952.75$

13. $\lim_{x \to \infty} \frac{x}{x \log_2 x} = \lim_{x \to \infty} \frac{\ln 2}{\ln x} = 0 \Rightarrow n \log_2 n$ grows faster than n, $\lim_{x \to \infty} \frac{x}{x^2} = \lim_{x \to \infty} \frac{1}{x} = 0 \Rightarrow n^2$ grows faster than n, $\lim_{x \to \infty} \frac{x}{x\left(\log_2 x\right)^2} = (\ln 2)^2 \left(\lim_{x \to \infty} \frac{1}{(\ln x)^2}\right) = 0 \Rightarrow n\left(\log_2 n\right)^2$ grows faster than n

$\therefore$ n is the best choice

15. If $\lim_{x \to \infty} \frac{f(x)}{g(x)} = L \neq 0$, then $\lim_{x \to \infty} \frac{g(x)}{f(x)} = \frac{1}{L}$ and by definition $f = O(g)$ and $g = O(f)$.

$\lim_{h \to 0} \frac{\left|E_s\right|}{h^4} \le \lim_{h \to 0} \frac{(b-a)\,M_2}{180} = \frac{(b-a)\,M_2}{180} \Rightarrow \left|E_s\right|$ is $O(h^4)$, when $h \to 0 \Rightarrow n \to \infty \Rightarrow \lim_{n \to \infty} \left|\frac{E_s}{E_t}\right| =$

$\lim_{n \to \infty} \left(\frac{(b-a)^5 M_2}{180\, n^4} \; \frac{12\, n^2}{(b-a)^3 M_1}\right) = \left(\frac{M_2\,(b-a)^2}{15\, M_1}\right)\left(\lim_{n \to \infty} \frac{1}{n^2}\right) = 0 \Rightarrow E_t$ grows faster than $E_s$

## 7.7 THE INVERSE TRIGONOMETRIC FUNCTIONS

1. a) $\frac{\pi}{4}$ b) $\frac{\pi}{3}$ c) $\frac{\pi}{6}$
3. a) $\frac{\pi}{6}$ b) $\frac{\pi}{4}$ c) $\frac{\pi}{3}$
5. a) $\frac{\pi}{3}$ b) $\frac{\pi}{4}$ c) $\frac{\pi}{6}$
7. a) $\frac{\pi}{4}$ b) $\frac{\pi}{6}$ c) $\frac{\pi}{3}$
9. a) $\frac{\pi}{4}$ b) $\frac{\pi}{3}$ c) $\frac{\pi}{6}$
11. a) $\frac{\pi}{4}$ b) $\frac{\pi}{6}$ c) $\frac{\pi}{3}$

13. $\cos \alpha = \frac{\sqrt{3}}{2}$, $\tan \alpha = \frac{1}{\sqrt{3}}$, $\sec \alpha = \frac{2}{\sqrt{3}}$, $\csc \alpha = 2$

15. $\sin\left(\cos^{-1}\frac{\sqrt{2}}{2}\right) = \sin\left(\frac{\pi}{4}\right) = \frac{1}{\sqrt{2}}$

17. $\sec\left[\cos^{-1}\left(\frac{1}{2}\right)\right] = \sec\left(\frac{\pi}{3}\right) = 2$

19. $\csc\left(\sec^{-1}(2)\right) = \csc\left(\frac{\pi}{3}\right) = \frac{2}{\sqrt{3}}$

21. $\cos\left(\cot^{-1} 1\right) = \cos\left(\frac{\pi}{4}\right) = \frac{1}{\sqrt{2}}$

23. $\cot\left(\cos^{-1} 0\right) = \cot\left(\frac{\pi}{2}\right) = 0$

25. $\tan\left(\sec^{-1} 1\right) = \tan\left(\cos^{-1} 1\right) = 0$

27. $\sin^{-1}(1) - \sin^{-1}(-1) = \frac{\pi}{2} - \left(-\frac{\pi}{2}\right) = \pi$

29. $\sec^{-1}(2) - \sec^{-1}(-2) = \cos^{-1}\left(\frac{1}{2}\right) - \cos^{-1}\left(-\frac{1}{2}\right) = \frac{\pi}{3} - \frac{2\pi}{3} = -\frac{\pi}{3}$

31. $\cos(\sin^{-1} 0.8) = 0.6$

33. $\cos^{-1}\left(-\sin\left(\frac{\pi}{6}\right)\right) = \cos^{-1}\left(-\frac{1}{2}\right) = \frac{2\pi}{3}$

35. $\lim_{x \to 1^-} \sin^{-1} x = \frac{\pi}{2}$

37. $\lim_{x \to \infty} \tan^{-1} x = \frac{\pi}{2}$

39. $\lim_{x \to \infty} \sec^{-1} x = \frac{\pi}{2}$

41. $\lim_{x \to \infty} \csc^{-1} x = \lim_{x \to \infty} \sin^{-1}\left(\frac{1}{x}\right) = 0$

43. The angle $\alpha$ = the large angle between the wall and the right end of the blackboard – the small angle between the left end of the blackboard and the wall = $\cot^{-1}\left(\frac{x}{15}\right) - \cot^{-1}\left(\frac{x}{3}\right)$

45. $V = \left(\frac{1}{3}\right)\pi r^2 h = \left(\frac{1}{3}\right)\pi(3\sin\theta)^2(3\cos\theta) = 9\pi\left(\cos\theta - \cos^3\theta\right)$, where $0 \le \theta \le \frac{\pi}{2} \Rightarrow$ $V' = -9\pi(\sin\theta)(1 - 3\cos^2\theta)$, $V' = 0 \Rightarrow \sin\theta = 0$ or $\cos\theta = \pm\frac{1}{\sqrt{3}} \Rightarrow$ the critical points are: 0, $\cos^{-1}\left(\frac{1}{\sqrt{3}}\right)$, and $\cos^{-1}\left(-\frac{1}{\sqrt{3}}\right)$; but $\cos^{-1}\left(-\frac{1}{\sqrt{3}}\right)$ is not in the domain. When $\theta = 0$, we have a minimum while when $\theta = \cos^{-1}\left(\frac{1}{\sqrt{3}}\right) \approx 54.7°$ we have a maximum.

47. $\cot^{-1}(2) = \frac{\pi}{2} - \tan^{-1}(2) \approx 26.57°$, $\sec^{-1}(1.5) = \sec^{-1}\left(\frac{3}{2}\right) = \cos^{-1}\left(\frac{2}{3}\right) \approx 48.19°$, $\csc^{-1}\left(\frac{3}{2}\right) = \sin^{-1}\left(\frac{2}{3}\right) \approx 41.81°$

## 7.8 DERIVATIVES OF INVERSE TRIGONOMETRIC FUNCTIONS; RELATED INTEGRALS

1. $y = \cos^{-1}\left(x^2\right) \Rightarrow \frac{dy}{dx} = \frac{-2x}{\sqrt{1-x^4}}$

3. $y = 5\tan^{-1}(3x) \Rightarrow \frac{dy}{dx} = \frac{15}{1+9x^2}$

5. $y = \sin^{-1}\left(\frac{x}{2}\right) \Rightarrow \frac{dy}{dx} = \frac{1}{\sqrt{4-x^2}}$

7. $y = \sec^{-1}(5x) \Rightarrow \frac{dy}{dx} = \frac{1}{|x|\sqrt{25x^2-1}}$

9. $y = \csc^{-1}\left(x^2+1\right) \Rightarrow \frac{dy}{dx} = \frac{-2x}{\left(x^2+1\right)\sqrt{x^4+2x^2}}$

11. $y = \csc^{-1}\left(x^{1/2}\right) + \sec^{-1}\left(x^{1/2}\right) \Rightarrow \frac{dy}{dx} = \frac{-\frac{1}{2}x^{-1/2}}{\sqrt{x}\sqrt{x-1}} + \frac{\frac{1}{2}x^{-1/2}}{\sqrt{x}\sqrt{x-1}} = 0$

13. $y = \cot^{-1}\left((x-1)^{1/2}\right) \Rightarrow \frac{dy}{dx} = \frac{-1}{2x\sqrt{x-1}}$

15. $y = \left(x^2-1\right)^{1/2} - \sec^{-1}(x) \Rightarrow \frac{dy}{dx} = \left(\frac{1}{2}\right)\left(x^2-1\right)^{-1/2}(2x) - \frac{1}{|x|\sqrt{x^2-1}} = \frac{x|x|-1}{|x|\sqrt{x^2-1}}$

17. $y = 2x\tan^{-1}(x) - \ln\left(x^2+1\right) \Rightarrow \frac{dy}{dx} = 2\tan^{-1}(x) + \frac{2x}{1+x^2} - \frac{2x}{x^2+1} = 2\tan^{-1}(x)$

19. $\int_0^{1/2} \frac{1}{\sqrt{1-x^2}}\,dx = \left[\sin^{-1}(x)\right]_0^{1/2} = \frac{\pi}{6}$

21. $\int_{\sqrt{2}}^{2} \frac{1}{x\sqrt{x^2-1}}\,dx = \left[\sec^{-1}(x)\right]_{\sqrt{2}}^{2} = \frac{\pi}{12}$

23. $\int_{-1}^{0} \frac{4}{1+x^2}\,dx = 4\left[\tan^{-1}(x)\right]_{-1}^{0} = \pi$

25. $\frac{1}{2}\int_0^{\sqrt{2}/2} \frac{2x}{\sqrt{1-x^4}}\,dx = \left(\frac{1}{2}\right)\left[\sin^{-1}\left(x^2\right)\right]_0^{\sqrt{2}/2} = \frac{\pi}{12}$

27. $\int_{1/\sqrt{3}}^{1} \frac{2}{2x\sqrt{4x^2-1}}\,dx = \left[\sec^{-1}(2x)\right]_{1/\sqrt{3}}^{1} = \frac{\pi}{6}$

29. $\int_0^1 \frac{4x}{\sqrt{4-x^4}}\,dx = 2\int_0^1 \frac{x}{\sqrt{1-\left(x^2/2\right)^2}}\,dx = 2\left[\sin^{-1}\left(\frac{x^2}{2}\right)\right]_0^1 = \frac{\pi}{3}$

31. $\int_{\sqrt[4]{2}}^{\sqrt{2}} \frac{2x}{x^2\sqrt{x^4-1}}\,dx = \int_{\sqrt{2}}^{2} \frac{1}{u\sqrt{u^2-1}}\,du = \left[\sec^{-1}(u)\right]_{\sqrt{2}}^{2} = \frac{\pi}{12}$, where $u = x^2$

33. $\int_1^{\sqrt{3}} \frac{2}{\left(1+x^2\right)\tan^{-1}(x)}\,dx = 2\int_{\pi/4}^{\pi/3} \frac{1}{u}\,du = 2[\ln u]_{\pi/4}^{\pi/3} = 2\left(\ln\left(\frac{\pi}{3}\right) - \ln\left(\frac{\pi}{4}\right)\right) = 2\ln\left(\frac{4}{3}\right)$,

where $u = \tan^{-1}(x)$

35. $\int_2^4 \frac{1}{2x\sqrt{x-1}}dx = \int_{\sqrt{2}}^2 \frac{2u}{2u^2\sqrt{u^2-1}}du = \int_{\sqrt{2}}^2 \frac{1}{u\sqrt{u^2-1}}du = \left[\sec^{-1}(u)\right]_{\sqrt{2}}^2 = \frac{\pi}{12}$, where $u^2 = x$

37. $\lim_{x \to 0} \frac{\sin^{-1}(x)}{x} = \lim_{x \to 0} \frac{1}{\sqrt{1-x^2}} = 1$

39. $\lim_{x \to 0} \frac{\tan^{-1}(x)}{x} = \lim_{x \to 0} \frac{\frac{1}{1+x^2}}{1} = 1$

41. $V = \pi \int_{-\sqrt{3}/3}^{\sqrt{3}} \left(\frac{1}{\sqrt{1+x^2}}\right)^2 dx = \pi \int_{-\sqrt{3}/3}^{\sqrt{3}} \frac{1}{1+x^2} dx = \pi\left[\tan^{-1}(x)\right]_{-\sqrt{3}/3}^{\sqrt{3}} = \frac{\pi^2}{2}$

43. $\alpha(x) = \cot^{-1}\left(\frac{x}{15}\right) - \cot^{-1}\left(\frac{x}{3}\right)$, $x > 0 \Rightarrow \alpha'(x) = \frac{-15}{225+x^2} + \frac{3}{9+x^2} = \frac{-15(9+x^2)+3(225+x^2)}{(225+x^2)(9+x^2)}$;
solving $\alpha'(x) = 0 \Rightarrow -135 - 15x^2 + 675 + 3x^2 = 0 \Rightarrow x = 3\sqrt{5}$, $\alpha'(x) > 0$ when $0 < x < 3\sqrt{5}$ and $\alpha'(x) < 0$ for $x > 3\sqrt{5} \Rightarrow$ at $3\sqrt{5}$ ft there is a maximum

45. $\theta(x) = \pi - \cot^{-1}(x) - \cot^{-1}(2-x) \Rightarrow \theta'(x) = \frac{4(1-x)}{(1+x^2)(1+(2-x)^2)}$, solving $\theta'(x) = 0 \Rightarrow x = 1$;
$\theta'(x) > 0$ for $0 < x < 1$ and $\theta'(x) < 0$ for $x > 1 \Rightarrow$ at $x = 1$ there is a maximum of $\theta(1) = \frac{\pi}{2}$

47. $\frac{dy}{dx} = \frac{1}{x\sqrt{x^2-1}}$ at $(2,\pi) \Rightarrow y = \sec^{-1}|x| + C = \cos^{-1}\left|\frac{1}{x}\right| + C$ at $(2,\pi) \Rightarrow \pi - \frac{\pi}{3} = C \Rightarrow C = \frac{2\pi}{3} \Rightarrow$
$y = \sec^{-1}|x| + \frac{2\pi}{3}$, but $x > 0 \Rightarrow y = \sec^{-1}x + \frac{2\pi}{3}$

49. $\frac{dy}{dx} = \frac{-1}{\sqrt{1-x^2}}$ at $(\sqrt{2}/2, \pi/2) \Rightarrow y = \cos^{-1}(x) + C$ at $(\sqrt{2}/2, \pi/2) \Rightarrow C = -\frac{\pi}{4} \Rightarrow y = \cos^{-1}(x) - \frac{\pi}{4}$

51. The Calculus Tool Kit yields 0.643517104.

53. Yes, $\sin^{-1}(x)$ and $\cos^{-1}(x)$ differ by the constant $\frac{\pi}{2}$.

55. $y = \cos^{-1}u \Rightarrow \cos y = u \Rightarrow (-\sin y)\, y' = u' \Rightarrow y' = -\frac{u'}{\sin y}$. Whereas $0 < y < \pi$, we have

$\sin y = \sqrt{1-\cos^2 y} \Rightarrow y' = -\frac{u'}{\sqrt{1-u^2}}$

57. $y = \cot^{-1}u = \frac{\pi}{2} - \tan^{-1}u \Rightarrow y' = -\frac{u'}{1+u^2}$

## 7.9 HYPERBOLIC FUNCTIONS

1. $\cosh x = \frac{5}{4}, \tanh x = -\frac{3}{5}, \coth x = -\frac{5}{3}, \text{sech } x = \frac{4}{5}, \text{csch} = -\frac{4}{3}$

3. $\sinh x = \frac{8}{15}, \tanh x = \frac{8}{17}, \coth x = \frac{17}{8}, \text{sech } x = \frac{15}{17}, \text{csch} = \frac{15}{8}$

5. $2\cosh(\ln x) = 2\left(\frac{e^{\ln x} + e^{-\ln x}}{2}\right) = x + \frac{1}{x}$

7. $\cosh(5x) + \sinh(5x) = \frac{e^{5x} + e^{-5x}}{2} + \frac{e^{5x} - e^{-5x}}{2} = e^{5x}$

9. $\cosh(3x) - \sinh(3x) = \frac{e^{3x} + e^{-3x}}{2} - \frac{e^{3x} - e^{-3x}}{2} = e^{-3x}$

11. a) $\sinh(2x) = \sinh(x + x) = \sinh(x)\cosh(x) + \cosh(x)\sinh(x) = 2\sinh(x)\cosh(x)$

    b) $\cosh(2x) = \cosh(x + x) = \cosh(x)\cosh(x) + \sinh(x)\sin(x) = \cosh^2(x) + \sinh^2(x)$

13. $y = \sinh(3x) \Rightarrow \frac{dy}{dx} = 3\cosh(3x)$

15. $y = 2\tanh\left(\frac{x}{2}\right) \Rightarrow \frac{dy}{dx} = \text{sech}^2\left(\frac{x}{2}\right)$

17. $y = \ln(\text{sech}(x)) \Rightarrow \frac{dy}{dx} = -\frac{\text{sech}(x)\tanh(x)}{\text{sech}(x)} = -\tanh(x)$

19. $y = \ln(\text{csch}(x) + \coth(x)) \Rightarrow \frac{dy}{dx} = \frac{-\text{csch}(x)\coth(x) - \text{csch}^2(x)}{\text{csch}(x) + \coth(x)} = -\text{csch}(x)$

21. $y = \left(\frac{1}{2}\right)\ln|\tanh(x)| \Rightarrow \frac{dy}{dx} = \frac{\text{sech}^2 x}{2\tanh x} = \frac{\cosh(x)}{2\sinh(x)\cosh^2(x)} = \frac{1}{2\sin(x)\cosh(x)} = \frac{1}{\sinh(2x)} = \text{csch}(2x)$

23. a) $y = \cosh^2(x) \Rightarrow \frac{dy}{dx} = 2\cosh(x)\sinh(x) = \sinh(2x)$

    b) $y = \sinh^2(x) \Rightarrow \frac{dy}{dx} = 2\sinh(x)\cosh(x) = \sinh(2x)$

    c) $y = \left(\frac{1}{2}\right)\cosh(2x) \Rightarrow \frac{dy}{dx} = \left(\frac{1}{2}\right)(\sinh(2x))(2) = \sinh(2x)$

25. $y = \sinh^{-1}(2x) \Rightarrow \frac{dy}{dx} = \frac{2}{\sqrt{1 + 4x^2}}$

27. $y = (1 - x)\tanh^{-1}(x) \Rightarrow \frac{dy}{dx} = (-1)\tanh^{-1}(x) + (1 - x)\left(\frac{1}{1 - x^2}\right) = \frac{1}{1 + x} - \tanh^{-1}(x)$

29. $y = x\,\text{sech}^{-1}(x) \Rightarrow \frac{dy}{dx} = \text{sech}^{-1}(x) + \frac{(x)(-1)}{x\sqrt{1 - x^2}} = \text{sech}^{-1}(x) - \frac{1}{\sqrt{1 - x^2}}$

31. $y = \sinh^{-1}(\tan(x)) \Rightarrow \frac{dy}{dx} = \frac{\sec^2 x}{\sqrt{1 + \tan^2 x}} = \frac{\sec^2 x}{\sec x} = \sec x$

33. $y = \tanh^{-1}(\sin x)$, where $-\frac{\pi}{2} < x < \frac{\pi}{2} \Rightarrow \frac{dy}{dx} = \frac{\cos x}{1 - \sin^2 x} = \frac{\cos x}{\cos^2 x} = \sec x$

35. $y = \text{sech}^{-1}(\sin x)$, where $0 < x < \frac{\pi}{2} \Rightarrow \frac{dy}{dx} = \frac{-\cos x}{(\sin x)\sqrt{1 + \sin^2 x}} = \frac{-1}{\sin x} = -\csc x$

37. $\int_{-1}^{1} \cosh(5x)\,dx = \frac{1}{5}\int_{-1}^{1} \cosh(5x)\,5\,dx = \left[\frac{\sinh(5x)}{5}\right]_{-1}^{1} = \left(\frac{2}{5}\right)\sinh 5$

39. $\int_{-3}^{3} \sinh(x)\,dx = 0$, $\sinh(x)$ is odd

41. $\int_0^{1/2} 4e^x \cosh(x)\,dx = \int_0^{1/2} 4e^x\left(\frac{e^x + e^{-x}}{2}\right)dx = \int_0^{1/2} e^{2x}(2) + 2\,dx = \left[e^{2x} + 2x\right]_0^{1/2} = e$

43. $\int_1^2 \frac{\cosh(\ln x)}{x}\,dx = [\sinh(\ln x)]_1^2 = \frac{3}{4}$

45. $\int_0^{\ln 3} \operatorname{sech}^2(x)\,dx = [\tanh(x)]_0^{\ln 3} = \frac{4}{5}$

47. $2\int_0^4 \frac{\cosh(\sqrt{x})}{2\sqrt{x}}\,dx = 2\left[\sinh\left(\sqrt{x}\right)\right]_0^4 = 2\sinh(2)$

51. $\cosh^{-1}\left(\frac{5}{3}\right) = \ln\left(\frac{5}{3} + \sqrt{\frac{25}{9} - 1}\right) = \ln 3$

53. $\tanh^{-1}\left(\frac{1}{2}\right) = \left(\frac{1}{2}\right)\ln\left(\frac{\frac{3}{2}}{\frac{1}{2}}\right) = \left(\frac{1}{2}\right)\ln 3 = \ln\sqrt{3}$

55. $\coth^{-1}\left(\frac{5}{4}\right) = \left(\frac{1}{2}\right)\ln\left(\frac{\frac{9}{4}}{\frac{1}{4}}\right) = \left(\frac{1}{2}\right)\ln 9 = \ln 3$

57. $\operatorname{sech}^{-1}\left(\frac{3}{5}\right) = \ln\left(\frac{1 + \sqrt{1 - 9/25}}{3/5}\right) = \ln 3$

59. $\operatorname{csch}^{-1}\left(\frac{5}{12}\right) = \ln\left(12/5 + \frac{\sqrt{1 + 25/144}}{5/12}\right) = \ln 5$

61. a) $\int_0^1 \frac{1}{\sqrt{1 + x^2}}\,dx = \left[\sinh^{-1}(x)\right]_0^1 = \sinh^{-1}(1) - \sinh^{-1}(0) = \sinh^{-1}(1)$

b) $\sinh^{-1}(1) = \ln(1 + \sqrt{2})$

63. a) $\int_{5/4}^{5/3} \frac{1}{\sqrt{x^2 - 1}}\,dx = \left[\cosh^{-1}(x)\right]_{5/4}^{5/3} = \cosh^{-1}\left(\frac{5}{3}\right) - \cosh^{-1}\left(\frac{5}{4}\right)$

b) $\cosh^{-1}\left(\frac{5}{3}\right) - \cosh^{-1}\left(\frac{5}{4}\right) = \ln\left(5/3 + \sqrt{25/9 - 1}\right) - \ln\left(5/4 + \sqrt{25/16 - 1}\right) = \ln\left(\frac{3}{2}\right)$

65. a) $\int_{5/4}^{2} \frac{1}{1 - x^2}\,dx = \left[\coth^{-1}(x)\right]_{5/4}^{2} = \coth^{-1}(2) - \coth^{-1}\left(\frac{5}{4}\right)$

b) $\coth^{-1}(2) - \coth^{-1}\left(\frac{5}{4}\right) = \left(\frac{1}{2}\right)\left[\ln(3) - \ln\left(\frac{9/4}{1/4}\right)\right] = \left(\frac{1}{2}\right)\ln\left(\frac{1}{3}\right)$

67. a) $\int_1^2 \frac{1}{x\sqrt{4 + x^2}}\,dx = -\frac{1}{2}\int_{1/2}^{1} \frac{-1}{u\sqrt{1 + u^2}}\,du = -\frac{1}{2}\left[\operatorname{csch}^{-1}(u)\right]_{1/2}^{1} =$

$\left(\frac{1}{2}\right)\left(\operatorname{csch}^{-1}\left(\frac{1}{2}\right) - \operatorname{csch}^{-1}(1)\right)$, where $u = \frac{x}{2}$

b) $\left(\frac{1}{2}\right)\left(\operatorname{csch}^{-1}\left(\frac{1}{2}\right) - \operatorname{csch}^{-1}(1)\right) = \left(\frac{1}{2}\right)\left[\ln\left(2 + \frac{\sqrt{5/4}}{1/2}\right) - \ln(1 + \sqrt{2})\right] = \left(\frac{1}{2}\right)\ln\left(\frac{2 + \sqrt{5}}{1 + \sqrt{2}}\right)$

69. $V = \pi\int_0^2 \cosh^2(x) - \sinh^2(x)\,dx = \pi\int_0^2 1\,dx = 2\pi$

71. $M_x = \int_{-\ln\sqrt{3}}^{\ln\sqrt{3}} \frac{\text{sech}^2(x)}{2}\,dx = \frac{1}{2}[\tanh(x)]_{-\ln\sqrt{3}}^{\ln\sqrt{3}} = \left(\frac{1}{2}\right)\left(\frac{1}{2} - \left(-\frac{1}{2}\right)\right) = \frac{1}{2}$, $M = \int_{-\ln\sqrt{3}}^{\ln\sqrt{3}} \text{sech}(x)\,dx =$

$\left[\sin^{-1}(\tanh(x))\right]_{-\ln\sqrt{3}}^{\ln\sqrt{3}} = \frac{\pi}{3} \Rightarrow \overline{y} = \frac{M_x}{M} = \frac{3}{2\pi}$, and $\overline{x} = 0$, by symmetry

73. Let $E(x) = \frac{f(x) + f(-x)}{2}$ and $O(x) = \frac{f(x) - f(-x)}{2}$. $E(x) + O(x) = \frac{f(x) + f(-x)}{2} + \frac{f(x) - f(-x)}{2} =$

$\frac{2\,f(x)}{2} = f(x)$. $E(-x) = \frac{f(-x) + f(-(-x))}{2} = \frac{f(x) + f(-x)}{2} = E(x) \Rightarrow E(x)$ is even.

$O(-x) = \frac{f(-x) - f(-(-x))}{2} = -\frac{f(x) - f(-x)}{2} = -O(x) \Rightarrow O(x)$ is odd. Consequently, $f(x)$ can be written as a sum of an even and odd function.

75. a) $m\frac{dv}{dt} = mg - kv^2 \Rightarrow \frac{m\frac{dv}{dt}}{mg - kv^2} = 1 \Rightarrow \frac{\frac{1}{g}\frac{dv}{dt}}{1 - \frac{kv^2}{mg}} = 1 \Rightarrow \frac{\sqrt{\frac{k}{mg}}\,dv}{1 - \left(v\sqrt{\frac{k}{mg}}\right)^2} = \sqrt{\frac{kg}{m}}\,dt \Rightarrow$

$\int \frac{\sqrt{\frac{k}{mg}}\,dv}{1 - \left(v\sqrt{\frac{k}{mg}}\right)^2}\,dv = \int \sqrt{\frac{kg}{m}}\,dt \Rightarrow \tanh^{-1}\left(\sqrt{\frac{k}{mg}}\,v\right) = \sqrt{\frac{kg}{m}}\,t + C \Rightarrow$

$v = \sqrt{\frac{mg}{k}}\tanh\left(\sqrt{\frac{gk}{m}}\,t\right) + C$ and $v(0) = 0 \Rightarrow C = 0 \therefore v = \sqrt{\frac{mg}{k}}\tanh\left(\sqrt{\frac{gk}{m}}\,t\right)$

b) $\lim_{t\to\infty} v = \lim_{t\to\infty} \sqrt{\frac{mg}{k}}\tanh\left(\sqrt{\frac{gk}{m}}\,t\right) = \sqrt{\frac{mg}{k}} \lim_{t\to\infty} \tanh\left(\sqrt{\frac{gk}{m}}\,t\right) = \sqrt{\frac{mg}{k}}(1) = \sqrt{\frac{mg}{k}}$

c) $\sqrt{\frac{160}{0.005}} = \sqrt{\frac{1600000}{5}} = \frac{400}{\sqrt{5}} = 80\sqrt{5} \approx 178.89$ ft/sec

77. $y = 10\cosh\left(\frac{x}{10}\right) \Rightarrow y' = \sinh\left(\frac{x}{10}\right) \Rightarrow (1 + (y')^2 = 1 + \sinh^2\left(\frac{x}{10}\right) = \cosh^2\left(\frac{x}{10}\right) \Rightarrow$

$L = \int_{-10\ln 10}^{10\ln 10} \sqrt{1 + (y')^2}\,dx = \int_{-10\ln 10}^{10\ln 10} \left|\cosh\left(\frac{x}{10}\right)\right| dx = \left[10\sinh\left(\frac{x}{10}\right)\right]_{-10\ln 10}^{10\ln 10} = 99$

79. $S = 2\pi\int_0^{\ln 8} 2\cosh\left(\frac{x}{2}\right)\sqrt{1 + \sinh^2\left(\frac{x}{2}\right)}\,dx = 4\pi\int_0^{\ln 8} \cosh^2\left(\frac{x}{2}\right)dx =$

$4\pi\int_0^{\ln 8} \frac{2\cosh(x) + 1}{2}\,dx = 2\pi[x + 2\sinh(x)]_0^{\ln 8} = \left(\ln(64) + \frac{63}{4}\right)\pi$

81. a) Let the point located at $(\cosh x, 0)$ be called T. $A(u)$ = area of the triangle $\Delta$ OTP – the area under the curve $y = \sqrt{x^2 - 1}$ from A to T $\Rightarrow A(u) = \frac{1}{2}\cosh u \sinh u - \int_1^{\cosh u} \sqrt{x^2 - 1}\,dx$.

b) $A(u) = \frac{1}{2}\cosh u \sinh u - \int_1^{\cosh u} \sqrt{x^2 - 1}\,dx \Rightarrow A'(u) = \frac{1}{2}\left(\cosh^2 u + \sinh^2 u\right) -$

$\sqrt{\cosh^2 u - 1}(\sinh u) = \frac{1}{2}\cosh^2 u + \frac{1}{2}\text{snh}^2 u - \sinh^2 u = \frac{1}{2}\left(\cosh^2 u - \sinh^2 u\right) = \left(\frac{1}{2}\right)(1) = \frac{1}{2}$

c) $A'(u) = \frac{1}{2} \Rightarrow A(u) = \frac{u}{2} + C$, from part a we have $A(0) = 0 \Rightarrow C = 0$ and $A(u) = \frac{u}{2} \Rightarrow u = 2A$

## PRACTICE EXERCISES

1. $y = \ln(\cos x) \Rightarrow y' = \frac{-\sin x}{\cos x} = -\tan x$

3. $y = \ln\left(\cos^{-1} x\right) \Rightarrow y' = \frac{\frac{-1}{\sqrt{1-x^2}}}{\cos^{-1} x} = \frac{-1}{\sqrt{1-x^2}\,\cos^{-1} x}$

5. $y = \log_2 x^2 = \frac{2\ln x}{\ln 2} \Rightarrow \frac{dy}{dx} = \frac{2}{(\ln 2)x}$

7. $y = e^{\tan^{-1} x} \Rightarrow \frac{dy}{dx} = e^{\tan^{-1} x}\left(\frac{1}{1+x^2}\right)$

9. $y = 8^{-x} \Rightarrow \ln y = -(\ln 8)x \Rightarrow \frac{y'}{y} = -\ln 8 \Rightarrow \frac{dy}{dx} = y' = -8^{-x}(\ln 8)$

11. $y = \sin^{-1}\left(\sqrt{1-x}\right) = \sin^{-1}\left((1-x)^{1/2}\right) \Rightarrow \frac{dy}{dx} = \frac{(1/2)(1-x)^{-1/2}(-1)}{\sqrt{1-(1-x)}} = \frac{-1}{2\sqrt{x-x^2}}$

13. $y = \tan^{-1}(\tan(2x)) \Rightarrow y' = \frac{2\sec^2(2x)}{1+\tan^2(2x)} = 2$

15. $y = x\tan^{-1}(x) - (1/2)\ln x \Rightarrow \frac{dy}{dx} = \tan^{-1}(x) + x\left(\frac{1}{1+x^2}\right) - \left(\frac{1}{2}\right)\left(\frac{1}{x}\right) = \tan^{-1}(x) + \frac{x}{1+x^2} - \frac{1}{2x}$

17. $y = 2\sqrt{x-1}\,\sec^{-1}\sqrt{x} = 2(x-1)^{1/2}\sec^{-1}(x)^{1/2} \Rightarrow \frac{dy}{dx} =$

$2\left[\left(\frac{1}{2}\right)(x-1)^{-1/2}\sec^{-1}(x)^{1/2} + (x-1)^{1/2}\left(\frac{(1/2)x^{-1/2}}{\sqrt{x}\sqrt{x-1}}\right)\right] = \frac{\sec^{-1}\sqrt{x}}{\sqrt{x-1}} + \frac{1}{x}$

19. $y = \frac{2\left(x^2+1\right)}{\sqrt{\cos 2x}} \Rightarrow \ln y = \ln 2 + \ln\left(x^2+1\right) - (1/2)\ln(\cos 2x) \Rightarrow \frac{y'}{y} = \frac{2x}{x^2+1} - \left(\frac{1}{2}\right)\left(\frac{(\sin 2x)2}{\cos 2x}\right) =$

$y' = \frac{2\left(x^2+1\right)}{\sqrt{\cos 2x}}\left[\frac{2x}{x^2+1} + \tan 2x\right]$

21. $y = \left[\frac{(x+5)(x-1)}{(x-2)(x+3)}\right]^5 \Rightarrow \ln y = 5[\ln(x+5) + \ln(x-1) - \ln(x-2) - \ln(x+3)] \Rightarrow$

$\frac{y'}{y} = 5\left[\frac{1}{x+5} + \frac{1}{x-1} - \frac{1}{x-2} - \frac{1}{x-3}\right] \Rightarrow y' = 5\left[\frac{(x+5)(x-1)}{(x-2)(x+3)}\right]^5\left[\frac{1}{x+5} + \frac{1}{x-1} - \frac{1}{x-2} - \frac{1}{x-3}\right]$

23. $y = \left(1+x^2\right)e^{\tan^{-1} x} \Rightarrow \ln y = \ln\left(1+x^2\right) + \left(\tan^{-1}(x)\right)(\ln e) \Rightarrow \frac{y'}{y} = \frac{2x}{1+x^2} + \frac{1}{1+x^2} \Rightarrow$

$y' = \left(\left(1+x^2\right)e^{\tan^{-1} x}\right)\left(\frac{2x+1}{1+x^2}\right) = (2x+1)\,e^{\tan^{-1} x}$

25. $y = x - \coth x \Rightarrow y' = 1 + \operatorname{csch}^2 x = \coth^2 x$

27. $y = \ln(\operatorname{csch} x) + x\coth x \Rightarrow y' = \frac{-\operatorname{csch} x\coth x}{\operatorname{csch} x} + \coth x - x\operatorname{csch}^2 x = -x\operatorname{csch}^2 x$

29. $y = \sinh^{-1}(\tanh x) \Rightarrow y' = \frac{\operatorname{sech}^2 x}{\sqrt{1-\tanh^2 x}} = \operatorname{sech} x$

31. $y = \sqrt{1+x^2}\,\sinh^{-1} x \Rightarrow y' = \frac{x}{\sqrt{1+x^2}}\sinh^{-1} x + \sqrt{1+x^2}\,\frac{1}{\sqrt{1+x^2}} = \frac{x\sinh^{-1} x}{\sqrt{1+x^2}} + 1$

33. $y = 1 - \tanh^{-1}\left(x^{-1}\right), |x| > 1 \Rightarrow y' = \left(\frac{-1}{1 - x^{-2}}\right)\left(-x^{-2}\right) = \frac{1}{x^2 - 1}$

35. $y = \operatorname{sech}^{-1}(\cos 2x),\ 0 < x < \pi/4 \Rightarrow y' \frac{-(-\sin 2x)(2)}{(\cos 2x)\sqrt{1 - \cos^2 2x}} = \frac{2 \sin 2x}{(\cos 2x)(\sin 2x)} = 2 \sec 2x$

37. $\int_{-1}^{1} \frac{1}{3x - 4}\,dx = \frac{1}{3}[\ln|3x - 4|]_{-1}^{1} = -\frac{\ln 7}{3}$

39. $\int_0^{\pi} \tan\left(\frac{\pi}{3}\right) dx = -3 \int_0^{\pi} \frac{-\frac{1}{3}\sin\frac{x}{3}}{\cos\frac{x}{3}}\,dx = -3\left[\ln\left|\cos\frac{x}{3}\right|\right]_0^{\pi} = \ln 8$

41. $\int_0^4 \frac{2x}{x^2 - 25}\,dx = \left[\ln\left|x^2 - 25\right|\right]_0^4 = \ln\left(\frac{9}{25}\right)$

43. $\int_0^{\pi/4} \frac{\sec x \tan x + \sec^2 x}{\sec x + \tan x}\,dx = \left[\ln|\sec x + \tan x|\right]_0^{\pi/4} = \ln(\sqrt{2} + 1)$

45. $\int_1^8 \frac{\log_4 x}{x}\,dx = \frac{1}{\ln 4}\int_1^8 (\ln x)\left(\frac{1}{x}\right) dx = \frac{1}{2\ln 4}\left[(\ln x)^2\right]_1^8 = \frac{9 \ln 2}{4}$

47. $\int_0^1 x\, 3^{x^2}\,dx = \frac{1}{2\ln 3}\int_0^1 e^{(\ln 3)x^2}(2\ln 3)x\,dx = \frac{1}{2\ln 3}\left[e^{(\ln 3)x^2}\right]_0^1 = \frac{1}{2\ln 3}\left[3^{x^2}\right]_0^1 = \frac{1}{\ln 3}$

49. $\int_{-1/2}^{1/2} \frac{3}{\sqrt{1 - x^2}}\,dx = \left[3\sin^{-1} x\right]_{-1/2}^{1/2} = \pi$

51. $\int_{-1}^{1} \frac{1}{1 + x^2}\,dx = \left[\tan^{-1} x\right]_{-1}^{1} = \frac{\pi}{2}$

53. $\int_{1/2}^{3/4} \frac{1}{\sqrt{x}\sqrt{1 - x}}\,dx = \int_{\sqrt{1/2}}^{\sqrt{3}/2} \frac{2u}{u\sqrt{1 - u^2}}\,du = 2\int_{\sqrt{1/2}}^{\sqrt{3}/2} \frac{1}{\sqrt{1 - u^2}}\,du = 2\left[\sin^{-1} u\right]_{\sqrt{1/2}}^{\sqrt{3}/2} = \frac{\pi}{6}$, where $u = \sqrt{x}$

55. $\int_0^{\ln 2} 4e^x \cosh x\,dx = 4\int_0^{\ln 2} e^x\left(\frac{e^x + e^{-x}}{2}\right) dx = \int_0^{\ln 2} 2e^{2x} + 2\,dx = \left[e^{2x} + 2x\right]_0^{\ln 2} = 3 + \ln 4$

57. $\int_{-\ln 3}^{\ln 3} 3\sqrt{\cosh 2x + 1}\,dx = \int_{-\ln 3}^{\ln 3} 3\sqrt{\cosh^2 x + \sinh^2 x + \cosh^2 x - \sinh^2}\,dx =$

$3\sqrt{2}\int_{-\ln 3}^{\ln 3} \cosh x\,dx = 3\sqrt{2}[\sinh x]_{-\ln 3}^{\ln 3} = 8\sqrt{2}$

59. $\int_2^4 10 \operatorname{csch}^2 x \coth x\,dx = -10\int_2^4 (\operatorname{csch} x)(-\operatorname{csch} x \coth x)\,dx = -5\left[\operatorname{csch}^2 x\right]_2^4 = 5\left[\operatorname{csch}^2 2 - \operatorname{csch}^2 4\right]$

61. a) $\int_0^{\pi/2} \frac{\sin x}{\sqrt{1 + \cos^2 x}}\,dx = -\left[\sinh^{-1}(\cos x)\right]_0^{\pi/2} = \sinh^{-1}(1)$

b) $\sinh^{-1}(1) = \ln(1 + \sqrt{2})$

63. a) $\int_{1/5}^{1/2} \frac{4\tanh^{-1}x}{1-x^2}dx = 4\left[\frac{(\tanh^{-1}x)^2}{2}\right]_{1/5}^{1/2} = 2\left[(\tan^{-1}(1/2))^2 - (\tan^{-1}(1/5))^2\right]$

b) $2\left[(\tan^{-1}(1/2))^2 - (\tan^{-1}(1/5))^2\right] = 2\left[\left(\frac{1}{2}\ln\left(\frac{3/2}{1/2}\right)\right)^2 - \left(\frac{1}{2}\ln\left(\frac{6/5}{4/5}\right)\right)^2\right] = \ln\left(\frac{9}{2}\right)\ln\sqrt{2}$

65. a) $\int_{3/5}^{4/5} \frac{2\,\text{sech}^{-1}x}{x\sqrt{1-x^2}}dx = -2\int_{3/5}^{4/5} (\text{sech}^{-1}x)\left(\frac{-1}{x\sqrt{1-x^2}}\right)dx =$

$\left[(\text{sech}^{-1}x)^2\right]_{3/5}^{4/5} = \left(\text{sech}^{-1}\frac{3}{5}\right)^2 - \left(\text{sech}^{-1}\frac{4}{5}\right)^2$

b) $\left(\text{sech}^{-1}\frac{3}{5}\right)^2 - \left(\text{sech}^{-1}\frac{4}{5}\right)^2 = \left[\ln\left(\frac{1+\sqrt{16/25}}{3/5}\right)\right]^2 - \left[\ln\left(\frac{1+\sqrt{9/25}}{4/5}\right)\right]^2 = (\ln 6)\left(\ln\frac{3}{2}\right)$

67. $\left.\frac{df^{-1}}{dx}\right|_{x=2+\ln 2} = \left.\frac{1}{df/dx}\right|_{x=\ln 2} = \left.\frac{1}{e^x+1}\right|_{x=\ln 2} = \frac{1}{3}$

69. $K = \ln(5x) - \ln(3x) = \ln(5) + \ln(x) - \ln(3) - \ln(x) = \ln(5) - \ln(3) = \ln\left(\frac{5}{3}\right)$

71. $\int_1^e \frac{2\log_2 x}{x}dx = \frac{2}{\ln 2}\int_1^e (\ln x)\left(\frac{1}{x}\right)dx = \left[\frac{(\ln x)^2}{\ln 2}\right]_1^e = \frac{1}{\ln 2},\ \int_1^e \frac{2\log_4 x}{x}dx =$

$\frac{2}{2\ln 2}\int_1^e (\ln x)\left(\frac{1}{x}\right)dx = \left[\frac{(\ln x)^2}{2\ln 2}\right]_1^e = \frac{1}{2\ln 2} = \frac{1}{\ln 4}$, the first area is twice the second

73. $C(t) = C_o \exp\left[-\frac{\ln 2}{5700}t\right] \Rightarrow (0.1)C_o = C_o\exp\left[-\frac{\ln 2}{5700}t\right] \Rightarrow \ln(0.1) = -\frac{\ln 2}{5700}t \Rightarrow$

$t = -\frac{5700\ln(0.1)}{\ln 2} \approx 18935$ yr

75. $T - T_s = (T - T_s)e^{-kt} \Rightarrow 180 - 40 = (220-40)e^{-k/4}$, time in hours, $\Rightarrow k = -4\ln(7/9) = 4\ln(9/7) \Rightarrow$

$70 - 40 = (220-40)e^{-4\ln(9/7)t} \Rightarrow t = \frac{\ln 6}{4\ln(9/7)} \approx 1.78$ hr $\approx 107$ min, the total time $\Rightarrow$ the time it took

to cool from 180° F to 70° F was 107 − 15 = 92 min

77. $\lim_{t\to 0}\frac{t-\ln(1+2t)}{t^2} = \lim_{t\to 0}\frac{1-\frac{2}{1+2t}}{2t} = \lim_{t\to 0}\left[\frac{1}{2t} - \frac{1}{t(1+2t)}\right] = \lim_{t\to 0}\frac{2t-1}{2t(1+2t)} = -\infty$

79. $\lim_{x\to 0}\frac{x\sin x}{1-\cos x} = \lim_{x\to 0}\frac{\sin x + x\cos x}{\sin x} = \lim_{x\to 0}\frac{\cos x + \cos x - x\sin x}{\cos x} = 2$

81. $\lim_{x\to 0}\frac{2^{\sin x}-1}{e^x-1} = \lim_{x\to 0}\frac{2^{\sin x}(\ln 2)\cos x}{e^x} = \ln 2$

83. $\lim_{x\to\infty} x^{1/x}$, let $f(x) = x^{1/x} \Rightarrow \ln f(x) = \frac{\ln x}{x} \Rightarrow \lim_{x\to\infty}\frac{\ln x}{x} = \lim_{x\to\infty}\frac{1/x}{1} = 0 \Rightarrow \lim_{x\to\infty} x^{1/x} = e^0 = 1$

85. $\lim_{x\to\infty}\left(1+\frac{3}{x}\right)^x$, let $f(x) = \left(1+\frac{3}{x}\right)^x \Rightarrow \ln f(x) = \frac{\ln\left(1+3x^{-1}\right)}{x^{-1}} \Rightarrow \lim_{x\to\infty}\ln f(x) =$

$\lim_{x\to\infty}\frac{\ln\left(1+3x^{-1}\right)}{x^{-1}} = \lim_{x\to\infty}\frac{\frac{-3x^{-2}}{1+3x^{-1}}}{-x^{-2}} = \lim_{x\to\infty}\frac{3}{1+3/x} = 3 \Rightarrow \lim_{x\to\infty}\left(1+\frac{3}{x}\right)^x = e^3$

87. a) $\lim_{x \to \infty} \frac{x}{5x} = \frac{1}{5} \Rightarrow$ the growth rates are the same

b) $\lim_{x \to \infty} \frac{x+1/x}{x} = \lim_{x \to \infty} \frac{x^2+1}{x^2} = 1 \Rightarrow$ the growth rates are the same

c) $\lim_{x \to \infty} \frac{x^2+x}{x^2-x} = 1 \Rightarrow$ the growth rates are the same

89. a) $\lim_{x \to \infty} \frac{\frac{1}{x^2}+\frac{1}{x^4}}{\frac{1}{x^2}} = \lim_{x \to \infty} \frac{x^2+1}{x^2} = 1 \Rightarrow$ true

b) $\lim_{x \to \infty} \frac{\frac{1}{x^2}+\frac{1}{x^4}}{\frac{1}{x^4}} = \lim_{x \to \infty} x^2+1 = \infty \Rightarrow$ false

c) $\lim_{x \to \infty} \frac{\sqrt{x^2+1}}{x} = \sqrt{\lim_{x \to \infty} \frac{x^2+1}{x^2}} = \sqrt{1} = 1 \Rightarrow$ true

91. If $f(x) = \tan^{-1}x + \tan^{-1}\left(\frac{1}{x}\right)$, then $f'(x) = \frac{1}{1+x^2} + \frac{-x^{-2}}{1+x^{-2}} = 0 \Rightarrow f(x)$ is a constant. The constant is $f(1) = \frac{\pi}{2}$.

93. $\theta = \pi - \cot^{-1}\left(\frac{x}{60}\right) - \cot^{-1}\left(\frac{5}{3} - \frac{x}{30}\right)$, $0 < x < 50 \Rightarrow \theta' = \frac{1/60}{1+\left(\frac{x}{60}\right)^2} + \frac{-1/30}{1+\left(\frac{50-x}{30}\right)^2} =$

$30\left[\frac{2}{60^2+x^2} - \frac{1}{30^2+(50-x)^2}\right]$, solving $\theta' = 0 \Rightarrow x^2 - 200x + 3200 = 0 \Rightarrow x = 100 \pm 20\sqrt{17}$,

but $100 + 20\sqrt{17}$ is not in the domain, $\theta' > 0$ for $x < 20\left(5-\sqrt{17}\right)$, and $\theta' < 0$ for $20\left(5-\sqrt{17}\right) < x < 50 \Rightarrow$ at $x = 20\left(5-\sqrt{17}\right) \approx 17.54$ m is a maximum

95. $\frac{dy}{dx} = \frac{-1}{x\sqrt{1-x^2}} + \frac{x}{\sqrt{1-x^2}} \Rightarrow \int dy = \int \frac{-1}{x\sqrt{1-x^2}}\,dx + \int \frac{x}{\sqrt{1-x^2}}\,dx \Rightarrow$

$y = \operatorname{sech}^{-1}(x) - \sqrt{1-x^2} + C$ at $(1,0) \Rightarrow C = 0 \Rightarrow y = \operatorname{sech}^{-1}(x) - \sqrt{1-x^2}$

# CHAPTER 8

# TECHNIQUES OF INTEGRATION

## 8.1 BASIC INTEGRATION FORMULAS

1. $\int_0^1 \frac{16x}{\sqrt{8x^2+1}}\,dx = \int_1^9 u^{-1/2}\,du = \left[2u^{1/2}\right]_1^9 = 4$, where $u = 8x^2 + 1$

3. $\int_0^1 \frac{16x}{8x^2+2}\,dx = \int_2^{10} \frac{1}{u}\,du = [\ln u]_2^{10} = \ln 5$, where $u = 8x^2 + 2$

5. $\int 4x \tan x^2\,dx = 2\int \tan u\,du = -2\int \frac{-\sin u}{\cos u}\,dx = -2[\ln|\cos u|] + C = -2\left[\ln\left|\cos x^2\right|\right] + C$,

where $u = x^2$

7. $\int_{-\pi}^{\pi} \sec\left(\frac{x}{3}\right)\,dx = 3\int_{-\pi/3}^{\pi/3} \sec u\,du = 3[\ln|\sec u + \tan u|]_{-\pi/3}^{\pi/3} = \ln\left(2+\sqrt{3}\right)^6$, where $u = \frac{x}{3}$

9. $\int_{3\pi/2}^{7\pi/4} \csc(x-\pi)\,dx = \int_{\pi/2}^{3\pi/4} \csc u\,du = -[\ln|\csc u + \cot u|]_{\pi/2}^{3\pi/4} = -\ln\left(\sqrt{2}-1\right) = \ln\left(\sqrt{2}+1\right)$,

where $u = x - \pi$

11. $\int_0^{\sqrt{\ln 2}} e^{x^2}(2x)\,dx = \int_0^{\ln 2} e^u\,du = \left[e^u\right]_0^{\ln 2} = 1$, where $u = x^2$

13. $\int_{-1}^{0} 3^{(x+1)}\,dx = \int_0^1 3^u\,du = \left[\frac{3^u}{\ln 3}\right]_0^1 = \frac{2}{\ln 3}$, where $u = x + 1$

15. $\int_1^3 \frac{6}{\sqrt{y}(1+y)}\,dy = 12\int_1^{\sqrt{3}} \frac{1}{1+u^2}\,du = 12\left[\tan^{-1}(u)\right]_1^{\sqrt{3}} = \pi$, where $u = \sqrt{y}$

17. $\int_0^{1/6} \frac{1}{\sqrt{1-9x^2}}\,dx = \frac{1}{3}\int_0^{1/2} \frac{1}{\sqrt{1-u^2}}\,du = \frac{1}{3}\left[\sin^{-1}(u)\right]_0^{1/2} = \frac{\pi}{18}$, where $u = 3x$

19. $\int_{2/5\sqrt{3}}^{2/5} \frac{6}{x\sqrt{25x^2-1}}\,dx = 6\int_{2/\sqrt{3}}^{2} \frac{1}{u\sqrt{u^2-1}}\,du = 6\left[\sec^{-1}|u|\right]_{2/\sqrt{3}}^{2} = \pi$, where $u = 5x$

21. $\int \frac{1}{\sqrt{-x^2+4x-3}}\,dx = \int \frac{1}{\sqrt{1-(x-2)^2}}\,dx = \int \frac{1}{\sqrt{1-u^2}}\,du = \left[\sin^{-1}(u)\right] + C =$

$\left[\sin^{-1}(x-2)\right] + C$, where $u = x - 2$

23. $\int_1^2 \frac{8}{x^2-2x+2}\,dx = 8\int_1^2 \frac{1}{1+(x-1)^2}\,dx = 8\int_0^1 \frac{1}{1+u^2}\,du = 8\left[\tan^{-1}(u)\right]_0^1 = 2\pi$, where $u = x - 1$

25. $\int \frac{1}{(x+1)\sqrt{x^2+2x}}\,dx = \int \frac{1}{(x+1)\sqrt{(x+1)^2-1}}\,dx = \int \frac{1}{u\sqrt{u^2-1}}\,du =$

$\left[\sec^{-1}|u|\right] + C = \left[\sec^{-1}|x+1|\right] + C$, when $|x+1| > 1$ and $u = x + 1$

27. $\int_{\pi/4}^{3\pi/4} (\csc x - \cot x)^2\, dx = \int_{\pi/4}^{3\pi/4} 2\csc^2 x - 1 - 2\csc x \cot x\, dx =$

$[-2\cot(x) - x + 2\csc(x)]_{\pi/4}^{3\pi/4} = 4 - \frac{\pi}{2}$

29. $\int_{\pi/6}^{\pi/3} (\csc x - \sec x)(\sin x + \cos x)\, dx = \int_{\pi/6}^{\pi/3} \frac{\cos x}{\sin x} + \frac{-\sin x}{\cos x}\, dx = [\ln|\sin x| + \ln|\cos x|]_{\pi/6}^{\pi/3} = 0$

31. $\int_0^1 \frac{x}{x+1}\, dx = \int_0^1 1 - \frac{1}{x+1}\, dx = [x - \ln|x+1|]_0^1 = 1 - \ln 2$

33. $\int_{\sqrt{2}}^3 \frac{2x^3}{x^2 - 1}\, dx = \int_{\sqrt{2}}^3 2x + \frac{2x}{x^2 - 1}\, dx = \left[x^2 + \ln\left|x^2 - 1\right|\right]_{\sqrt{2}}^3 = 7 + \ln 8$

35. $\int_0^{\sqrt{3}/2} \frac{1 - x}{\sqrt{1 - x^2}}\, dx = \int_0^{\sqrt{3}/2} \left(\frac{1}{\sqrt{1 - x^2}} - \left(1 - x^2\right)^{-1/2}(x)\right) dx = \left[\sin^{-1}(x) + \sqrt{1 - x^2}\right]_0^{\sqrt{3}/2} = \frac{2\pi - 3}{6}$

37. $\int_0^{\pi/4} \frac{1 + \sin x}{\cos^2 x}\, dx = \int_0^{\pi/4} \sec^2 x + \sec x \tan x\, dx = [\tan x + \sec x]_0^{\pi/4} = \sqrt{2}$

39. $y = \ln(\cos x) \Rightarrow 1 + (y')^2 = 1 + \left(\frac{-\sin x}{\cos x}\right)^2 = 1 + \tan^2 x = \sec^2 x \Rightarrow L = \int_0^{\pi/3} \sqrt{1 + (y')^2}\, dx =$

$\int_0^{\pi/3} \sqrt{\sec^2 x}\, dx = \int_0^{\pi/3} \sec x\, dx = [\ln|\sec x + \tan x|]_0^{\pi/3} = \ln\left(2 + \sqrt{3}\right)$

41. $M_x = \int_{-\pi/4}^{\pi/4} \frac{\sec^2 x}{2}\, dx = \int_0^{\pi/4} \sec^2 x\, dx = [\tan x]_0^{\pi/4} = 1,\ M = 2\int_0^{\pi/4} \sec x\, dx = 2[\ln|\sec x + \tan x|]_0^{\pi/4} =$

$2\ln\left(\sqrt{2} + 1\right) = \ln\left(2\sqrt{2} + 3\right) \Rightarrow \overline{y} = \frac{M_x}{M} = \frac{1}{\ln\left(2\sqrt{2} + 3\right)}$, and $\overline{x} = 0$ by symmetry

43. $25\,[\ln|\sec x + \tan x|]_{30°}^{45°} = 25\ln\left(\frac{\sqrt{2} + 1}{\sqrt{3}}\right) \approx 8.30$ cm

45. $\int_0^{\pi/2} 3\sqrt{\sin x}\cos x\, dx = \int_0^1 3\sqrt{u}\, du = \left[2u^{3/2}\right]_0^1 = 2$, where $u = \sin x$

47. $\int_{-\pi}^0 \frac{\sin x}{2 + \cos x}\, dx = -\int_{-1}^1 \frac{1}{2 + u}\, du = -[\ln|2 + u|]_{-1}^1 = -\ln 3$, where $u = \cos x$

49. $\int_0^{1/4} \sec \pi x\, dx = \frac{1}{\pi}\int_0^{\pi/4} \sec u\, du = \frac{1}{\pi}[\ln|\sec u + \tan u|]_0^{\pi/4} = \frac{\ln\left(\sqrt{2} + 1\right)}{\pi}$, where $u = \pi x$

51. $\int_0^{\pi/3} e^{\tan x}\sec^2 x\, dx = \int_0^{\sqrt{3}} e^u\, du = \left[e^u\right]_0^{\sqrt{3}} = e^{\sqrt{3}} - 1$, where $u = \tan x$

53. $\int_1^4 \frac{2^{\sqrt{x}}}{2\sqrt{x}}\, dx = \int_1^2 2^u\, du = \left[\frac{2^u}{\ln 2}\right]_1^2 = \frac{2}{\ln 2}$, where $u = \sqrt{x}$

55. $\int_0^{\sqrt{3}/3} \frac{9}{1 + 9x^2}\, dx = 3\int_0^{\sqrt{3}} \frac{1}{1 + u^2}\, du = 3\left[\tan^{-1}(u)\right]_0^{\sqrt{3}} = \pi$, where $u = 3x$

57. $\int_0^{1/4} \frac{2}{\sqrt{1-4x^2}}\,dx = \int_0^{1/2} \frac{1}{\sqrt{1-u^2}}\,du = \left[\sin^{-1}(u)\right]_0^{1/2} = \frac{\pi}{6}$, where $u = 2x$

59. $\int_{1/\sqrt{2}}^{1} \frac{1}{x\sqrt{4x^2-1}}\,dx = \int_{\sqrt{2}}^{2} \frac{1}{u\sqrt{u^2-1}}\,du = \left[\sec^{-1}|u|\right]_{\sqrt{2}}^{2} = \frac{\pi}{12}$, where $u = 2x$

## 8.2 INTEGRATION BY PARTS

1. Let $v = x$ and $du = \sin x \Rightarrow dv = dx$ and $u = -\cos x$

$\int x \sin x\,dx = -x\cos x + \int \cos x\,dx = -x\cos x + \sin x + C$

$\int x\cos(2x)\,dx = \frac{x\sin(2x)}{2} - \frac{1}{2}\int \sin 2x\,dx = \frac{x\sin(2x)}{2} + \frac{\cos(2x)}{4} + C$

3. $\int x^2 \sin x\,dx = -x^2\cos x + 2x\sin x + 2\cos x + C$, tabular integration

| | | |
|---|---|---|
| $x^2$ | + | $\sin x$ |
| $2x$ | − | $-\cos x$ |
| $2$ | + | $-\sin x$ |
| $0$ | | $\cos x$ |

5. Let $v = \ln x$ and $du = x\,dx \Rightarrow dv = \frac{1}{x}$ and $u = \frac{x^2}{2}$

$\int_1^2 x\ln x\,dx = \left[\frac{x^2\ln x}{2}\right]_1^2 - \frac{1}{2}\int_1^2 x\,dx = \left[\frac{x^2\ln x}{2} - \frac{x^2}{4}\right]_1^2 = \ln(4) - \frac{3}{4}$

7. Let $v = \tan^{-1}(x)$ and $du = dx \Rightarrow dv = \frac{dx}{1+x^2}$ and $u = x$

$\int \tan^{-1}(x)\,dx = x\tan^{-1}(x) - \frac{1}{2}\int \frac{2x}{1+x^2}\,dx = x\tan^{-1}(x) - \ln\sqrt{1+x^2} + C$

9. Let $v = x$ and $du = \sec^2 x\,dx \Rightarrow dv = dx$ and $u = \tan x$

$\int x\sec^2 x\,dx = x\tan x + \int \frac{-\sin x}{\cos x}\,dx = x\tan x + \ln|\cos x| + C$

11. $\int x^3 e^x\,dx = \left(x^3 - 3x^2 + 6x - 6\right)e^x + C$, tabular integration

| | | |
|---|---|---|
| $x^3$ | + | $e^x$ |
| $3x^2$ | − | $e^x$ |
| $6x$ | + | $e^x$ |
| $6$ | − | $e^x$ |
| $0$ | | $e^x$ |

13. $\int (x^2 - 5x)\, e^x\, dx = \left[(x^2 - 5x) - (2x - 5) + (2)\right] e^x + C = (x^2 - 7x + 7)\, e^x + C$, tabular integration

$$\begin{array}{ccc} x^2-5x & + & e^x \\ 2x-5 & - & e^x \\ 2 & + & e^x \\ 0 & & e^x \end{array}$$

15. $\int x^5 e^x\, dx = (x^5 - 5x^4 + 20x^3 - 60x^2 + 120x - 120)\, e^x + C$, tabular integration

$$\begin{array}{ccc} x^5 & + & e^x \\ 5x^4 & - & e^x \\ 20x^3 & + & e^x \\ 60x^2 & - & e^x \\ 120x & + & e^x \\ 120 & - & e^x \\ 0 & & e^x \end{array}$$

17. $\int_0^{\pi/2} x^2 \sin 2x\, dx = \left[-\frac{x^2 \cos 2x}{2} + \frac{x \sin 2x}{2} + \frac{\cos 2x}{4}\right]_0^{\pi/2} = \frac{\pi^2 - 4}{8}$, tabular integration

$$\begin{array}{ccc} x^2 & + & \sin 2x \\ 2x & - & (-\cos 2x)/2 \\ 2 & + & (-\sin 2x)/4 \\ 0 & & (\cos 2x)/8 \end{array}$$

19. Let $v = \sec^{-1} x$ and $du = x\, dx \Rightarrow dv = \frac{dx}{x\sqrt{x^2 - 1}}$ and $u = \frac{x^2}{2}$

$\int_1^2 x \sec^{-1} x\, dx = \left[\frac{x^2 \sec^{-1} x}{2}\right]_1^2 - \frac{1}{4}\int_1^2 (x^2 - 1)^{-1/2}\, 2x\, dx =$

$\left[\frac{x^2 \sec^{-1} x}{2}\right]_1^2 - \frac{1}{2}\left[2(x^2 - 1)^{1/2}\right]_1^2 = \frac{2\pi}{3} - \frac{\sqrt{3}}{2}$

21. Let $u = e^x$ and $dv = \sin x\, dx \Rightarrow du = e^x dx$ and $v = -\cos x \Rightarrow \int e^x \sin x\, dx = -e^x \cos x + \int e^x \cos x\, dx$

Now let $u = e^x$ and $dv = \cos x\, dx \Rightarrow du = e^x dx$ and $v = \sin x \Rightarrow \int e^x \sin x\, dx = -e^x \cos x +$

$\int e^x \cos x\, dx = -e^x \cos x + e^x \sin x - \int e^x \sin x\, dx \Rightarrow 2\int e^x \sin x\, dx = -e^x \cos x + e^x \sin x + C \Rightarrow$

$\int e^x \sin x\, dx = \frac{1}{2}\left[-e^x \cos x + e^x \sin x\right] + C$

23. Let $u = e^{2x}$ and $dv = \cos 3x\,dx \Rightarrow du = 2e^{2x}dx$ and $v = \frac{\sin 3x}{3} \Rightarrow \int e^{2x}\sin 3x\,dx = \frac{e^{2x}\sin 3x}{3} -$

$\frac{2}{3}\int e^{2x}\sin 3x\,dx$, now let $u = e^{2x}$ and $dv = \sin 3x\,dx \Rightarrow du = 2e^{2x}dx$ and $v = -\frac{\cos 3x}{3} \Rightarrow$

$$\int e^{2x}\sin 3x\,dx = \frac{e^{2x}\sin 3x}{3} + \frac{2}{3}\left[\frac{e^{2x}\cos 3x}{3} - \int 2e^{2x}\frac{\text{co}3x}{3}\,dx\right] \Rightarrow$$

$$\frac{13}{9}\int e^{2x}\cos 3x\,dx = \frac{e^{2x}\sin 3x}{3} + \frac{2e^{2x}\cos 3x}{9} + C \Rightarrow \int e^{2x}\cos 3x\,dx = \frac{e^{2x}}{13}(3\sin 3x + 2\cos 3x) + C$$

25. a) $\int_0^{\pi} x\sin x\,dx = [-x\cos x + \sin x]_0^{\pi} = \pi$

b) $\int_{\pi}^{2\pi} x\sin x\,dx = [-x\cos x + \sin x]_{\pi}^{2\pi} = -3\pi$, but area is positive $\therefore 3\pi$

27. $V = 2\pi\int_0^1 x\,e^{-x}\,dx = 2\pi\left[-x\,e^{-x} - e^{-x}\right]_0^1 = 2\pi - \frac{4\pi}{e}$

29. $M_y = \int_0^{\pi} x(1 + x)\sin x\,dx = \left[\left(x + x^2\right)(-\cos x) + (1 + 2x)(\sin x) + (2)(\cos x)\right]_0^{\pi} = \pi^2 + \pi - 4$, use tabular integration

| | sign | |
|---|---|---|
| $x + x^2$ | + | $\sin x$ |
| $1 + 2x$ | − | $-\cos x$ |
| 2 | + | $-\sin x$ |
| 0 | | $\cos x$ |

## 8.3 TRIGONOMETRIC INTEGRALS

1. $\int_0^{\pi/2} \sin^5 x\,dx = \int_0^{\pi/2} \left(1-\cos^2 x\right)^2 \sin x\,dx = -\left[\cos x - \frac{2\cos^3 x}{3} + \frac{\cos^5 x}{5}\right]_0^{\pi/2} = \frac{8}{15}$

3. $\int_{-\pi/2}^{\pi/2} \cos^3 x\,dx = 2\int_0^{\pi/2} \left(1-\sin^2 x\right)\cos x\,dx = 2\left[\sin x - \frac{\sin^3}{3}\right]_0^{\pi/2} = \frac{4}{3}$, $\cos^3 x$ is even

5. $\int_0^{\pi/2} \sin^7 y\,dy = \int_0^{\pi/2} \left(1-\cos^2 y\right)^3 \sin y\,dy = \left[-\cos y + \cos^3 y - \frac{3\cos^5 y}{5} + \frac{\cos^7 y}{7}\right]_0^{\pi/2} = \frac{16}{35}$

7. $\int_0^{\pi} 8\sin^4 x\,dx = 8\int_0^{\pi} \left(\frac{1-\cos 2x}{2}\right)^2 dx = \int_0^{\pi} 3 - 4\cos 2x + \cos 4x\,dx =$

$\left[3x - 2\sin 2x + \frac{\sin 4x}{4}\right]_0^{\pi} = 3\pi$

9. $\int_{-\pi/4}^{\pi/4} 16\sin^2 x\cos^2 x\,dx = 4\int_{-\pi/4}^{\pi/4} (2\sin x\cos x)^2\,dx = 8\int_0^{\pi/4} (\sin 2x)^2\,dx =$

$4\int_0^{\pi/4} 1 - \cos 4x\,dx = 4\left[x - \frac{\sin 4x}{4}\right]_0^{\pi/4} = \pi$

11. $\int_0^{\pi/2} 35\sin^4 x\cos^3 x\,dx = \int_0^{\pi/2} 35\sin^4 x\left(1-\sin^2 x\right)\cos x\,dx = 35\left[\frac{\sin^5 x}{5} - \frac{\sin^7}{7}\right]_0^{\pi/2} = 2$

13. $\int_0^{\pi/4} 8\cos^3 2\theta\sin 2\theta\,d\theta = -4\int_0^{\pi/4} (\cos 2\theta)^3(-2\sin 2\theta)\,d\theta = -\left[(\cos 2\theta)^4\right]_0^{\pi/4} = 1$

15. $\int_0^{2\pi} \sqrt{\frac{1-\cos\theta}{2}}\,d\theta = \int_0^{2\pi} \left|\sin\frac{\theta}{2}\right| d\theta = 2\int_0^{2\pi} \left(\frac{1}{2}\right)\sin\frac{\theta}{2}\,d\theta = -2\left[\cos\frac{\theta}{2}\right]_0^{2\pi} = 4$

17. $\int_0^{\pi} \sqrt{1-\sin^2 t}\,dt = \int_0^{\pi} |\cos t|\,dt = \int_0^{\pi/2} \cos t\,dt - \int_{\pi/2}^{\pi} \cos t\,dt = 2$

19. $\int_{-\pi/4}^{\pi/4} \sqrt{1+\tan^2 x}\,dx = \int_{-\pi/4}^{\pi/4} |\sec x|\,dx = 2\int_0^{\pi/4} \sec x\,dx = 2\left[\ln|\sec x + \tan x|\right]_0^{\pi/4} = \ln\left(3+\sqrt{2}\right)$

21. $\int_0^{\pi/2} \theta\sqrt{1-\cos 2\theta}\,d\theta = \sqrt{2}\int_0^{\pi/2} \theta\sqrt{\frac{1-\cos 2\theta}{2}}\,d\theta = \sqrt{2}\int_0^{\pi/2} \theta\sin\theta\,d\theta =$

$\sqrt{2}\left[-\theta\cos\theta + \sin\theta\right]_0^{\pi/2} = \sqrt{2}$

23. Let $v = \sec x$ and $du = \sec^2 x\,dx \Rightarrow dv = \sec x\tan x\,dx$ and $u = \tan x$, $\int \sec^3 x\,dx =$

$\sec x\tan x - \int \sec x\left(\sec^2 x - 1\right) dx = \sec x\tan x - \int \sec^3 x\,dx + \int \sec x\,dx \Rightarrow \int \sec^3 x\,dx =$

$\frac{1}{2}(\sec x\tan x + \ln|\sec x + \tan x|) + C \;\therefore\; \int_{-\pi/3}^{0} 2\sec^3 x\,dx = [\sec x\tan x + \ln|\sec x + \tan x|]_{-\pi/3}^{0} =$

$2\sqrt{3} - \ln\left(2-\sqrt{3}\right)$

25. $\displaystyle\int_0^{\pi/4} \sec^4\theta\, d\theta = \int_0^{\pi/4} \left(1 + \tan^2\theta\right)\left(\sec^2\theta\right) d\theta = \left[\tan\theta + \frac{\tan^3\theta}{3}\right]_0^{\pi/4} = \frac{4}{3}$

27. $\displaystyle\int_{\pi/4}^{\pi/2} \csc^4\theta\, d\theta = \int_{\pi/4}^{\pi/2} \left(1 + \cot^2\theta\right) \csc^2\theta\, d\theta = -\left[\cot\theta + \frac{\cot^3\theta}{3}\right]_{\pi/4}^{\pi/2} = \frac{4}{3}$

29. $\displaystyle\int_0^{\pi/4} 4\tan^3 x\, dx = 4\int_0^{\pi/4} \left(\sec^2 x - 1\right) \tan x\, dx = 4\int_0^{\pi/4} \tan x \sec^2 x - \frac{\sin x}{\cos x}\, dx =$

$\displaystyle 4\left[\frac{\tan^2 x}{2} + \ln|\cos x|\right]_0^{\pi/4} = 2 - \ln 4$

31. $\displaystyle\int_{\pi/6}^{\pi/3} \cot^3 x\, dx = \int_{\pi/6}^{\pi/3} \left(\csc^2 x - 1\right)(\cot x)\, dx = \int_{\pi/6}^{\pi/3} \cot x\left(\csc^2 x\right) - \frac{\cos x}{\sin x}\, dx =$

$\displaystyle -\left[\frac{\cot^2 x}{2} + \ln|\sin x|\right]_{\pi/6}^{\pi/3} = \frac{4}{3} - \ln\sqrt{3}$

33. $\displaystyle\int_{-\pi}^{0} \sin 3x \cos 2x\, dx = \frac{1}{2}\int_{-\pi}^{0} \sin x + \sin 5x\, dx = -\frac{1}{2}\left[\cos x + \frac{\cos 5x}{5}\right]_{-\pi}^{0} = -\frac{6}{5}$

35. $\displaystyle\int_{-\pi}^{\pi} \sin 3x \sin 3x\, dx = \frac{1}{2}\int_{-\pi}^{\pi} 1 - \cos 6x\, dx = \frac{1}{2}\left[x - \frac{\sin 6x}{6}\right]_{-\pi}^{\pi} = \pi$

37. $\displaystyle\int_0^{\pi} \cos 3x \cos 4x\, dx = \frac{1}{2}\int_0^{\pi} \cos x + \cos 7x\, dx = \frac{1}{2}\left[\sin x + \frac{\sin 7x}{7}\right]_0^{\pi} = 0$

39. The integrands for a, b, c, e, g and i are odd and are integrated over a symmetric interval about zero, which implies that the corresponding integrals are zero. The integrands for d, f and h are nonnegative and even which implies that the corresponding integrals are not zero.

41. The integrals in parts d, f and h have even integrands. $\displaystyle\int_{-\pi+2}^{\pi/2} x \sin x\, dx = 2[-x\cos x + \sin x]_0^{\pi/2} = 2,$

$\displaystyle\int_{-\pi/2}^{\pi/2} \cos^3 x\, dx = 2\int_0^{\pi/2} \left(1 - \sin^2\right) \cos x\, dx = 2\left[\sin x - \frac{\sin^3 x}{3}\right]_0^{\pi/2} = \frac{4}{3},\ \int_{-\pi/2}^{\pi/2} \sin x \sin 2x\, dx =$

$\displaystyle 4\int_0^{\pi/2} \sin^2 x \cos x\, dx = 4\left[\frac{\sin^3 x}{3}\right]_0^{\pi/2} = \frac{4}{3}$

43. $\displaystyle\int \csc x\, dx = -\int \csc x \left(\frac{-\cot x - \cot x}{\csc x + \cot x}\right) dx = -\int \frac{-\csc x \cot x - \csc x}{\csc x + \cot x}\, dx = -\ln|\csc x + \cot x| + C$

## 8.4 TRIGONOMETRIC SUBSTITUTIONS

1. $\displaystyle\int_{-2}^{2}\frac{1}{4+x^2}\,dx = 2\int_0^2\frac{1}{2^2+x^2}\,dx = 2\left[\frac{1}{2}\tan^{-1}\frac{x}{2}\right]_0^2 = \frac{\pi}{4}$

3. $\displaystyle\int_0^{3/2}\frac{1}{\sqrt{9-x^2}}\,dx = \int_0^{\pi/6}\frac{3\cos\theta}{|3\cos\theta|}\,dx = \frac{\pi}{6}$

5. $\displaystyle\int\frac{1}{\sqrt{x^2-4}}\,dx = \int\frac{2\sec\theta\tan\theta}{\sqrt{4(\sec^2\theta-1)}}\,d\theta = \int\sec\theta\,d\theta = \ln|\sec\theta+\tan\theta| + C =$

$\ln\left|x+\sqrt{x^2-4}\right| + C$, where $x = 2\sec\theta$

7. $\displaystyle\int_0^{\sqrt{3}/2}\frac{2}{1+4y^2}\,dy = \int_0^{\sqrt{3}/2}\frac{2}{1+(2y)^2}\,dy = \left[\tan^{-1}2y\right]_0^{\sqrt{3}/2} = \frac{\pi}{3}$

9. $\displaystyle\int_0^{3\sqrt{2}/4}\frac{1}{\sqrt{9-4x^2}}\,dx = \frac{3}{2}\int_0^{\pi/4}\frac{\cos\theta}{3\cos\theta}\,d\theta = \frac{\pi}{8}$, where $2x = 3\sin\theta$

11. $\displaystyle\int_{1/\sqrt{3}}^{1}\frac{2}{z\sqrt{4z^2-1}}\,dz = 2\int_{1/\sqrt{3}}^{1}\frac{2}{2z\sqrt{(2z)^2-1}}\,dz = 2\left[\sec^{-1}|2z|\right]_{1/\sqrt{3}}^{1} = \frac{\pi}{3}$

13. $\displaystyle\int_0^2\frac{1}{\sqrt{4+x^2}}\,dx = \int_0^{\pi/4}\frac{2\sec^2\theta}{2\sec\theta}\,d\theta = \ln|\sec\theta+\tan\theta|\Big|_0^{\pi/4} = \ln\left(\sqrt{2}+1\right)$, where $x = 2\tan\theta$

15. $\displaystyle\int_1^2\frac{6}{\sqrt{4-(x-1)^2}}\,dx = \int_0^{\pi/6}\frac{6(2\cos\theta)}{2|\cos\theta|}\,d\theta = \pi$, where $x-1 = 2\sin\theta$

17. $\displaystyle\int_1^3\frac{1}{y^2-2y+5}\,dy = \int_1^3\frac{1}{(y-1)^2+2^2}\,dy = \frac{1}{2}\left[\tan^{-1}\left(\frac{y-1}{2}\right)\right]_1^3 = \frac{\pi}{8}$

19. $\displaystyle\int_1^{3/2}\frac{x-1}{\sqrt{2x-x^2}}\,dx = \int_1^{3/2}\frac{x-1}{\sqrt{1-(x-1)^2}}\,dx = \int_0^{1/2}\left(1-u^2\right)^{-1/2}u\,du =$

$\displaystyle-\left[1-u^2\right]_0^{1/2} = \frac{2-\sqrt{3}}{2}$, where $u = x-1$

21. $\displaystyle\int\frac{1}{\sqrt{x^2-2x}}\,dx = \int\frac{1}{\sqrt{(x-1)^2-1}}\,dx = \int\frac{\sec\theta\tan\theta}{\tan\theta}\,d\theta = \ln|\sec\theta+\tan\theta| + C =$

$\ln\left|(x-1)+\sqrt{x^2-2x}\right| + C$, where $x-1 = \sec\theta$

23. $\displaystyle\int_{-2}^{2}\frac{x+2}{\sqrt{x^2+4x+13}}\,dx = \int_{-2}^{2}\frac{x+2}{\sqrt{(x+2)^2+3^2}}\,dx = \int_0^{\arctan 4/3}\frac{3\tan\theta\,3\sec^2\theta}{3\sec\theta}\,d\theta =$

$3\left[\sec\theta\right]_0^{\arctan 4/3} = 2$, where $x+2 = 3\tan\theta$

25. $\displaystyle\int \frac{4x^2}{\left(1-x^2\right)^{3/2}}\,dx = \int \frac{4\sin^2\theta\cos\theta}{\cos^3\theta}\,d\theta = 4\int \sec^2\theta - 1\,d\theta = 4\tan\theta - 4\theta + C =$

$\displaystyle\frac{4x}{\sqrt{1-x^2}} - 4\sin^{-1}x + C$, where $x = \sin\theta$

27. $\displaystyle A = \int_0^3 \frac{\sqrt{9-x^2}}{3}\,dx = \frac{1}{3}\int_0^{\pi/2} 3\cos\theta\, 3\cos\theta\,d\theta = \frac{3}{2}\int_0^{\pi/2} 1+\cos 2\theta\,d\theta =$

$\displaystyle\frac{3}{2}\left[\theta + \frac{\sin 2\theta}{2}\right]_0^{\pi/2} = \frac{3\pi}{4}$, where $x = 3\sin\theta$

29. $\displaystyle A = \int_0^1 \frac{2}{x^2-4x+5}\,dx = \int_0^1 \frac{2}{(x-2)^2+1}\,dx = \int_{-2}^{-1} \frac{2}{1+u^2}\,du = 2\left[\tan^{-1}u\right]_{-2}^{-1} \approx 0.643501108$

## 8.5 RATIONAL FUNCTIONS AND PARTIAL FRACTIONS

1. $\displaystyle\frac{5x-13}{(x-3)(x-2)} = \frac{A}{x-3} + \frac{B}{x-2} = \frac{2}{x-3} + \frac{3}{x-2}$

3. $\displaystyle\frac{x+4}{(x+1)^2} = \frac{A}{x+1} + \frac{B}{(x+1)^2} = \frac{1}{x+1} + \frac{3}{(x+1)^2}$

5. $\displaystyle\frac{x+1}{x^2(x-1)} = \frac{A}{x} + \frac{B}{x^2} + \frac{C}{x-1} = \frac{-2}{x} + \frac{-1}{x^2} + \frac{2}{x-1}$

7. $\displaystyle\frac{x^2+8}{x^2-5x+6} = 1 + \frac{5x+2}{x^2-5x+6} = 1 + \frac{A}{x-3} + \frac{B}{x-2} = 1 + \frac{17}{x-3} + \frac{-12}{x-2}$

9. $\displaystyle\int_0^{1/2} \frac{1}{1-x^2}\,dx = \frac{1}{2}\int_0^{1/2} \frac{1}{1+x} + \frac{1}{1-x}\,dx = \frac{1}{2}\Big[\ln|1+x| - \ln|1-x|\Big]_0^{1/2} = \ln\sqrt{3}$

11. $\displaystyle\int \frac{x+4}{x^2+5x-6}\,dx = \frac{1}{7}\int \frac{2}{x+6} + \frac{5}{x-1}\,dx = \frac{1}{7}\Big[2\ln|x+6| + 5\ln|x-1|\Big] + C =$

$\displaystyle\frac{1}{7}\ln\left|(x+6)^2(x-1)^5\right| + C$

13. $\displaystyle\int_4^8 \frac{y}{y^2-2y-3}\,dy = \frac{1}{4}\int_4^8 \frac{3}{y-3} + \frac{1}{y+1}\,dy = \frac{1}{4}\Big[3\ln|y-3| + \ln|y+1|\Big]_4^8 = \ln\sqrt{15}$

15. $\displaystyle\int \frac{1}{t^3+t^2-2t}\,dt = \int \frac{-1/2}{t} + \frac{1/6}{t+2} + \frac{1/3}{t-1}\,dt = -\frac{1}{2}\ln|t| + \frac{1}{6}\ln|t+2| + \frac{1}{3}\ln|t-1| + C$

17. $\displaystyle\int \frac{x^3}{x^2+2x+1}\,dx = \int x - 2 + \frac{3}{x+1} - \frac{1}{(x+1)^2}\,dx = \frac{x^2}{2} - 2x + 3\ln|x+1| + \frac{1}{x+1} + C$

19. $\displaystyle\int \frac{1}{\left(x^2-1\right)^2}\,dx = \frac{1}{4}\int \frac{1}{x+1} + (x+1)^{-2} - \frac{1}{x-1} + (x-1)^{-2}\,dx =$

$\displaystyle\frac{1}{4}\Big[\ln|x+1| - (x+1)^{-1} - \ln|x-1| - (x-1)^{-1}\Big] + C = \frac{1}{4}\ln\left|\frac{x+1}{x-1}\right| - \frac{x}{2\left(x^2-1\right)} + C$

21. $\int_0^{2\sqrt{2}} \frac{x^3}{x^2+1}\,dx = \int_0^{2\sqrt{2}} x - \frac{x}{x^2+1}\,dx = \left[\frac{x^2}{2} - \frac{1}{2}\ln\left(x^2+1\right)\right]_0^{2\sqrt{2}} = 4 - \ln 3$

23. $\int_1^2 \frac{1}{y^3+y}\,dy = \int_1^2 \frac{1}{y} - \frac{y}{y^2+1}\,dy = \left[\ln|y| - \frac{1}{2}\ln\left(1+x^2\right)\right]_1^2 = \ln\left(\frac{2\sqrt{2}}{\sqrt{5}}\right)$

25. $\int_0^{\sqrt{3}} \frac{5x^2}{x^2+1}\,dx = \int_0^{\sqrt{3}} 5 - \frac{5}{1+x^2}\,dx = \left[5x - 5\tan^{-1}x\right]_0^{\sqrt{3}} = 5\sqrt{3} - \frac{5\pi}{3}$

27. $\int \frac{4x+4}{x^2\left(x^2+1\right)}\,dx = 4\int \frac{1}{x} + x^{-2} - \frac{x+1}{x^2+1}\,dx = 4\ln|x| - 4x^{-1} - 2\ln\left(x^2+1\right) - 4\tan^{-1}x + C =$

$\ln\left(\frac{x^4}{\left(x^2+1\right)^2}\right) - \frac{4}{x} - 4\tan^{-1}x + C$

29. $\int_{-1}^0 \frac{2x}{\left(x^2+1\right)(x-1)^2}\,dx = \int_{-1}^0 \frac{1}{(x-1)^2} - \frac{1}{1+x^2}\,dx = \left[-\frac{1}{x-1} - \tan^{-1}x\right]_{-1}^0 = \frac{2-\pi}{4}$

31. $\int_0^1 \frac{x^3+1}{x^2+1}\,dx = \int_0^1 x + \frac{1}{x^2+1} - \frac{x}{x^2+1}\,dx = \left[\frac{x^2}{2} + \tan^{-1}x - \frac{1}{2}\ln\left(x^2+1\right)\right]_0^1 = \frac{2+\pi-\ln 4}{4}$

33. $\int_0^1 \frac{y^2+2y+1}{\left(y^2+1\right)^2}\,dy = \int_0^1 \frac{1}{y^2+1} + \frac{2y}{\left(y^2+1\right)^2}\,dy = \left[\tan^{-1}y - \frac{1}{y^2+1}\right]_0^1 = \frac{\pi+2}{4}$

35. $\int_0^1 \frac{2r^3+3r^2+5r+2}{r^2+r+1}\,dr = \int_0^1 2r+1+\frac{2r+1}{r^2+r+1}\,dr = \left[r^2+r+\ln\left|r^2+r+1\right|\right]_0^1 = 2+\ln 3$

37. $V = \pi\int_{1/2}^{5/2}\left(\frac{3}{\sqrt{3x-x^2}}\right)^2 dx = 9\pi\int_{1/2}^{5/2}\frac{1}{x(3-x)}\,dx = 3\pi\int_{1/2}^{5/2}\frac{1}{x} + \frac{1}{(3-x)}\,dx =$

$3\pi\left[\ln|x| - \ln|3-x|\right]_{1/2}^{5/2} = 3\pi\ln 25$

39. a) $\frac{dx}{dt} = kx(N-x) \Rightarrow \int \frac{1}{x(N-x)}\,dx = \int k\,dt \Rightarrow \frac{1}{N}\int \frac{1}{x} + \frac{1}{N-x}\,dx = \int k\,dt \Rightarrow$

$\frac{1}{N}\left[\ln|x| - \ln|N-x|\right] = kt + C \Rightarrow \ln\left(\frac{x}{N-x}\right) = kNt + C$, for $x < N \Rightarrow \frac{x}{N-x} = A\,e^{kNt} \Rightarrow$

$x = \frac{NAe^{kNt}}{1+Ae^{kNt}}$, where $k = \frac{1}{250}$ and $N = 1000 \Rightarrow x = \frac{1000Ae^{4t}}{1+Ae^{4t}}$, where $t = 0$ and $x = 2 \Rightarrow$

$x = \frac{1000\,e^{4t}}{499+e^{4t}}$

b) Solving $500 = \frac{1000\,e^{4t}}{499+e^{4t}} \Rightarrow t = \frac{\ln 499}{4} \approx 1.55$ days

41. $\int_0^{\pi/2} \frac{1}{1+\sin x}dx = \int_0^1 \frac{1}{1+\frac{2z}{1+z^2}} \frac{2}{1+z^2}dz = 2\int_0^1 \frac{1}{1+2z+z^2}dz = 2\int_0^1 (z+1)^{-2}dz =$

$\left[\frac{-2}{z+1}\right]_0^1 = 1$, where $z = \tan\frac{x}{2}$

43. $\int \frac{1}{1-\sin x}dx = \int \frac{1}{1-\frac{2z}{1+z^2}} \frac{2}{1+z^2}dz = 2\int \frac{1}{(z-1)^2}dz = -\frac{2}{z-1} + C = \frac{2}{1-\tan\left(\frac{x}{2}\right)} + C$

45. $\int \frac{\cos x}{1-\cos x}dx = \int \frac{\frac{1-z^2}{1+z^2}\frac{2}{1+z^2}}{1-\frac{1-z^2}{1+z^2}}dz = \int \frac{1-z^2}{z^2\left(1+z^2\right)}dz = \int \frac{1}{z^2} - \frac{2}{1+z^2}dz =$

$-\frac{1}{z} - 2\tan^{-1}z + C = \frac{-1}{\tan\left(\frac{x}{2}\right)} - 2\tan^{-1}\left(\tan\left(\frac{x}{2}\right)\right) + C = -\cot\left(\frac{x}{2}\right) - x + C$

47. $\int \frac{1}{\sin x - \cos x}dx = \int \frac{\frac{2}{1+z^2}}{\frac{2z}{1+z^2} - \frac{1-z^2}{1+z^2}}dz = \int \frac{2}{(z+1)^2 - 2}dz = \int \frac{2}{u^2-2}du =$

$\frac{1}{\sqrt{2}}\int \frac{-1}{u+\sqrt{2}} + \frac{1}{u-\sqrt{2}}du = \frac{1}{\sqrt{2}}\ln\left|\frac{u-\sqrt{2}}{u+\sqrt{2}}\right| + C = \frac{1}{\sqrt{2}}\ln\left|\frac{z+1-\sqrt{2}}{z+1+\sqrt{2}}\right| + C =$

$\frac{1}{\sqrt{2}}\ln\left|\frac{\tan\left(\frac{x}{2}\right)+1-\sqrt{2}}{\tan\left(\frac{x}{2}\right)+1+\sqrt{2}}\right| + C$, where $u = z+1$ and $z = \tan\left(\frac{x}{2}\right)$

## 8.6 USING INTEGRAL TABLES

1. $\int_0^{\infty} e^{-x^2}dx = \frac{1}{2}\sqrt{\frac{\pi}{1}} = \frac{\sqrt{\pi}}{2}$, formula 140

$\frac{1}{2}\left(\frac{1}{2}\sin^{-1}(x) - \frac{1}{2}x\sqrt{1-x^2}\right) + C = \frac{x^2}{2}\cos^{-1}(x) + \frac{1}{4}\sin^{-1}(x) - \frac{1}{4}x\sqrt{1-x^2} + C$, formulas 33, 100

3. $\int_6^9 \frac{1}{x\sqrt{x-3}}dx = \frac{2}{\sqrt{3}}\left[\tan^{-1}\sqrt{\frac{x-3}{3}}\right]_6^9 = \frac{2}{\sqrt{3}}\left(\tan^{-1}\sqrt{2} - \frac{\pi}{4}\right)$, formula 13a

5. $\int \frac{1}{\left(9-x^2\right)^2}dx = \frac{x}{18\left(9-x^2\right)} + \frac{1}{18}\int \frac{1}{9-x^2}dx = \frac{x}{18\left(9-x^2\right)} + \frac{1}{108}\ln\left|\frac{x+3}{x-3}\right| + C$ , formulas 18,19

7. $\int_3^{11} \frac{1}{x^2\sqrt{7+x^2}}\,dx = -\left[\frac{\sqrt{7+x^2}}{7x}\right]_3^{11} = \frac{44-24\sqrt{21}}{231}$, formula 27

9. $\int_{-2}^{-\sqrt{2}} \frac{\sqrt{x^2-2}}{x}\,dx = \left[\sqrt{x^2-2} - \sqrt{2}\sec^{-1}\left|\frac{x}{\sqrt{2}}\right|\right]_{-2}^{-\sqrt{2}} = \sqrt{2}\left(\frac{\pi}{4}-1\right)$, formula 42

11. $\int \frac{1}{4+5\sin 2x}\,dx = -\frac{1}{6}\left[\ln\left|\frac{5+4\sin 2x+3\cos 2x}{4+5\sin 2x}\right|\right] + C$, formula 71

13. $\int x\sqrt{2x-3}\,dx = \frac{(2x-3)^{3/2}(x+1)}{5} + C$, formula 7

15. $\int_0^{\infty} x^{10}e^{-x}\,dx = \Gamma(11) = 10!$, formula 139

17. $\int_0^1 \sin^{-1}\sqrt{x}\,dx = 2\int_0^1 u\left(\sin^{-1}u\right)du = 2\left[\frac{u^2}{2}\sin^{-1}u\right]_0^1 - \frac{1}{2}\int_0^1 \frac{u^2}{\sqrt{1-u^2}}\,du =$

$\sin^{-1}1 - \left[\frac{1}{2}\sin^{-1}u - \frac{1}{2}u\sqrt{1-u^2}\right]_0^1 = \frac{\pi}{4}$, where $u = \sqrt{x}$ and by formulas 33, 99

19. $\int_0^{1/2} \frac{\sqrt{x}}{\sqrt{1-x}}\,dx = 2\int_0^{1/\sqrt{2}} \frac{u^2}{\sqrt{1-u^2}}\,du = \left[\sin^{-1}u - u\sqrt{1-u^2}\right]_0^{1/\sqrt{2}} = \frac{\pi-2}{4}$,

where $u = \sqrt{x}$ and by formula 33

21. $M = \int_0^3 \frac{1}{\sqrt{x+1}}\,dx = 2\left[(x+1)^{1/2}\right]_0^3 = 2$, $M_x = \frac{1}{2}\int_0^3 \frac{1}{x+1}\,dx = \frac{1}{2}\left[\ln|x+1|\right]_0^3 = \ln 2$,

$M_y = \int_0^3 \frac{x}{\sqrt{x+1}}\,dx = \sqrt{x+1}\left[\frac{2(x+1)}{3} - 2\right]_0^3 = \frac{8}{3}$, $\overline{x} = \frac{M_y}{M} = \frac{4}{3}$ and $\overline{y} = \frac{M_x}{M} = \frac{\ln 2}{2} = \ln\sqrt{2}$, formula 7

23. Let $y = -\frac{1}{a}\sqrt{\frac{2a-x}{x}} + C = -\frac{1}{a}\left(2ax^{-1}-1\right)^{1/2} + C \Rightarrow y' = -\frac{1}{2a}\left(2axz^{-1}-1\right)^{-1/2}\left(-2ax^{-2}\right) =$

$\frac{|x|}{x^2\sqrt{2ax-x^2}} = \frac{1}{x\sqrt{2ax-x^2}}$, when $x > 0$ and $\frac{-1}{x\sqrt{2ax-x^2}}$, when $x < 0$.

Therefore, $\int \frac{1}{x\sqrt{2ax-x^2}}\,dx = \frac{1}{a}\sqrt{\frac{2a-x}{x}} + C$, when $x > 0$

25. $\int \frac{x}{(ax+b)^2}\,dx = \frac{1}{a^2}\int \frac{u-b}{u^2}\,du = \frac{1}{a^2}\int \frac{1}{u} - bu^{-2}\,du = \frac{1}{a^2}\left[\ln|u| + bu^{-1}\right] + C =$

$\frac{1}{a}\left[\ln|ax+b| + \frac{b}{ax+b}\right] + C$, where $u = ax+b$

27. $\int_{-\pi}^{\pi} \cos^4 x\,dx = 2\int_0^{\pi} \cos^4 x\,dx = 2\left(\left[\frac{\left(\cos^3 x\right)(\sin x)}{4}\right]_0^{\pi} + \frac{3}{4}\int_0^{\pi} \cos^2 x\,dx\right) =$

$\frac{3}{2}\left(\left[\frac{\cos x\ \sin x}{2}\right]_0^{\pi} + \frac{1}{2}\int_0^{\pi} dx\right) = \frac{3\pi}{4}$, formula 61

29. $\int_0^{\pi} \sin^4 x\,dx = -\left[\frac{\sin^3 x \cos x}{4}\right]_0^{\pi} + \frac{3}{4}\int_0^{\pi} \sin^2 x\,dx = \frac{3}{4}\left(\left[-\frac{\sin x \cos x}{2}\right]_0^{\pi} + \frac{1}{2}\int_0^{\pi} dx\right) = \frac{3\pi}{8}$, formula 6

31. $\int_0^{\pi/8} \tan^3 2x\,dx = \left[\frac{\tan^2 2x}{4}\right]_0^{\pi/8} + \int_0^{\pi/8} \tan 2x\,dx = \frac{1}{4} + \frac{1}{2}\left[\ln|\cos 2x|\right]_0^{\pi/8} = \frac{1 - \ln 2}{4}$, formula 86

33. $\int_{\pi/4}^{3\pi/4} \cot^4 x\,dx = \left[-\frac{\cot^3 x}{3}\right]_{\pi/4}^{3\pi/4} - \int_{\pi/4}^{3\pi/4} \cot^2 x\,dx = \frac{2}{3} - [-\cot x]_{\pi/4}^{3\pi/4} - \int_{\pi/4}^{3\pi/4} dx = \frac{3\pi - 8}{6}$, formula 87

35. $\int_{-\pi/3}^{\pi/3} \sec^4 x\,dx = 2\int_0^{\pi/3} \sec^4 x\,dx = 2\left(\left[\frac{\sec^2 x \tan x}{3}\right]_0^{\pi/3} + \frac{2}{3}\int_0^{\pi/3} \sec^2 x\,dx\right) =$

$2\left[\frac{\sec^2 x \tan x}{3} + \frac{2}{3}\tan x\right]_0^{\pi/3} = 4\sqrt{3}$, formula 92

37. $\int_{\pi/4}^{\pi/2} \csc^4 x\,dx = \left[-\frac{\csc^2 x \cot x}{3}\right]_{\pi/4}^{\pi/2} + \frac{2}{3}\int_{\pi/4}^{\pi/2} \csc^2 x\,dx = \left[-\frac{\csc^2 x \cot x}{3} - \frac{2}{3}\cot x\right]_{\pi/4}^{\pi/2} = \frac{4}{3}$, formula 93

39. $\int 16x^3(\ln x)^2\,dx = 16\left(\frac{x^4(\ln x)^2}{4} - \frac{1}{2}\int x^3(\ln x)\,dx\right) = 16\left(\frac{x^4(\ln x)^2}{4} - \frac{1}{2}\left(\frac{x^4(\ln x)}{4} - \frac{1}{4}\int x^3\,dx\right)\right)$, where

$\int x^m(\ln x)^n\,dx = \frac{x^{m+1}(\ln x)^n}{m+1} - \frac{n}{m+1}\int x^m(\ln x)^{n-1}\,dx \therefore \int_1^3 16x^3(\ln x)^2\,dx =$

$\left[16\left(\frac{x^4(\ln x)^2}{4} - \frac{x^4(\ln x)}{8} + \frac{x^4}{32}\right)\right]_1^3 = 324(\ln 3)^2 - 162(\ln 3) + 40$

41. $\int_0^1 \left(x^2 + 1\right)^{-3/2} dx = \int_0^{\pi/4} \left((\tan u)^2 + 1\right)^{-3/2} \sec^2 u\,du = \int_0^{\pi/4} \cos u\,du = [\sin u]_0^{\pi/4} =$

$\frac{1}{\sqrt{2}}$, where $x = \tan u$

43. $\int_0^{3/5} \frac{1}{\left(1 - x^2\right)^3}\,dx = \int_0^{\arcsin(3/5)} \frac{\cos u}{\cos^6 u}\,du = \int_0^{\arcsin(3/5)} \sec^5 u\,du$, where $x = \sin u \Rightarrow$

$\int_0^{3/5} \frac{1}{\left(1 - x^2\right)^3}\,dx = \left[\frac{\sec^3 u \tan u}{4} + \frac{3 \sec u \tan u}{8} + \frac{3}{8}\ln|\sec u + \tan u|\right]_0^{\arcsin(3/5)} =$

$\frac{735 + 384 \ln 2}{1024}$, formula 92

45. $y = -\frac{\sin^{n-1} x \cos x}{n} + \frac{n-1}{n}\int \sin^{n-2} x\,dx \Rightarrow y' = -\frac{(n-1)\sin^{n-2} x \cos^2 c + \sin^n x}{n} + \frac{n-1}{n}\sin^{n-2} x =$

$\frac{(n-1)\sin^n x + \sin^n x}{n} = \sin^n x \Rightarrow$ that the formula is true

47. a) Let $u = x^n$ and $dv = \sin x\,dx \Rightarrow du = n\,x^{n-1}\,dx$ and $v = -\cos x$

$\int x^n \sin x\,dx = -x^n \cos x + n\int x^{n-1} \cos x\,dx$

b) Let $u = x^n$ and $dv = \sin ax\,dx \Rightarrow du = n\,x^{n-1}\,dx$ and $v = -\frac{\cos ax}{a}$

$\int x^n \sin ax\,dx = -\frac{x^n \cos ax}{a} + \frac{n}{a}\int x^{n-1} \cos ax\,dx$

## 8.7 IMPROPER INTEGRALS

1. $\int_0^{\infty} \frac{1}{x^2+1}\,dx = \lim_{t \to \infty} \left[\tan^{-1}\right]_0^t = \left(\lim_{t \to \infty} \tan^{-1} t\right) - \tan^{-1} 0 = \frac{\pi}{2}$

3. $\int_{-1}^{1} \frac{1}{x^{2/3}}\,dx = \int_{-1}^{0} x^{-2/3}\,dx + \int_0^1 x^{-2/3}\,dx = \lim_{t \to 0^-} \int_{-1}^{t} x^{-2/3}\,dx + \lim_{t \to 0^+} \int_t^1 x^{-2/3}\,dx =$

$\lim_{t \to 0^-} \left[3x^{1/3}\right]_{-1}^{t} + \lim_{t \to 0^+} \left[3x^{1/3}\right]_t^1 = (0) - (-3) + (3) - (0) = 6$

5. $\int_0^4 \frac{1}{\sqrt{4-x}}\,dx = \lim_{t \to 4^-} \int_0^t (4-x)^{-1/2}\,dx = = -2 \lim_{t \to 4^-} \left[(4-t)^{1/2} - 2\right] = 4$

7. $\int_0^1 \frac{1}{x^{0.999}}\,dx = \lim_{t \to 0^+} \int_t^1 x^{-0.999}\,dx = 1000 \lim_{t \to 0^+} \left(1^{0.001} - t^{0.001}\right) = 1000$

9. $\frac{2}{x^2 - x} = -\frac{2}{x} + \frac{2}{x-1} \Rightarrow \int_2^{\infty} \frac{2}{x^2 - x}\,dx = \lim_{t \to \infty} \int_2^t -\frac{2}{x} + \frac{2}{x-1}\,dx = 2 \lim_{t \to \infty} \left[\ln\left|\frac{x-1}{x}\right|\right]_2^t = \ln 4$

11. $\int_1^{\infty} \frac{1}{\sqrt{x}}\,dx = \lim_{t \to \infty} \int_1^t x^{-1/2}\,dx = 2 \lim_{t \to \infty} \left[\sqrt{x}\right]_1^t = \infty$, diverges

13. $\int_1^{\infty} \frac{1}{x^3+1}\,dx$ converges, for $\lim_{t \to \infty} \frac{\frac{1}{x^3+1}}{\frac{1}{x^3}} = 1$, exercise 12 and Theorem 2

15. $\int_0^{\infty} \frac{1}{x^{3/2}+1}\,dx = \int_0^1 \frac{1}{x^{3/2}+1}\,dx + \int_1^{\infty} \frac{1}{x^{3/2}+1}\,dx$; the first integral is finite and the second is dominated by $\int_1^{\infty} \frac{1}{x^{3/2}}\,dx$ which converges, $\therefore \int_0^{\infty} \frac{1}{x^{3/2}+1}\,dx$ converges

17. $\int_0^{\pi/2} \tan x\,dx = -\lim_{t \to \pi/2^-} \int_0^t -\frac{\sin x}{\cos x}\,dx = -\lim_{t \to \pi/2^-} \left(\ln|\cos t| - \ln 1\right) = \infty$, diverges

19. $\int_{-1}^{1} \frac{1}{x^{2/5}}\,dx = 2 \lim_{t \to 0^+} \int_t^1 x^{-2/5}\,dx = \frac{10}{3} \lim_{t \to 0^+} \left(t^{3/5} - (-1)^{3/5}\right) = \frac{10}{3}$, converges

21. For $x > 2 \Rightarrow x^2 > x - 1 \Rightarrow x > \sqrt{x-1} \Rightarrow \frac{1}{x} < \frac{1}{\sqrt{x-1}}$ and $\int_2^{\infty} \frac{1}{x}\,dx$ diverges $\Rightarrow \int_2^{\infty} \frac{1}{\sqrt{x-1}}\,dx$

also diverges by Theorem 1

23. $\int_0^2 \frac{1}{1-x^2}dx = \frac{1}{2}\int_0^2 \frac{1}{1+x}dx + \frac{1}{2}\int_0^2 \frac{1}{1-x}dx = \frac{1}{2}\int_0^2 \frac{1}{1+x}dx + \frac{1}{2}\int_0^1 \frac{1}{1-x}dx + \frac{1}{2}\int_1^2 \frac{1}{1-x}dx$, from example 2 we have that $\int_0^1 -\frac{1}{x}dx$, diverges and $\lim_{t \to \infty} \frac{-1/x}{1/(1-x)} = 1$ $\therefore$ by Theorem 2, $\int_0^1 \frac{1}{1-x}dx$ diverges $\Rightarrow \int_0^2 \frac{1}{1-x^2}dx$ diverges

25. $\int_0^\infty \frac{1}{x^3}dx = \int_0^1 \frac{1}{x^3}dx + \int_1^\infty \frac{1}{x^3}dx$, $\int_0^1 \frac{1}{x^3}dx = \lim_{t \to 0^+} \int_t^1 x^{-3}dx = -\frac{1}{2}\lim_{t \to 0^+}\left(1 - \frac{1}{t^2}\right) = \infty$, diverges, $\lim_{t \to \infty} \frac{\frac{1}{\sqrt{x^6+1}}}{\frac{1}{x^3}} = \sqrt{\lim_{t \to \infty} \frac{x^6}{x^6+1}} = 1$; by Theorem 2, $\int_0^\infty \frac{1}{\sqrt{x^6-1}}dx$ diverges

27. $\int_0^\infty x^2 e^{-x}dx = \lim_{t \to \infty} \int_0^t x^2 e^{-x}dx = \lim_{t \to \infty} \left[-\left(x^2+2x+2\right)e^{-x}\right]_0^t =$
$-\lim_{t \to \infty}\left[\left(\frac{t^2}{e^t} + \frac{2t}{e^t} + \frac{2}{e^t}\right) - 2\right] = 2$, converges

29. $\frac{1}{x} \le \frac{2+\cos x}{x}$ and $\int_\pi^\infty \frac{1}{x}dx$ diverges $\Rightarrow \int_\pi^\infty \frac{2+\cos x}{x}dx$ diverges by Theorem 1

31. $\int_0^\infty \frac{1}{\sqrt{x+5}}dx = \lim_{t \to \infty}\int_0^t (x+5)^{-1/2}dx = 2\lim_{t \to \infty}\left(\sqrt{t+5} - \sqrt{5}\right) = \infty$, diverges

33. $\int_2^\infty \frac{2}{x^2-1}dx = \lim_{t \to \infty}\int_2^t \frac{1}{x-1} - \frac{1}{x+1}dx = \lim_{t \to \infty}\left[\ln\left|\frac{x-1}{x+1}\right|\right]_2^t = \lim_{t \to \infty}\left(\ln\left|\frac{t-1}{t+1}\right| - \ln\left(\frac{1}{3}\right)\right) =$
$-\ln\left(\frac{1}{3}\right) = \ln 3$, converges

35. $\frac{1}{x} \le \frac{1}{\ln x}$ and $\int_1^\infty \frac{1}{x}dx$ diveges $\Rightarrow \int_1^\infty \frac{1}{\ln x}dx$ diverges by Theorem 1

37. $\int_1^\infty \frac{1}{e^x}dx = \lim_{t \to \infty}\int_1^t e^{-x}dx = \lim_{t \to \infty}\left[-e^{-x}\right]_1^t = 1 \Rightarrow \int_1^\infty \frac{1}{e^x}dx$ converges and $\lim_{t \to \infty}\frac{\frac{1}{e^x - 2^x}}{\frac{1}{e^x}} =$
$\lim_{t \to \infty}\frac{e^x}{e^x - 2^x} = \lim_{t \to \infty}\frac{1}{1-(2/e)^t} = 1 \Rightarrow \int_1^\infty \frac{1}{e^x - 2^x}dx$ converges by Theorem 2

39. $\int_0^{\infty} \frac{1}{\sqrt{x+x^4}}\,dx = \int_0^{1} \frac{1}{\sqrt{x+x^4}}\,dx + \int_1^{\infty} \frac{1}{\sqrt{x+x^4}}\,dx$; $\frac{1}{\sqrt{x+x^4}} < \frac{1}{\sqrt{x}}$ and $\int_0^1 \frac{1}{\sqrt{x}}\,dx = 2$,

converges from exercise 2 $\Rightarrow \int_0^{1} \frac{1}{\sqrt{x+x^4}}\,dx$ converges by Theorem 1 ; $\frac{1}{\sqrt{x+x^4}} < \frac{1}{x^2}$ and

$\int_1^{\infty} \frac{1}{x^2}\,dx$ converges from remarks following example 5 $\Rightarrow \int_1^{\infty} \frac{1}{\sqrt{x+x^4}}\,dx$ converges by

Theorem 1; $\therefore \int_0^{\infty} \frac{1}{\sqrt{x+x^4}}\,dx$ must also converge

41. $\int_3^{\infty} e^{-3x}\,dx = -\frac{1}{3} \lim_{t \to \infty} \left[e^{-3x}\right]_3^t = -\frac{1}{3} \lim_{t \to \infty} \left(\frac{1}{e^{3t}} - \frac{1}{e^9}\right) = \frac{1}{3\,e^9} \approx 0.000041136$; from the

Calculus Tool Kit we get 0.88617257

43. $\int_1^{\infty} \frac{1}{x^p}\,dx = \lim_{t \to \infty} \int_1^{t} \frac{1}{x^p}\,dx = \lim_{t \to \infty} \left[\frac{x^{(1-p)}}{1-p}\right]_1^t = \left(\frac{1}{1-p}\right) \lim_{t \to \infty} \left(\frac{1}{t^{(p-1)}} - 1\right)$

$\therefore \int_1^{\infty} \frac{1}{x^p}\,dx = \frac{1}{p-1}$ when $p > 1$ and diverges when $p < 1$

45. $A = \lim_{t \to \infty} \int_0^{t} e^{-x}\,dx = -\lim_{t \to \infty} \left[\frac{1}{e^x}\right]_0^t = -\lim_{t \to \infty} \left(\frac{1}{e^t} - 1\right) = 1$

47. $V = 2\pi \int_0^{\infty} x\,e^{-x}\,dx = 2\pi$; see exercise 46.

49. $\int_0^{\pi/2} (\sec x - \tan x)\,dx = \lim_{t \to (\pi/2)^-} \int_0^{t} \left(\sec x - \frac{\sin x}{\cos x}\right) dx =$

$\lim_{t \to (\pi/2)^-} \left[\ln|\sec x + \tan x| + \ln|\cos x|\right]_0^t = \lim_{t \to (\pi/2)^-} \ln\left|\frac{\sec t + \tan t}{\sec t}\right| = \lim_{t \to (\pi/2)^-} \ln|1 + \sin t| = \ln 2$

## PRACTICE EXERCISES

1. $\int_0^{\pi/2} \frac{\cos x}{\sqrt{1+\sin x}}\,dx = \int_0^{\pi/2} (1+\sin x)^{-1/2}\cos x\,dx = 2\left[(1+\sin x)^{1/2}\right]_0^{\pi/2} = 2\sqrt{2}-2$

3. $\int_{-1}^{1} \frac{2y}{y^4+1}\,dy = 0$; integrand is odd.

5. $\int_0^{\sqrt{2}/2} \frac{\sin^{-1}x}{\sqrt{1-x^2}}\,dx = \frac{1}{2}\left[\left(\sin^{-1}x\right)^2\right]_0^{\sqrt{2}/2} = \frac{\pi^2}{32}$

7. $\int_{\pi/4}^{\pi/3} \frac{1}{2\sin x\cos x}\,dx = \int_{\pi/4}^{\pi/3} \csc(2x)\,dx = -\frac{1}{2}\left[\ln|\csc 2x + \cot 2x|\right]_{\pi/4}^{\pi/3} = \frac{\ln\sqrt{3}}{2}$

9. $\int \frac{x+4}{x^2+1}\,dx = \frac{1}{2}\int \frac{2x}{x^2+1}\,dx + 4\int \frac{1}{1+x^2}\,dx = \frac{1}{2}\ln\left(x^2+1\right) + 4\tan^{-1}x + C$

11. Let $v = \ln x$ and $du = x^2\,dx \Rightarrow dv = \frac{1}{x}\,dx$ and $u = \frac{x^3}{3}$

$\int x^2 \ln x\,dx = \frac{x^3}{3}\ln x - \frac{1}{3}\int x^3 \frac{1}{x}\,dx = \frac{x^3}{3}\ln x - \frac{1}{3}\int x^2\,dx = \frac{x^3}{3}\ln x - \frac{x^3}{9} + C$

13. $\int x^5 \sin x\,dx = -x^5\cos x + 5x^4\sin x + 20x^3\cos x - 60x^2\sin x - 120x\cos x + 120\sin x + C$,

tabular integration

| | | |
|---|---|---|
| $x^5$ | + | $\sin x$ |
| $5x^4$ | − | $-\cos x$ |
| $20x^3$ | + | $-\sin x$ |
| $60x^2$ | − | $\cos x$ |
| $120x$ | + | $\sin x$ |
| $120$ | − | $-\cos x$ |
| $0$ | | $-\sin x$ |

15. Let $u = e^x$ and $dv = \cos 2x\,dx \Rightarrow du = e^x\,dx$ and $v = \frac{\sin 2x}{2}$

$\int e^x\cos 2x\,dx = \frac{e^x\sin 2x}{2} - \frac{1}{2}\int e^x\sin 2x\,dx$

let $u = e^x$ and $dv = \sin 2x\,dx \Rightarrow du = e^x\,dx$ and $v = \frac{-\cos 2x}{2}$

$\int e^x\sin 2x\,dx = -\frac{e^x\cos 2x}{2} + \frac{1}{2}\int e^x\cos 2x\,dx$

$\therefore \int e^x\cos 2x\,dx = \frac{e^x\sin 2x}{2} - \frac{1}{2}\left[-\frac{e^x\cos 2x}{2} + \frac{1}{2}\int e^x\cos 2x\,dx\right] \Rightarrow$

$\frac{5}{4}\int e^x\cos 2x\,dx = \frac{e^x\sin 2x}{2} + \frac{e^x\cos 2x}{4} + C \Rightarrow \int e^x\cos 2x\,dx = \frac{2e^x\cos 2x}{5} + \frac{e^x\cos 2x}{5} + C$

17. $\int \sin^3 y\, dy = \int \left(1 - \cos^2 y\right) \sin y\, dy = \int \sin y - \cos^2 y \sin y\, dy = -\cos y + \frac{\cos^3 y}{3} + C$

19. $\int \sin^4 x \cos^2 x\, dx = \int \sin^4 x \left(1 - \sin^2 x\right) dx = \int \sin^4 x - \sin^6 x\, dx = \int \sin^4 x\, dx - \int \sin^6 x\, dx =$

$\int \sin^4 x\, dx + \frac{\sin^5 x \cos x}{6} - \frac{5}{6}\int \sin^4 x\, dx = \frac{1}{6}\int \sin^4 x\, dx + \frac{\sin^5 x \cos x}{6} =$

$\frac{1}{6}\left[-\frac{\sin^3 x \cos x}{4} + \frac{3x}{8} - \frac{3 \sin 2x}{16}\right] + \frac{\sin^5 x \cos x}{6} + C =$

$-\frac{\sin^3 x \cos x}{24} + \frac{x}{16} - \frac{\sin 2x}{32} + \frac{\sin^5 x \cos x}{6} + C$, reduction formula 60

21. $\int_0^{\pi} \sqrt{\frac{1 + \cos 2x}{2}}\, dx = \int_0^{\pi} |\cos x|\, dx = \int_0^{\pi/2} \cos x\, dx - \int_{\pi/2}^{\pi} \cos x\, dx = [\sin x]_0^{\pi/2} - [\sin x]_{\pi/2}^{\pi} = 2$

23. $\int_0^{\pi/3} \tan^3 t\, dt = \int_0^{\pi/3} \left(\sec^2 t - 1\right) \tan t\, dt = \int_0^{\pi/3} (\sec t)(\sec t \tan t) - \frac{\sin t}{\cos t}\, dt =$

$\left[\frac{\sec^2 t}{2} + \ln|\cos t|\right]_0^{\pi/3} = \frac{3}{2} - \ln 2$

25. $\int_0^3 \frac{1}{\left(16 + z^2\right)^{3/2}}\, dz = \frac{1}{16}\int_0^{\arctan 3/4} \cos\theta\, d\theta = \left[\frac{\sin\theta}{16}\right]_0^{\arctan 3/4} = \frac{3}{80}$, where $z = 4\tan\theta$

27. $\int \frac{1}{x^2\sqrt{1 - x^2}}\, dx = \int \csc^2\theta\, d\theta = -\cot\theta + C = -\frac{\sqrt{1 - x^2}}{x} + C$, where $x = \sin\theta$

29. $\int_{5/4}^{5/3} \frac{12}{\left(x^2 - 1\right)^{3/2}}\, dx = 12\int_{\text{arcsec } 5/4}^{\text{arcsec } 5/3} \csc\theta \cot\theta\, d\theta = \left[-12 \csc\theta\right]_{\text{arcsec } 5/4}^{\text{arcsec } 5/3} = 5$, where $x = \sec\theta$

31. $\int_{1/3}^{1} \frac{3}{9x^2 - 6x + 5}\, dx = \int_{1/3}^{1} \frac{3}{(3x - 1)^2 + 2^2}\, dx = \frac{1}{2}\left[\tan^{-1}\left(\frac{3x - 1}{2}\right)\right]_{1/3}^{1} = \frac{\pi}{8}$

33. $\int_0^1 \frac{1}{(x + 1)\sqrt{x^2 + 2x}}\, dx = \int_0^1 \frac{1}{(x + 1)\sqrt{(x + 1)^2 - 1}}\, dx = \int_1^2 \frac{1}{u\sqrt{u^2 - 1}}\, dx =$

$\left[\sec^{-1}|u|\right]_1^2 = \frac{\pi}{3}$, where $u = x + 1$

35. $\int_2^6 \frac{x^3 + x^2}{x^2 + x - 2}\, dx = \int_2^6 x + \frac{4/3}{x + 2} + \frac{2/3}{x - 1}\, dx = \left[\frac{x^2}{2} + \frac{4}{3}\ln|x + 2| + \frac{2}{3}\ln|x - 1|\right]_2^6 = \frac{48 + \ln 400}{3}$

37. $\int \frac{x}{(x - 1)^2}\, dx = \int \frac{1}{u} + u^{-2}\, du = \ln u - \frac{1}{u} + C = \ln|x - 1| - \frac{1}{x - 1} + C$, where $u = x - 1$

39. $\int \frac{4}{x^3 + 4x}\, dx = \int \frac{1}{x} - \frac{x}{x^2 + 4}\, dx = \ln|x| - \frac{1}{2}\int \frac{2x}{x^2 + 4}\, dx = \ln|x| - \frac{1}{2}\ln\left|x^2 + 4\right| + C =$

$\ln\left|\frac{x}{\sqrt{x^2 + 4}}\right| + C$

41. $\int_3^{\infty} \frac{2}{x^2 - 2x}\,dx = = \lim_{t \to \infty} \int_3^t -\frac{1}{x} + \frac{1}{x-2}\,dx = = \lim_{t \to \infty} \left[\ln\left|\frac{x-2}{x}\right|\right]_3^t =$

$\lim_{t \to \infty} \left(\ln\left|\frac{t-2}{t}\right| - \ln\left(\frac{1}{3}\right)\right) = \ln 3$

43. Let $v = (\ln x)^2$ and $du = dx \Rightarrow dv = 2(\ln x)\frac{1}{x}\,dx$ and $u = x$, $\int (\ln x)^2\,dx = x(\ln x)^2 - 2\int \ln x\,dx$,

let $v = \ln x$ and $du = dx \Rightarrow dv = \frac{1}{x}\,dx$ and $u = x$, $\int \ln x\,dx = x\ln x - \int dx = x\ln x - x + C$

$\therefore \int (\ln x)^2\,dx = x(\ln x)^2 - 2(x\ln x - x) + C = x(\ln x)^2 - 2x\ln x + 2x + C$

$M_x = 2\int_1^e \frac{1-(\ln x)^2}{2}\,dx = \int_1^e 1 - (\ln x)^2\,dx = e - 1 - \int_1^e (\ln x)^2\,dx =$

$e - 1 - \left[x(\ln x)^2 - 2x\ln x + 2x\right]_1^e = 1$

45. $V = 2\pi\int_0^1 x\left(3x\sqrt{1-x}\right)dx = 6\pi\int_0^1 u^{1/2} - 2u^{3/2} + u^{5/2}\,du = 6\pi\left[\frac{2}{3}u^{3/2} - \frac{4}{5}u^{5/2} + \frac{2}{7}u^{7/2}\right]_0^1 = \frac{32\pi}{35}$,

where $u = 1 - x$

47. $\frac{dx}{dt} = kx(a-x) \Rightarrow \frac{1}{ak}\left(\frac{1}{x} + \frac{1}{a-x}\right)dx = dt \Rightarrow \int \frac{1}{ak}\left(\frac{1}{x} + \frac{1}{a-x}\right)dx = \int dt \Rightarrow \frac{1}{ak}\ln\left|\frac{x}{a-x}\right| = t + C$,

$x = x_0$ when $t = 0 \Rightarrow \frac{x}{a-x} = Ae^{akt}$, where $A = \frac{x_0}{a-x_0}$; solving $\frac{x}{a-x} = Ae^{akt}$ for x when $A = \frac{x_0}{a-x_0}$

yields $x = \frac{x_0 a e^{akt}}{a - x_0 + x_0 e^{akt}}$

49. $V = \pi \lim_{t \to 0^+} \int_t^1 (\ln x)^2\,dx = \pi \lim_{t \to 0^+} \left[x(\ln x)^2 - 2x\ln x + 2x\right]_t^1 = \pi \lim_{t \to 0^+} \left(2 - t(\ln t)^2 + 2t\ln t - 2t\right) =$

$\pi\left[\lim_{t \to 0^+}\left(\frac{1/t}{-1/t^2}\right) + 2\right] = 2\pi$; see exercise 43.

# APPENDICES

## APPENDIX A-2 PROOFS OF THE LIMIT THEOREMS IN CHAPTER 2

1. Let $\lim_{x \to x_0} f_1(x) = L_1$, $\lim_{x \to x_0} f_2(x) = L_2$, $\lim_{x \to x_0} f_3(x) = L_3$. Then $\lim_{x \to x_0} (f_1(x) + f_2(x)) = L_1 + L_2$ by Theorem 1. Thus $\lim_{x \to x_0}(f_1(x) + f_2(x) + f_3(x)) = \lim_{x \to x_0}((f_1(x) + f_2(x)) + f_3(x)) = (L_1 + L_2) + L_3 = L_1 + L_2 + L_3$.

   Suppose functions $f_1(x), f_2(x), f_3(x), \ldots, f_n(x)$ have limits $L_1, L_2, L_3, \ldots, L_n$ as $x \to x_0$.
   Prove $\lim_{x \to x_0}\left(f_1(x) + f_2(x) + f_3(x) + \cdots + f_n(x)\right) = L_1 + L_2 + L_3 + \cdots + L_n$, n a positive integer.

   Step 1: For n = 1, we are given that $\lim_{x \to x_0} f_1(x) = L_1$.

   Step 2: Assume $\lim_{x \to x_0} \left(f_1(x) + f_2(x) + f_3(x) + \cdots + f_k(x)\right) = L_1 + L_2 + L_3 + \cdots + L_k$ for some k.
   Then $\lim_{x \to x_0} \left(f_1(x) + f_2(x) + f_3(x) + \cdots + f_k(x) + f_{k+1}(x)\right) =$
   $\lim_{x \to x_0} \left(\left(f_1(x) + f_2(x) + f_3(x) + \cdots + f_k(x)\right) + f_{k+1}(x)\right) =$
   $\left(L_1 + L_2 + L_3 + \cdots + L_k\right) + L_{k+1} = L_1 + L_2 + L_3 + \cdots + L_k + L_{k+1}$.

   $\therefore$ by Steps 1 and 2 and mathematical induction, $\lim_{x \to x_0} \left(f_1(x) + f_2(x) + f_3(x) + \cdots + f_n(x)\right)$
   $= L_1 + L_2 + L_3 + \cdots + L_n$

3. Given $\lim_{x \to x_0} x = x_0$. Then $\lim_{x \to x_0} x^n = \lim_{x \to x_0} \left(x \bullet x \bullet x \bullet \cdots \bullet x\right) = x_0 \bullet x_0 \bullet x_0 \bullet \cdots \bullet x_0 = x_0^n$ by Exercise 2.
   (n factors) (n factors)

5. $$\lim_{x \to x_0} \frac{f(x)}{g(x)} = \frac{\lim_{x \to x_0} f(x)}{\lim_{x \to x_0} g(x)} \text{ (by Theorem 1) } = \frac{f(x_0)}{g(x_0)} \text{ if } g(x_0) \neq 0 \text{ (by Exercise 4).}$$

## APPENDIX A-4 MATHEMATICAL INDUCTION

1. Step 1: For n = 1, $|x_1| = |x_1| \leq |x_1|$.
   Step 2: Assume $|x_1 + x_2 + \cdots + x_k| \leq |x_1| + |x_2| + \cdots + |x_k|$ for some positive integer k.
   Then $|x_1 + x_2 + \cdots + x_k + x_{k+1}| = |(x_1 + x_2 + \cdots + x_k) + x_{k+1}|$
   $\leq |x_1 + x_2 + \cdots + x_k| + |x_{k+1}|$ (by the triangle inequality)
   $\leq |x_1| + |x_2| + \cdots + |x_k| + |x_{k+1}|$. $\therefore$ $|x_1 + x_2 + \cdots + x_n| \leq |x_1| + |x_2| + \cdots + |x_n|$ for all positive integers n by Steps 1 and 2 and mathematical induction.

3. Step 1: For $n = 1$, $\frac{d}{dx}(x) = 1 = 1 \cdot x^0$.

   Step 2: Assume $\frac{d}{dx}(x^k) = kx^{k-1}$ for some positive integer k.

   Then $\frac{d}{dx}(x^{k+1}) = \frac{d}{dx}(x^k x) = \frac{d}{dx}(x^k)x + x^k \frac{d}{dx}(x) = kx^{k-1}x + 1 \cdot x^k = kx^k + x^k = (k + 1)x^k = (k + 1)x^{(k+1)-1}$.

   $\therefore \frac{d}{dx}(x^n) = nx^{n-1}$ for all positive integers n by Steps 1 and 2 and mathematical induction.

5. Step 1: For $n = 1$, $\frac{2}{3^1} = \frac{2}{3} = 1 - \frac{1}{3^1}$.

   Step 2: Assume $\frac{2}{3^1} + \frac{2}{3^2} + \cdots + \frac{2}{3^k} = 1 - \frac{1}{3^k}$ for some positive integer k.

   Then $\frac{2}{3^1} + \frac{2}{3^2} + \cdots + \frac{2}{3^k} + \frac{2}{3^{k+1}} = 1 - \frac{1}{3^k} + \frac{2}{3^{k+1}} = 1 - \frac{3^{k+1}}{3^k 3^{k+1}} + \frac{2(3^k)}{3^k 3^{k+1}} = 1 - \left(\frac{3^{k+1} - 2(3^k)}{3^k 3^{k+1}}\right) =$
   $1 - \left(\frac{3^k(3 - 2)}{3^k 3^{k+1}}\right) = 1 - \frac{1}{3^{k+1}}$. $\therefore \frac{2}{3^1} + \frac{2}{3^2} + \cdots + \frac{2}{3^n} = 1 - \frac{1}{3^n}$ for all positive integers n by Steps 1 and 2 and mathematical induction.

7.

| n | 1 | 2 | 3 | 4 | 5 | 6 |
|---|---|---|---|---|---|---|
| $2^n$ | 2 | 4 | 8 | 16 | 32 | 64 |
| $n^2$ | 1 | 4 | 9 | 16 | 25 | 36 |

   Step 1: For $n = 5$, $2^5 = 32 > 25 = 5^2$.

   Step 2: Assume $2^k > k^2$ for some $k \geq 5$, k a positive integer.

   Then $2^{k+1} = 2^k(2) > 2k^2$. Now $k \geq 5 \Rightarrow k - 1 \geq 4 \Rightarrow (k - 1)^2 \geq 16 \Rightarrow (k - 1)^2 - 2 \geq 14 \Rightarrow$
   $(k - 1)^2 - 2 > 0$. Then $k^2 - 2k + 1 - 2 > 0 \Rightarrow k^2 - 2k - 1 > 0 \Rightarrow k^2 > 2k + 1 \Rightarrow k^2 + k^2 > k^2 + 2k + 1 \Rightarrow$
   $2k^2 > (k + 1)^2$.

   $\therefore 2^{k+1} > (k + 1)^2$. Thus $n^n > n^2$ for positive integers $n \geq 5$ by Steps 1 and 2 and mathematical induction.

9. Step 1: For $n = 1$, $1^2 = \frac{1(1 + 1)(2(1) + 1)}{6}$.

   Step 2: Assume $1^2 + 2^2 + 3^2 + \cdots + k^2 = \frac{k(k + 1)(2k + 1)}{6}$ for some positive integer k.

   Then $1^2 + 2^2 + 3^2 + \cdots + k^2 + (k + 1)^2 = \frac{k(k + 1)(2k + 1)}{6} + (k + 1)^2 \Rightarrow$

   $1^2 + 2^2 + 3^2 + \cdots + k^2 + (k + 1)^2 = \frac{k(k + 1)(2k + 1) + 6(k + 1)^2}{6} \Rightarrow$

   $1^2 + 2^2 + 3^2 + \cdots + k^2 + (k + 1)^2 = \frac{(k + 1)(k(2k + 1) + 6(k + 1))}{6} \Rightarrow$

   $1^2 + 2^2 + 3^2 + \cdots + k^2 + (k + 1)^2 = \frac{(k + 1)(2k^2 + 7k + 6)}{6} \Rightarrow$

   $1^2 + 2^2 + 3^2 + \cdots + k^2 + (k + 1)^2 = \frac{(k + 1)(k + 2)(2k + 3)}{6} \Rightarrow$

   $1^2 + 2^2 + 3^2 + \cdots + k^2 + (k + 1)^2 = \frac{(k + 1)((k + 1) + 1)(2(k + 1) + 1)}{6}$.

   $\therefore$ by Steps 1 and 2 and mathematical induction, $1^2 + 2^2 + 3^2 + \cdots + n^2 = \frac{n(n + 1)(2n + 1)}{6}$ for all positive integers n.

11. a) Step 1: For n = 1, $\sum_{k=1}^{1} (a_k + b_k) = a_1 + b_1 = \sum_{k=1}^{1} a_k + \sum_{k=1}^{1} b_k$.

Step 2: Assume $\sum_{k=1}^{j} (a_k + b_k) = \sum_{k=1}^{j} a_k + \sum_{k=1}^{j} b_k$ for some positive integer j.

Then $\sum_{k=1}^{j+1} (a_k + b_k) = \left( \sum_{k=1}^{j} (a_k + b_k) \right) + (a_{j+1} + b_{j+1}) = \sum_{k=1}^{j} a_k + \sum_{k=1}^{j} b_k + a_{j+1} + b_{j+1} =$

$\left( \sum_{k=1}^{j} a_k \right) + a_{j+1} + \left( \sum_{k=1}^{j} b_k \right) + b_{j+1} = \sum_{k=1}^{j+1} a_k + \sum_{k=1}^{j+1} b_k$.

$\therefore$ by Steps 1 and 2 and mathematical induction, $\sum_{k=1}^{n} (a_k + b_k) = \sum_{k=1}^{n} a_k + \sum_{k=1}^{n} b_k$ for all positive integers n.

b) Step 1: For n = 1, $\sum_{k=1}^{1} (a_k - b_k) = a_1 - b_1 = \sum_{k=1}^{1} a_k - \sum_{k=1}^{1} b_k$.

Step 2: Assume $\sum_{k=1}^{j} (a_k - b_k) = \sum_{k=1}^{j} a_k - \sum_{k=1}^{j} b_k$ for some positive integer j.

Then $\sum_{k=1}^{j+1} (a_k - b_k) = \left( \sum_{k=1}^{j} (a_k - b_k) \right) + (a_{j+1} - b_{j+1}) = \sum_{k=1}^{j} a_k - \sum_{k=1}^{j} b_k + a_{j+1} - b_{j+1} =$

$\left( \sum_{k=1}^{j} a_k \right) + a_{j+1} - \left( \left( \sum_{k=1}^{j} b_k \right) + b_{j+1} \right) = \sum_{k=1}^{j+1} a_k - \sum_{k=1}^{j+1} b_k$.

$\therefore$ by Steps 1 and 2 and mathematical induction, $\sum_{k=1}^{n} (a_k - b_k) = \sum_{k=1}^{n} a_k - \sum_{k=1}^{n} b_k$ for all positive integers n.

c) Step 1: For n = 1, $\sum_{k=1}^{1} ca_k = ca_1 = c \sum_{k=1}^{1} a_k$.

Step 2: Assume $\sum_{k=1}^{j} ca_k = c \sum_{k=1}^{j} a_k$ for some positive integer j.

11. c) (Continued)

Then $\sum_{k=1}^{j+1} ca_k = \sum_{k=1}^{j} ca_k + ca_{j+1} = c\left(\sum_{k=1}^{j} a_k\right) + ca_{j+1} = c\left(\left(\sum_{k=1}^{j} a_k\right) + a_{j+1}\right) = c\sum_{k=1}^{j+1} a_k.$

∴ by Steps 1 and 2 and mathematical induction, $\sum_{k=1}^{n} ca_k = c\sum_{k=1}^{n} a_k$ for all positive integers n.

d) Step 1: For n = 1, $\sum_{k=1}^{1} a_k = a_1 = c = 1 \cdot c.$

Step 2: Assume $\sum_{k=1}^{j} a_k = jc$ for some positive integer j.

Then $\sum_{k=1}^{j+1} a_k = \sum_{k=1}^{j} a_k + a_{j+1} = jc + c = (j + 1)c.$

∴ by Steps 1 and 2 and mathematical induction, $\sum_{k=1}^{n} a_k = nc$ for all positive integers n and where $a_k = c$, a constant.